The Gentianaceae - Volume 1: Characterization and Ecology

Gentiana tibetica King. (Photograph A. Mikuła)

Jan J. Rybczyński · Michael R. Davey
Anna Mikuła
Editors

The Gentianaceae - Volume 1: Characterization and Ecology

Springer

Editors
Jan J. Rybczyński
Anna Mikuła
Polish Academy of Sciences
Botanical Garden-Center for Biological Diversity Conservation
Warsaw
Poland

Michael R. Davey
Plant and Crop Sciences Division
School of Biosciences
University of Nottingham
Loughborough
UK

ISBN 978-3-642-54009-7 ISBN 978-3-642-54010-3 (eBook)
DOI 10.1007/978-3-642-54010-3
Springer Heidelberg New York Dordrecht London

Library of Congress Control Number: 2014931384

Printed on acid-free paper

Springer is part of Springer Science+Business Media (www.springer.com)

This volume is dedicated by the editors to their spouses

Preface

The Gentianaceae, or Gentian family, is worldwide in distribution with approximately 100 genera and about 1,800 species that include monocarpic and perennial herbs, shrubs, trees, and lianes, with terrestrial and epiphytic representatives. The plants are diverse in habit, the majority being herbaceous. The tropics are the main source of new species of the Gentianaceae. *Gentiana* (360 species), *Gentianella* (250 species), and *Swertia* (135 species) are the three largest genera; members of the family are protected by law. Several species are important pharmacologically because of their secondary metabolites, as some of the compounds have a broad spectrum of biological activity.

Initial contacts with Gentians often occur during childhood when skin is protected from bacterial infection by *Gencjana* (Polish) or *Violetum Gentianae* (Latin), while children suffering from chickenpox are also painted with *Violetum* to counteract infection by *Herpes virus varicellae*. The importance of Gentians escalated in the 1980s when several studies at the plant level focused on the vegetative propagation of species, such as *Gentiana lutea* and *G. cruciata*, following recognition of the secondary products synthesized by some members of this genus. Research into Gentians, especially in Poland, was stimulated further by publication of the "Red Book" of the Polish Flora. This volume included reference to numerous Gentians and Gentianellas, with the need for their multiplication and reintroduction into the wild because of destructive over-collection of wild material for pharmaceutical use, combined with loss of natural habitats. Some species are now rare and endangered. Variation in plant habit, especially flower morphology and pigmentation, also make members of the Gentianaceae attractive for outdoor and indoor cultivation. The establishment of the Web site "Gentiana.pl" supplemented the earlier reference site "Gentiana Research Network" established by Dr. Lena Struwe at Rutgers University, New Brunswick, USA. Much deliberation, contacts at the scientific level and discussions with colleagues at Springer, resulted in the compilation of these two volumes on Gentians. Volume 1 includes contributions on the characterization of this family of plants, while Volume 2 is devoted to aspects of biotechnology and their applications.

Volume 1: Characterization and Ecology

Volume 1, comprising 12 chapters, centers upon the characterization and ecology of the Gentianaceae, with some emphasis on the application of molecular and cytological approaches in relation to taxonomy. The first three chapters consider classification of this family of plants, with Chap. 2 reviewing research progress since the earlier revision of the Gentianaceae in 2002. This revision resulted in reclassification of some plants and the naming of new genera. Chapter 3 provides the most comprehensive report to date of the systematics of South American Neotropical woody members of the Gentians, with discussion of the use of cytological and molecular technologies to facilitate classification. Other reviews (Chaps. 4, 5) include details of the Gentianaceae in The Ukraine and Balkan Peninsula, with discussion of the taxonomy of representative species in these regions. Floral pigmentation in members of this family has been a topic of investigation for many years, with the key biochemical steps that result in the diversity of flower colors found in Gentians being summarized in Chap. 6. Other aspects of this diverse, interesting group of plants include the cytology of European species (Chap. 7), and a historical account of the importance of Gentians in herbal medicines, with links to evolution and classification (Chap. 8). Analysis of gene expression in overwintering buds is presented as an approach with which to study several aspects of plant taxonomy, phenotypic characteristics, phylogeography, and pedigree (Chap. 9). Two (Chaps. 10, 11) indicate the importance of Gentians in India in terms of their exploitation as herbal-based medicines, but emphasize the need for conservation to negate the loss of germplasm from natural habitats resulting from random harvesting. Finally, Chap. 12 presents evidence for the importance of fungi from the Phylum Glomeromycota in developing arbuscular mycorrhizal associations with the roots of members of the Gentianaceae. The role of such associations in plant growth and development is also discussed. Volume 1 of the Gentianaceae provides a general, broad-based foundation for more biotechnological approaches that are considered in Volume 2.

Volume 2: Biotechnology and Applications

The Gentianaceae includes species which are popular as ornamentals in the form of cut flowers and pot plants, with market demands necessitating improvement in flower quality, particularly characteristics such as inflorescence longevity. Micropropagation has become a routine procedure for multiplication of horticultural genera, including *Blakstonia*, *Centaurium*, *Genetiana*, *Gentianella*, and *Swertia*, with seedlings being the most common source of explants for plant propagation in vitro. Although organogenesis is the main route of plant regeneration, somatic embryogenesis is also a pathway in routine use for plant multiplication. These approaches are discussed in detail in Chaps. 1–6. Embryogenic cultures, such as cell suspensions, are an excellent source of protoplasts for gene transfer by somatic hybridization and cybridization. The relevance of the latter technologies (Chap. 7) is that they generate nuclear and cytoplasmic combinations normally unavailable to plant breeders through conventional sexual hybridization. Techniques presented in Volume 2 also include the generation of haploid and dihaploid plants from cultured anthers, and the genetic variation that may arise from tissue and organ culture (Chaps. 8, 9). Subsequent chapters discuss the molecular breeding of Gentians, particularly gene transfer by transformation, with associated genetic analyses (Chap. 10). Molecular markers facilitate breeding and cultivar identification. Vegetative propagation to generate genetically uniform populations and, conversely, manipulations to increase genetic variability, often rely upon cryopreservation as a common technology for long-term storage of relevant germplasm (Chap. 11). Other reviews consider the postharvest physiology of Gentian flowers (Chap. 12), and the biosynthesis of secondary metabolites, including antimalarial compounds (Chaps. 13–18). Modification of secondary metabolites has application in human health protection. Interestingly, the beauty of Gentian flowers and the pharmaceutical value of the plants have been the reasons for the special interest in the Gentianaceae since ancient times.

These two volumes should serve as key references for persons from a wide range of disciplines, including students and staff of universities and institutes, as well as professional gardeners and plant hobbyists.

Jan J. Rybczyński
Michael R. Davey
Anna Mikuła

Contents

Contributors

T. S. Baburaj Regional Agricultural Research Station, Kerala University of Agriculture, Vellanikkara, Kerala 680656, India, e-mail: tsbraj@gmail.com

N. M. Drobyk 'Volodymir Hnatiuk Ternopil' National Pedagogical University, 2 M. Kryvonis Str., Ternopil 46027, Ukraine, e-mail: drobyk.n@gmail.com

J. R. Grant Institut de Biologie, Université de Neuchâtel, Rue Emile-Argand 11, 2000 Neuchâtel, Switzerland, e-mail: jason.grant@unine.ch

T. Hikage Hachimantai City Floricultural Research and Developmental Center, Hachimantai, Iwate 028-7592, Japan

S. Jha Department of Botany, University of Calcutta, 35 B. C. Road, Kolkata 700019, India

T. B. Jha Plant Biotechnology Laboratory, Department of Botany, Presidency University, 86/1 College Street, Kolkata 700073, India, e-mail: tbjha2000@yahoo.co.in

V. A. Kunakh Institute of Molecular Biology and Genetics, National Academy of Sciences of Ukraine, Akad. Zabolotnogo Str. 150, Kyiv 03680, Ukraine, e-mail: Kunakh@imbg.org.ua

G. Mansion Botanical Garden in Dalhem, Berlin, Germany, e-mail: g.mansion@bgdm.org

V. M. Mel'nyk Institute of Molecular Biology and Genetics, National Academy of Sciences of Ukraine, Akad. Zabolotnogo Str. 150, Kyiv 03680, Ukraine, e-mail: v.m.melnyk@imbg.org.ua

E. Młodzińska Department of Plant Physiology, Institute of Experimental Biology, ul. Kanonia 6/8, 50-328, Wrocław, Poland, e-mail: ewamloda@biol.uni.wroc.pl

C. Narayanankutty Agricultural Research Station, Kerala University of Agriculture, Vellanikkara, Kerala 680656, India, e-mail: cnkutty@gmail.com

J. S. Pringle Royal Botanical Gardens, P.O. Box 399, Hamilton, ON L8N 3H8, Canada, e-mail: jpringle@rbg.c

T. Samaddar Department of Botany, University of Calcutta, 35 B. C. Road, Kolkata 700019, India

N. M. Shiyan National Herbarium of Ukraine, M.G. Kholodny Institute of Botany of the NAS of Ukraine, 2, Tereshhenkivska Str., Kyiv 01601, Ukraine, e-mail: herbarium_kw@ukr.net

V. Stenovic Faculty of Biology, The Institute of Botany and Botanical Garden "Jevremovac", University of Belgrade, 43 Takovska str, Belgrade, Serbia, e-mail: vstev@bio.bg.ac.rs

L. Struwe Department of Ecology, Evolution and Natural Resources, Cook College, Rutgers University, 237 Foran Hall, 59 Dudley Road, New Brunswick, USA, e-mail: struwe@aesop.rutgers.edu

Z. Sýkorová Department of Mycorrhizal Symbiose, Institute of Botany, Academy of Sciences of the Czech Republic, 252 43 Pruhonice, Czech Republic, e-mail: Zuzana.Sykorova@ibot.cas.cz

K-I. Tsutsumi Cryobiofrontier Research Center, Iwate University, Morioka, Iwate 020-8550, Japan, e-mail: kentsu@iwate-u.ac.jp

M. O. Twardovska Institute of Molecular Biology and Genetics, National Academy of Sciences of Ukraine, Akad. Zabolotnogo Str. 150, Kyiv 03680, Ukraine

S. Unniampurath Agricultural Research Station, Kerala University of Agriculture, Vellanikkara, Kerala 680656, India, e-mail: sreeu2005@gmail.com

Chapter 1
Morphological Characteristics of the Family Gentianaceae

James S. Pringle

Abstract The Gentianaceae or Gentian family is worldwide in distribution, with 1,800 or so species in about 100 genera. The plants are diverse in habit, although the majority are herbaceous. The leaves of most species are opposite, with the margins generally entire; stipules are absent in most genera. The perianths are biseriate, actinomorphic or somewhat zygomorphic, sympetalous, and usually synsepalous. The stamens are epipetalous, isomerous with the petals, and the pistil is solitary, bicarpellate, with a superior ovary containing several to many ovules.

1.1 Introduction

The plant family Gentianaceae comprises about 100 genera and 1,800 species, native to all continents except Antarctica. It is the type family of the order Gentianales, with morphological and molecular evidence indicating that it is most closely related to the Loganiaceae, Gelsemiaceae, and Apocynaceae (Struwe et al. 1994, 2002; Backlund et al. 2000; Stephens 2001; Albert and Struwe 2002; Mészáros et al. 2002). The tribal and subtribal classification accepted here follows that of Struwe et al. (2002). In the past, the genera constituting the subtribe Potaliinae were often placed in the Loganiaceae, but molecular evidence strongly supports their inclusion in the Gentianaceae (Struwe et al. 2002). The Menyanthaceae, formerly included in the Gentianaceae, are now generally recognized as a distinct family, and placed in the Asterales.

Species of the Gentianaceae occur in diverse vegetation zones and plant communities, including tropical forests and savannahs, temperate woodlands and prairies, and arctic and alpine regions, as well as in littoral and ruderal plant communities. Most grow in mesic habitats. Some are found in wetlands or on tidal flats, but none is

J. S. Pringle (✉)
Royal Botanical Gardens, P.O. Box 399, Hamilton, ON L8N 3H8, Canada
e-mail: jpringle@rbg.ca

J. J. Rybczyński et al. (eds.), *The Gentianaceae - Volume 1: Characterization and Ecology*, DOI: 10.1007/978-3-642-54010-3_1,

truly aquatic or marine. True desert species do not exist in the family, although that status is approached by some species in the tribe Chironieae, and many species grow in relatively dry savannahs or fell-fields. A few tropical species are epiphytes (Gilg 1895; Struwe et al. 2002). Many are cultivated as ornamentals.

The largest genus is *Gentiana* L., with about 400 species. This genus is nearly worldwide in distribution, but absent from low-altitude tropical regions and all but northernmost Africa and all but eastern Australia. It reaches its greatest diversity in China, where numerous species are present in the high mountains. The second-largest genus is *Gentianella* Moench, with about 300 species, similar in the extent of its distribution, but with its greatest diversity in the Andes of South America. Another large genus is *Sebaea* R. Br., with about 100 species in Asia, Africa, and Australasia. *Swertia* L., if widely circumscribed, includes 125–150 species, but its appropriate generic delimitation remains in question (Ho and Pringle 1996; Albert and Struwe 2002).

1.2 Habit

The majority of the species in the Gentianaceae are herbs or subshrubs, although many of the tropical species are shrubs or small to medium-sized trees (to ca. 10 m in most arborescent species, to 35 m in some species of the subtribe Potaliinae), and a few are lianas. The herbaceous species may be annual, winter-annual, biennial, pliestesial (long-lived but monocarpic), or perennial. Most are erect or decumbent, but a few are trailing or twining vines. Both sympodial and, less commonly, monopodial growth occurs among the perennial species. Most herbaceous perennial species develop an erect, often branched caudex. Some perennial and some otherwise monocarpic species produce new crowns from elongating rhizomes, spreading roots, or stems that root at proximal nodes, forming colonies. The plants of most species are glabrous; aside from axillary colleters, the indument of the vegetative parts, when present, consists of papillae or simple or rarely uniseriate, non-glandular or rarely glandular hairs, which may be present on the stems but absent from the leaves. Spiny stems occur only in some species of *Anthocleista* Afzel. ex R. Br. Phyllodia, bulbs, and highly modified, succulent stems do not occur in the family Gentianaceae (Wood and Weaver 1982; Struwe et al. 1994, 2002; Mészáros et al. 2002).

Persistent tap roots are common. Many perennial species produce secondary roots, which are often stout, and lack root hairs (Gilg 1895). Endomycorrhizae of the order Glomales (Zygomycotina) are generally present (Imhof 1999; Cameron and Bolin 2010). Most species are chlorophyllous, but several genera have greatly reduced leaves and root systems. *Voyria* Aubl., *Voyriella* (Miq.) Miq., and some species of *Exacum* L. (formerly segregated as *Cotylanthera* Blume) are completely mycoheterotrophic, lacking chlorophyll, with minute, scale-like leaves and coralline or "bird's-nest" root systems (Imhof 1999).

1.3 Leaves

Phyllotaxy is opposite or less often whorled, or rarely with some or all leaves, alternate. Basal leaves are often present. The nodes are usually unilacunar or trilacunar, occasionally multilacunar. Leaves are simple and unlobed, sessile, or less often petiolate; distinctly petiolate leaves are common among the tropical, woody species, but are absent among species of the temperate and Arctic-alpine zones. Primary venation (except in species with narrowly linear or very small leaves) is usually basally or suprabasally acrodromous, or pinnate and camptodromous, especially in woody genera. Secondary venation is diverse, often more or less brochidodromous and sometimes with an intramarginal vein, or eucamptodromous. In nearly all species, the margins are entire (except for the projecting distal corners of marginal cell walls, which in some species form minute teeth or cilia); in the few exceptions the margins are shallowly crenulate or serrulate. Raised lines often connect the bases of the opposite or whorled leaves, but in many other species, sometimes within the same genus, such lines are absent, with the leaf margins often being decurrent as longitudinal ridges on the stems. In some species, the bases of the opposite or whorled leaves are connate and strongly ascending, forming a sheath around the stem. The blades of the distal or all leaf-pairs are perfoliate in some species in the tribe Chironieae. Other than lateral expansions of the petioles, which occur in several genera, stipules are absent except in the subtribe Potaliinae and some species of *Tachia* Aubl., in which they form low ocreae, and in *Macrocarpaea zophoflora* R. E. Weaver & J. R. Grant, in which they are interpetiolar and deciduous. Colleters are often present in the leaf axils (Gilg 1895; Metcalfe and Chalk 1950; Struwe et al. 1994, 2002; Mészáros et al. 2002).

The leaves are dorsiventral or, in a few small genera, isobilateral without a distinct palisade layer. The stomates are anomocytic or anisocytic, on the abaxial or less often both leaf surfaces. Some species of *Voyria* lack stomates. Kranz leaf anatomy does not occur in the family (Metcalfe and Chalk 1950; Struwe et al. 1994, 2002).

1.4 Inflorescences

The inflorescences are terminal and lateral. They are basically cymose, but many variations occur. The cymes range from dichasial to partly or wholly monochasial and, in some species, are racemoid, umbelloid, capitate, or grouped in thyrses or verticillasters, or the inflorescence may consist only of one flower. The flowers may be pedicellate or sessile, erect, horizontal, or nodding. They may or may not be subtended by paired bracts (Gilg 1895; Struwe et al. 1994, 2002; Mészáros et al. 2002).

1.5 Flowers

1.5.1 General

The flowers are hypogynous. In the majority of genera, the flower is actinomorphic, except often for some dorsal compression of the pistil. Slightly to moderately (but not strongly) zygomorphic flowers occur in some tropical and subtropical genera, mostly in the tribe Helieae. Zygomorphy in such flowers is usually more pronounced in the corolla than in the calyx, and most strongly expressed in the sexual parts (Gilg 1895; Wood and Weaver 1982; Struwe et al. 1994, 2002; Mészáros et al. 2002).

1.5.2 Perianths

The perianth, with the exception, presumably, in *Obolaria* L., is biseriate. In most genera, the perianth (or corolla) is tetramerous or pentamerous. Those of a few genera (or of sections, subsections, or individual species) are 6- to 16-merous; those of *Pycnosphaera* Gilg are trimerous. The sepals and petals are usually isomerous, but in a few genera, the lobes of the corolla are more numerous than those of the calyx.

The calyx is differentiated from the corolla, and is generally smaller, green and more or less leaf-like in texture, or occasionally coriaceous or nearly hyaline. Although usually inconspicuous, in a few genera the calyx forms a significant part of the floral display, notably in *Potalia* Aubl. in which, with the pedicels, it is yellow to orange, although not larger than the corolla. In *Prepusa* Mart., it is greenish yellow, rose-violet, or red and conceals much of the corolla. The corolla is usually actinomorphic, occasionally more or less irregular in its lobing, variably spathaceous, or slightly zygomorphic. In most genera the sepals are united proximally, but in a few genera they are completely separate. Aestivation of the calyx lobes is imbricate or valvate. In *Crawfurdia* Wall. and most sections of *Gentiana*, an adaxial invagination of the calyx tube forms an intracalycular membrane, so that the lobes appear to diverge abaxially from the tube slightly below its summit. *Obolaria* lacks a calyx (structures in that genus interpreted here as subtending bracts are sometimes regarded as a two-part calyx). Colleters, or less often paired scales or, mostly in woody species, discoid glands, are frequently present adaxially near the base of the calyx tube (Gilg 1895; Wood and Weaver 1982; Metcalfe and Chalk 1950; Struwe et al. 1994; Mészáros et al. 2002).

The corollas are petaloid and usually conspicuous, highly diverse in shape, with tubular, salverform, campanulate, and rotate corollas all being frequent. Corollas are also diverse in color, often with guidelines or "eyes," mostly in the tribe Chironieae and *Gentiana* section *Chondrophyllae* Bunge. In all species, the corolla is gamopetalous, although the depth of lobing varies greatly. Aestivation of the corolla is usually contorted, but is imbricate in a few genera, and proximally valvate in *Aripuana* (Struwe et al. 2002), *Jaeschkea* Kurz, and *Tachia schomburgkiana* Benth. In many species in the tribe Gentianeae, there is a fringe of vascularized or non-vascularized trichomes or a pair of deeply fimbriate scales near the base of each corolla lobe, or a more or less continuous fringe at or near the summit of the tube. In some genera in the subtribe Swertiinae, fringes of trichomes and/or tubular or scale-like prolongations of the rim surround the openings of discoid glands on the adaxial corolla surface (Gilg 1895; Wood and Weaver 1982; Struwe et al. 1994, 2002; Ho and Pringle 1996; Maas and Ruyters 1986; Mészáros et al. 2002).

1.5.3 Androecia, Gynoecia, and Nectaries

The stamens are in one cycle, isomerous and alternate with the corolla lobes, and are inserted either on the corolla tube or in the sinuses between the corolla lobes. Filaments are usually present; the anthers are sessile in some species of *Voyria* and nearly so in a few other genera. The filaments are mostly slender and often flattened; those of some genera or species are ± abruptly wider near the base. In most species, the stamens are equal, with the filaments straight or nearly so, forming an actinomorphic androecium. In some genera or species therein, mostly in the subtribe Chironiinae, all of the stamens are deflexed to one side, away from the stigma during the early floral stages, and later become erect. In some tropical species, mostly in the tribe Helieae, the stamens are declinate, forming a zygomorphic androecium, and are sometimes also coiled circinnately distally. In a few genera or individual species, also mostly tropical, especially in the subtribe Canscorinae, the stamens are unequal, and differ in filament length and/or level of insertion on the corolla tube. The anthers may differ in size, and one or two stamens may be reduced, having a sterile anther or lacking an anther. In most genera, the filaments are free above the level of insertion, but in some cases they are connected by a corona, which is often fringed between the filaments, or corona scales may be alternate with the filaments. The filaments are fused proximally forming a tube in the subtribe Potaliinae. The anthers are tetrasporangiate and dithecal, dorsifixed or basifixed, and free or more or less connate in some genera. In other genera, they are appendaged, sometimes with glands at the base or apex and in some cases they coil helically or circinnately upon dehiscence. Dehiscence is by longitudinal rimae, usually introrse, rarely latrorse or extrorse. In *Exacum*, dehiscence is restricted to the distal portion of the anther, forming one pore per anther in the chlorophyllous species, in which the thecae are fused distally, or two pores per anther in the achlorophyllous species (Gilg 1895; Lindsey 1940; Struwe et al. 1994, 2002; Stephens 2001).

Except in the achlorophyllous taxa, the pollen grains are small to medium-sized, 18–65 μm in the longest dimension, trinucleate in most genera, but binucleate in some. They are most often tricolporate, but in some genera are tetracolporate, pentacolporate, pericolporate, or triporate. The pores are round or nearly so. When shed in monads, as in the majority of genera, the pollen grains are usually prolate, occasionally spheroid. The pollen is shed in tetrads in most of the genera in the tribe Helieae and in several genera in other tribes, and in polyads in a few genera in the tribe Helieae. Except in some achlorophyllous species, both the exine and intine are well developed, the exine consisting of nexine and sexine, the sexine of endexine and ectexine. The sexine is striate, striato-reticulate, or reticulate in most genera, perforate in a few, and nearly smooth in *Halenia* Borkh.; projections of diverse forms occur in the tribe Helieae. The tectum is generally supported by columellae. Granules and/or other subtectal processes are sometimes also present. Pollen tends to be smaller in the achlorophyllous taxa. Pollen of most *Voyria* species is 11–15 μm in diameter, 1- or 2(-6)-porate, with the exine smooth and not distinctly stratified (Nilsson 2002).

The pistil is solitary, bicarpellate, and syncarpous at least below the distal portion of the style. A style may be present or absent. If present, it is solitary, usually not cleft below the stigmas or stigmatic lobes, but shallow to rather deeply bifid in some genera or species within some genera. The style is erect in most species, but in some species in the tribe Chironieae it is deflexed away from the stamens during the early stages of the flower, later becoming erect. It is usually declinate in species with more or less zygomorphic corollas and androecia. When the style is not bifid the stigmas may be solitary or paired. When solitary, they may be capitate, peltate, funnel-form, bilobed, or in *Sebaea*, bottle brush-like. They range when paired from orbicular to oblong or strap-shaped, and may become increasingly recurved as the flowers age. In genera with relatively long, narrow stigmas, the latter may remain straight or, along with the style branches, coil helically before and/or after receptivity. In most species of *Lomatogonium* A. Br. and to some extent in *Bartonia* Muhl. ex Willd., the stigmas are decurrent along the sutures of the ovary. The stigmatic surface is wet and usually papillate, sometimes nearly smooth. In genera with stout and/or poorly differentiated styles, the latter is generally persistent in the fruit; in species with slender styles, the style is often completely or partially deciduous (Gilg 1895; Wood and Weaver 1982; Bertin and Newman 1993; Struwe et al. 1994; Stephens 2001).

The ovary is superior. It may be sessile or stipitate; a stipe, when present often elongates in the fruit. Both carpels are ovuliferous. In most genera, the ovary is unilocular with parietal but often deeply inserted placentation. In some genera, the ovaries are proximally or completely bilocular with axile placentation, and in the achlorophyllous species of *Exacum* they are unilocular with partially free-central placentation. The ovules are few to very many, usually sessile and anatropous, tenuinucellate, and unitegmic, but ategmic in *Voyria* subgenus *Leiphaimos* (Schltdl. & Cham.) V. A. Albert & Struwe and the achlorophyllous species of *Exacum*. The embryo sac is of the *Polygonum* type (Gilg 1895; Wood and Weaver 1982; Struwe et al. 1994, 2002; Shamrov 1996; Stephens 2001; Bouman et al. 2002; Mészáros et al. 2002).

The flowers of most species produce nectar, except in many genera of the tribe Chironieae, and most also produce abundant pollen. The nectaries may be borne on the ovary or ovarian stipe, may form a ring at the base of the ovary, or may be epipetalous. When epipetalous, they may be on the adaxial corolla surface, in spurs in the majority of *Halenia* species, as clusters of nectariferous hairs in *Faroa* Welw., or in some genera of the subtribe Swertiinae, in pits within the corolline tissue. There may be one or two per corolla lobe, with the pits opening to the adaxial corolla surface immediately adaxial to the nectary or distal to it. In species in which the nectaries do not form a continuous ring, they are usually isomerous with or twice as many as the corolla lobes, but only two (if any), borne on the ovary, are present in *Voyria*. In *Orphium* E. Mey. and a few other genera, glandular discs between the calyx and corolla also appear to function as nectaries. Abaxial calycine and extrafloral nectaries are common in the tribe Helieae. In some species, mostly in the tribe Chironieae, pollen is the only "reward." Most species of *Exacum* L. produce "buzz flowers," in which pollen is released from the anthers by resonant vibration produced by the action of visiting bees (Wood and Weaver 1982; Struwe et al. 1994, 2002; Stephens 2001; Mészáros et al. 2002).

1.6 Pollination and Breeding Systems

Most species of the Gentianaceae are self-compatible, but pollination usually requires external agents. Most species in temperate regions are pollinated primarily by bees, although flies and other insects are also important, especially in *Sabatia* Adans., *Swertia*, and closely related genera. Butterflies, moths, hummingbirds, and bats are also important pollinators, especially among the tropical species. Wind-pollination does not occur in the family. Geitonogamous pollination is frequent, as the flowers of most species are long-lived and those on any one plant may open over a long period (Gilg 1895; Stephens 2001; Struwe et al. 2002).

Herkogamy, with the style being deflected to one side and the stamens deflected in the opposite direction, occurs in several genera. The style and stamens later become erect, allowing autogamy if xenogamous or geitonogamous pollination has not occurred previously (Bertin and Newman 1993).

In most species, the flowers are bisexual and monomorphic, usually protandrous and predominantly xenogamous. Most species in the tribe Saccifoliae and a few in *Sebaea*, although having all flowers bisexual, are heterostylous. *Veratrilla* (Baill.) Franch. is dioecious, and some species of *Gentianella* in the Southern Hemisphere are gynodioecious, andromonoecious, gynomonoecious, or trimonoecious. Species with obligately or usually autogamous flowers also occur mostly in the subtribe Chironiinae, and in many other species the flowers are facultatively autogamous, often with the stigma tardily coming into contact with the anthers. Agamospermy is believed to occur in achlorophyllous species of *Exacum* (Gilg 1895; Filippa and Barboza 2006; Freitas and Sagima 2009).

Fig. 1.1 The diversity of flowers of the species of the family Gentianaceae from North America. **a** *Frasera albicaulis* var. *nitida* (Benth.) C. L. Hitchc. in California, USA, showing nectary-pit openings and corona scales. *Photo* Norman Jensen: Copyright Allen L. Jensen with permission. **b** *Obolaria virginica* L., a strongly mycotrophic but chlorophyllous species, in Tennessee, USA. **c** *Gentiana andrewsii* Griseb., in Ontario, Canada, a species with permanently closed corollas. **d** *Gentiana newberryi* A. Gray var. *newberryi*. in Oregon, USA. *Photo* Tanya Harvey, with permission

Fig. 1.2 The diversity of the species of the family Gentianaceae from South America. **a** *Gentianella gilgiana* (Reimers) Fabris ex J. S. Pringle. Peru. A species of high altitudes in the Andes. Note the actinomorphic flowers, isomerous calyx and corolla, gamopetalous but deeply lobed corolla, and epipetalous stamens alternating with the corolla lobes. Photo by Victor Villalobos, with permission. **b** *Symbolanthus mathewsii* (Griseb.) Gilg. Peru. A shrubby species with a slightly zygomorphic corolla. Photo by Jason R. Grant, with permission. **c** *Macrocarpaea subsessilis* Weaver & J. R. Grant. Ecuador. A shrubby species with slightly zygomorphous flowers with pale corollas that remain open at night, pollinated primarily by bats. Photo by Jason R. Grant, with permission. **d** *Voyria acuminata* Benth. French Guiana. Species is mycoheterotrophic, lacking chlorophyll. Photo by Carole Gracie; © 2013 Carol Gracie, with permission

1.6.1 Fruits

The calyx is persistent in fruit in all but two genera and some species in a third, usually not expanding, but in some tropical genera becoming indurate and/or accrescent. The corolla is marcescent in most genera. The fruits generally remain attached, but fall from the plants in a few tropical species. In most genera, the fruits are septicidal, bivalvate capsules, diverse in wall thickness, but in a few genera they are irregularly dehiscent or indehiscent capsules or berries (Gilg 1895).

1.6.2 Seeds and Dispersal

The seeds are diverse in shape, nonarillate, with the seed coat membranous or crustaceous. Seed-coat surfaces are diverse at maturity, often reticulate, sometimes winged or with papillose projections. In most genera, the seeds are small and light and, especially those with winged seed coats, are largely wind-dispersed. In some species in the tribe Gentianeae, elongation of the ovarian stipe elevates the capsule above the marcescent corolla and renders it more susceptible to breeze action. Rain-washed and melt-water from snowfields are also significant in seed dispersal, especially in alpine species of *Gentiana* and *Gentianella*. The capsules of some species of *Gentiana* section *Chondrophyllae* probably function as splash-cups. Rainwash is also significant in genera with indehiscent dry fruits, from which the seeds are released when the walls decay, and in species of *Curtia* Cham. & Schltdl. and *Voyria*, with fruits that dehisce after having fallen. *Gentianopsis detonsa* subsp. *victorinii* (Fernald) Lammers, and probably some other taxa, have seeds that float. Projections on the seed coats of some species may function in exozoochorous dispersal. The seeds are dispersed endozoochorically in the genera with baccate fruits. Most of the achlorophyllous species produce minute "dust seeds," dispersed by wind or rain-wash. The embryo is small and cylindric or conical in the autotrophic species and minute or undifferentiated in the achlorophyllous species. The endosperm is copious, fleshy or oily, and nuclear in the autotrophic taxa, scant and in some species cellular in the achlorophyllous taxa. Detached vegetative disseminules such as bulblets do not occur in the family (Gilg 1895; Struwe et al. 1994, 2002; Stephens 2001; Bouman et al. 2002; Mészáros et al. 2002).

1.7 Conclusions

Morphological and molecular evidence indicates that the Gentianaceae is a natural group. Although many species are exceptional in one or more respects, the family is largely characterized morphologically by the combination of opposite, simple leaves with entire margins, a biseriate, an actinomorphic or nearly actinomorphic perianth, synsepalous and sympetalous and more or less persistent in fruit, epipetalous stamens, isomerous with the corolla lobes and alternating with them. There is a single bicarpelate pistil with a superior, unilocular ovary containing several to many ovules (Figs. 1.1, 1.2).

References

Albert VA, Struwe L (2002) Gentianaceae in context. In: Struwe L, Albert VA (eds) Gentianaceae: systematics and natural history. Cambridge University Press, Cambridge, pp 1–20

Backlund M, Oxelman B, Bremer B (2000) Phylogenetic relationships within the Gentianales based on *ndh*F and on *rbc*L sequences, with particular reference to the Loganiaceae. Amer J Bot 67:1029–1043

Bertin RI, Newman CM (1993) Dichogamy in angiosperms. Bot Rev 59:112–152

Bouman F, Cobb L, Devente N, Goethals V, Maas PJM, Smets E (2002) The seeds of Gentianaceae. In: Struwe L, Albert VA (eds) Gentianaceae: systematics and natural history. Cambridge University Press, Cambridge, pp 498–572

Cameron DD, Bolin JF (2010) Isotypic evidence of partial mycoheterotrophy in the Gentianaceae: *Bartonia virginica* and *Obolaria virginica* as case studies. Amer J Bot 97:1272–1277

Filippa EM, Barboza GE (2006) Gentianaceae Juss. Flora Fanerogámica Argentina 102. Programa PROFLORA-CONICET, Córdoba, Argentina

Freitas L, Sagima M (2009) Floral biology and mechanisms of spontaneous self-pollination in five neotropical species of Gentianaceae. Bot J Lin Soc 160:357–368

Gilg E (1895) Gentianaceae. In: Engler A, Prantl K (eds) Die Natürlichen Pflanzenfamilien, vol 4. Verlag von Wilhelm Engelmann, Leipzig, pp 50–108

Ho TN, Pringle JS (1996) Gentianaceae. In: Flora of China Editorial Committee, Flora of North America North of Mexico, vol 16. Science Press, Beijing, China and Missouri Botanical Garden, St. Louis, MO, USA. pp 1–139

Imhof S (1999) Root morphology, anatomy and mycotrophy of the achlorophyllous *Voyria aphylla* (Jacq.) Pers. (Gentianaceae). Mycorrhiza 9:33–39

Lindsey AA (1940) Floral anatomy in the Gentianaceae. Amer J Bot 27:640–652

Maas PJM, Ruyters P (1986) *Voyria* and *Voyriella* (saprophytic Gentianaceae). Flora Neotropica Monograph No. 41. The New York Botanical Garden, New York, USA

Mészáros S, De Laet J, Goethals V, Smets E, Nilsson S (2002) Cladistics of Gentianaceae: a morphological approach. In: Struwe L, Albert VA (eds) Gentianaceae: systematics and natural history. Cambridge University Press, Cambridge, pp 310–376

Metcalfe CR, Chalk L (1950) Anatomy of the dicotyledons, 1st edn. Clarendon Press, Oxford

Nilsson S (2002) Gentianaceae: review of palynology. In: Struwe L, Albert VA (eds) Gentianaceae: systematics and natural history. Cambridge University Press, Cambridge, pp 377–497

Shamrov II (1996) Ovule development and significance of its features for Gentianaceae systematics. Opera Bot Belg 7:113–118

Stephens PF (2001) Angiosperm phylogeny website. Version 12, July 2012 and updates. www.mobot.org/MOBOT/research/APweb. Accessed 25 Oct 2012

Struwe L, Albert VA, Bremer B (1994) Cladistics and family level classification of the Gentianales. Cladistics 10:175–206

Struwe L, Kadereit JW, Klackenberg J, Nilsson S, Thiv M, von Hagen KB, Albert VA (2002) Systematics, character evolution, and biogeography of Gentianaceae, including a new tribal and subtribal classification. In: Struwe L, Albert VA (eds) Gentianaceae: Systematics and Natural History. Cambridge University Press, Cambridge, pp 21–309
Wood CE Jr, Weaver RE Jr (1982) The genera of Gentianaceae in the southeastern United States. J Arnold Arboretum Harvard Univ 65:441–487

Chapter 2
Classification and Evolution of the Family Gentianaceae

Lena Struwe

Abstract This chapter reviews research progress and resulting changes in classification in the Gentian family since the worldwide revision in 2002. Currently, the Gentianaceae includes 99 genera and approximately 1,736 species. The tribal classification still stands, but there have been some important changes of genus delimitations based on new evolutionary work. This includes reclassifications of poly- or paraphyletic genera (e.g., *Canscora*, *Centaurium*, *Fagraea*, *Sebaea*) that have led to the description of new or resurrected genera such as *Cyrtophyllum*, *Duplipetala*, *Exochaenium*, *Gyrandra*, *Lagenias*, *Limahlania*, *Klackenbergia*, *Phyllocyclus*, *Picrophloeus*, *Schenkia*, *Utania*, and *Zeltnera*. New genera have been discovered in South America, including *Roraimaea* and *Yanomamua*. Some genera were incorporated into others to preserve monophyly (*Cotylanthera* into *Exacum*, *Wurdackanthus* into *Symbolanthus*). *Bisgoeppertia* has been moved to the Potalieae. Unsolved generic delimitation problems remain in Gentianeae-Swertiinae and Helieae. The placement of the enigmatic mycoheterotroph *Voyria* is still uncertain, but it is likely an isolated, basally placed branch in the family. Recent biogeographic studies that address large-scale distribution patterns, vicariance events, and the significance of these new results are reviewed in this chapter, as are examples of evolutionary research progress within each tribe.

L. Struwe (✉)
Department of Ecology, Evolution, and Natural Resources, Rutgers University,
237 Foran Hall, 59 Dudley Road, New Brunswick, NJ 08901, USA
e-mail: struwe@aesop.rutgers.edu

L. Struwe
Department of Plant Biology and Pathology, Rutgers University, 237 Foran Hall,
59 Dudley Road, New Brunswick, NJ 08901, USA

J. J. Rybczyński et al. (eds.), *The Gentianaceae - Volume 1: Characterization and Ecology*, DOI: 10.1007/978-3-642-54010-3_2,

2.1 Introduction

By the early 1990s, almost no phylogenetic family-level work had been done on gentians, which changed very quickly with the development of PCR, and rapid and efficient DNA sequencing, paired with computational advantages in speed, algorithms, and computer hardware. This revolution in evolutionary research occurred throughout all kingdoms, and for many plant families we had the first preliminary family-level phylogenies constructed by the late 1990s, gentians included. International collaborations sped up the process and, for the gentians, the primary laboratories involved in this early cladistic work were those of Joachim Kadereit and his students in Germany (Bernhard von Hagen, Mike Thiv), Philippe Küpfer's research group in Switzerland (Philippe Chassot, Jason Grant, Guilhem Mansion, Yong-Ming Yuan), and the work at The New York Botanical Garden in USA by myself and Victor Albert with Katherine (Gould) Mathews, Jason Grant, and other visiting gentian researchers.

About a decade ago, the book *Gentianaceae: Natural History and Systematics* was published (Struwe and Albert 2002). It was the first comprehensive volume on gentian systematics ever published in English, as well as the first detailed book in any language on this taxonomic topic since Grisebach's treatment in Latin about 150 years earlier (Grisebach 1839). There were 17 contributors to the book, and the main 290-page chapter was a new, multi-authored family-level classification and description of the Gentianaceae genera based on monophyly and new phylogenetic data (Struwe et al. 2002). Before then, Gilg's (1895) tribal and subtribal gentian classification was used mostly in taxonomy. The book also included important chapters, for example, on gentian pollen by palynologist Siwert Nilsson (who sadly passed away shortly after the book was published), seed anatomy and morphology, morphological cladistics, and phytochemistry (Bouman et al. 2002; Jensen and Shripsema 2002; Mészáros et al. 2002; Nilsson 2002).

This chapter outlines progress since the last book was published in 2002, so it primarily reviews and discusses papers published during 2001–2012. The number of scientists focusing their efforts on this not just beautiful, but incredibly exciting and interesting family, has increased steadily over the last decade and great strides have been made in understanding the evolution of the family both when it comes to phylogenetic and biogeographic relationships and changes in morphological, anatomical, and chemical diversity (the latter will be addressed in other chapters). In general, the tribal classification has held up, with only some changes for a few genera that earlier were placed tentatively based on non-DNA data, or necessary reclassifications of genera. There are still some taxonomic areas that need attention and a changed classification, and this will be discussed here at the end of this chapter.

2.2 Large-Scale Classification, Species Number Estimates, and Phylogenetic Progress

Based on molecular systematics research starting in the mid-1990s, the Gentianaceae is now known to have seven major monophyletic clades, corresponding to six tribes in the current classification, namely the Saccifolieae, Exaceae, Chironieae, Gentianeae, Helieae, and Potalieae (Fig. 2.1). The mycotrophic genus *Voyria* was placed as *incertae sedis* in 2002 and in this classification, but recent data indicates that it should be recognized as its own tribe (Merckx, "personal communication"). Three of the tribes have monophyletic subtribes. Chironieae includes the three subtribes Chironieae (mostly temperate or subtropical, cosmopolitan), Coutoubeinae (neotropical), and Canscorinae (paleotropical). Cosmopolitan and mostly temperate Gentianeae are divided into two subtribes, Swertiinae and Gentianinae. The tribe Potalieae has three subtribes, namely Faroinae (mostly tropical Africa), Lisianthiinae (neotropical), and Potaliinae (pantropical). Of the remaining tribes, Saccifolieae and Helieae are strictly neotropical, while Exaceae is strictly paleotropical. The mycoheterotrophic genus *Voyria* forms an early branch in the family, and might be classified soon as its own tribe (see discussion below).

The phylogenetic study that was the basis for the classification of Struwe et al. (2002) was published in the same book chapter as the classification. Since then no overarching family-level study has been published, but many tribal level phylogenies have been added and expanded our knowledge on relationships within tribes and subtribes. Major trends and results from these studies are discussed here, and new data are also reviewed under each tribe below.

The most basally positioned clade in the Gentianaceae is the tribe Saccifolieae (Fig. 2.1), which represents the first diverging lineage in the family and is also taxonomically the smallest one with only 5 genera and 18 species. All are restricted to continental South America (mostly on the Guayana and Brazilian Shields; Table 2.1). The next diverging clade in the phylogeny is the tribe Exaceae, with 184 (mostly) herbaceous species from Africa, Madagascar, and Southern Asia in 8 genera. Diverging above the Exaceae is the clade containing the tribe Chironieae, a morphologically variable group with 26 genera containing about 161 mostly herbaceous species.

Sister to the Chironieae in the family phylogeny is a large clade with the three most species-rich tribes; the Gentianeae (974 species; 56 % of species in the family), Helieae (218 species), and Potalieae (163 species; see Table 2.1 for statistics). These three tribes form a trichotomy in most phylogenetic studies (or Gentianeae is placed as the more basally positioned clade but with very low branch support; Struwe et al. 2002). Additional sequencing has not clarified these relationships further, despite the use of additional markers (Molina and Struwe, "unpublished"), so it is possible that the divergence of these three clades happened relatively suddenly, or that major lineages at the base of each clade might have become extinct and thereby have limited our possibilities to reconstruct the actual order of diversification events at that point in time.

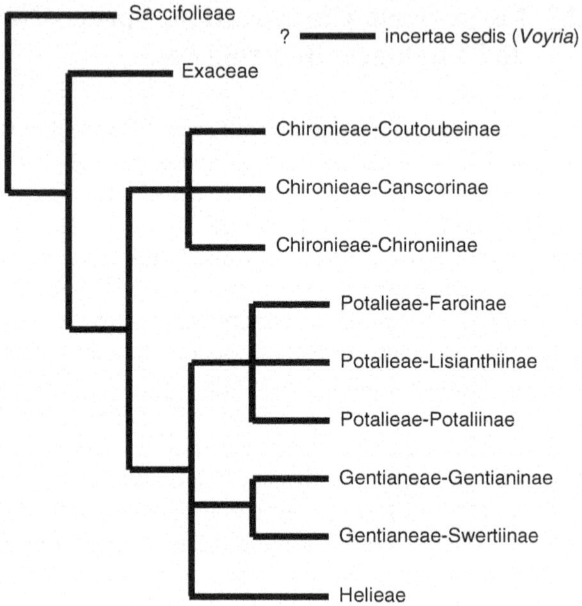

Fig. 2.1 Overview of the major clades of the Gentianaceae. Each tribe and subtribe is supported as monophyletic by having substantial branch support in several phylogenetic analyses. The position of *Voyria* is uncertain, and it might form its own branch (and tribe) close to the base of the Gentianaceae

Table 2.1 Number and percentages of genera and species within each tribe (and *Voyria*) as part of the total in the Gentianaceae

	Number of species	% of species	Number of genera	% of genera
Saccifolieae	18	1	5	5
Exaceae	184	11	8	8
Voyria	18	1	1	1
Chironieae	161	9	26	26
Gentianeae	974	56	18	18
Potalieae	163	9	18	18
Helieae	218	13	23	23
Total Gentianaceae	1,736		99	

This pattern of more species in the relatively more recently diverged clades is not unusual, and can also be seen in clades such as the milkweeds (Apocynaceae-Asclepiadoideae). Some of the more species-rich genera most likely reflect relatively recent speciation influenced by mountain orogenies, island formation, and/or habitat fragmentation due to climate change, sea level changes, or tectonics (*Symbolanthus* and *Macrocarpaea* in the Andes; *Gentiana*, *Gentianella*, and *Swertia* in several mountain areas worldwide; *Fagraea* in Southeast Asia. Many of these diversification and biogeographic hypotheses still remain to be tested for gentians.

There are large discrepancies in the number of estimated species in many genera of the Gentianeae with, for example, *Gentianella* being estimated to have 150–275 species by current gentian researchers. *Gentiana* (with its estimated 360 species) most likely includes many Chinese species that are synonymous with other Asian or European species. The lack of global revisions of these large, nearly cosmopolitan, genera hinders a full taxonomic and evolutionary understanding of these groups. Another example that has been resolved is *Halenia*, which was estimated to have 80–100 species, but only has 39 species recognized in its first global revision (Bernhard von Hagen, "personal communication"). In contrast to *Halenia*, estimates of species numbers in other genera have increased after new revisions, such as *Fagraea*, *Macrocarpaea*, and *Symbolanthus*, highlighting the historical tendency of lumping species together in these groups. In this discussion, estimates have been made of species numbers based on most recent information from current researchers. The synonymization of species within some genera has led to the fact that the estimated numbers of species in the family has been relatively stagnant, despite the discovery of new species. Such additions have been outweighed by recent revisions that resolved taxonomic inflation and included not just one country, but the total distribution.

2.3 New Generic Placements and Additions in the Classification

In general, the classification from 2002 has held up well, and the only required only changes needed are (1) para- or polyphyletic genera that have been divided into monophyletic units, (2) a few genera classified based on morphology that since have been sequenced and changed placement on their DNA-based result, and (3) two newly described genera for newly discovered taxa. The updated generic classification is outlined in Table 2.2. These are the generic taxonomic changes since the 2002 classification (see highlights below for discussion):

- The paraphyletic genus *Sebaea* (Exaceae) has been reclassified into four monophyletic genera, namely *Exochaenium*, *Klackenbergia*, *Lagenias* and *Sebaea* sensu stricto. *Klackenbergia* is a newly described genus.
- *Cotylanthera* (Exaceae) has been included in *Exacum*.
- *Centaurium* sensu lato (Chironieae-Chironiinae) was a paraphyletic clade and was therefore split into the four monophyletic genera *Centaurium* sensu stricto, *Gyrandra*, *Schenkia*, and *Zeltnera*. Of these, *Zeltnera* is a newly described genus.
- The genus *Duplipetala* was described as a segregate of the paraphyletic *Canscora* sensu lato in Chironieae-Canscoriinae, and *Canscora* sensu stricto is still accepted.
- The genus *Metagentiana* (Gentianeae-Gentianiinae) was described as a segregate from the largest genus in the family, *Gentiana*, but turned out to be polyphyletic.
- *Wurdackanthus* (Helieae) was included in *Symbolanthus*, since it formed a paraphyletic grade toward *Symbolanthus*.
- The genus *Roraimaea* (Helieae) was described from northern South America.

- The monotypic genus *Yanomamua* (Helieae) was described from Brazil.
- The genus *Bisgoeppertia* was moved from Chironieae-Chironiinae to Potalieae-Lisianthiinae based on new molecular data.
- The genus *Fagraea* sensu lato (Potalieae-Potaliinae) is now described as five distinct genera, i.e., *Fagraea* sensu stricto, *Cyrtophyllum*, *Limahlania*, *Picrophloeus* and *Utania* based on molecular and morphological data. *Limahlania* is a newly described monotypic genus.
- The genus *Voyria* forms an independent clade placed basally in the family, and will eventually be classified as a seventh tribe.

2.4 New Biogeographic and Age Findings

Tribes and subtribes of gentians are mostly restricted to the Neotropics or Paleotropics or temperate areas worldwide, except for a few late dispersals into new territories. This pattern is even more obvious at the generic level, with very few genera present on more than one continent (Table 2.2). The center of diversity for the Gentianaceae can be measured in several ways. Species diversity reaches its maximum in East Asia, with Gentianaceae in the *Flora of China* listing 20 genera and more than 400 species just within China's borders (Ho and Pringle 1995), leading China to be probably the most species-rich country in the world with regards to gentians. However, nearly all of these species are in the tribe Gentianeae, and 248 of them are in the genus *Gentiana* (Ho and Pringle 1995).

In contrast to species diversity, the greatest evolutionary diversity occurs in South America, which is home to 47 genera (36 of which are endemic) and 5 of 6 tribes, as well as the unclassified genus *Voyria* (Albert and Struwe 2002). Saccifolieae, the most basally placed tribe in the phylogeny, is strictly from South and Central America, which supports the fact that South America is the ancestral, or at least part of the ancestral distribution, of the family. There is no strong support for China or a larger part of temperate Asia as part of the ancestral distribution of the gentian family.

Despite the high species diversity in China, many of the basal gentian clades are either absent from East Asia, or only have a few weedy, widespread species present such as *Exacum tetragonum* and *Sebaea microphylla*. This is a good example of how the number of species (i.e., 'center of species diversity') does not translate well into "center of origin" for a larger taxonomic group.

A family-wide biogeographic study analyzing the ancestral geographic origin for the gentians is still lacking, but progress has been made on biogeographical studies at tribal levels. Long distance dispersal of strongly disjunct taxa within transatlantic genera is the likely explanation for the presence of *Schultesia* on both sides of the Atlantic, but for potentially older genera, the case is less clear (*Neurotheca* and *Voyria*; Renner 2004).

Table 2.2 Updated tribal and subtribal classification of the Gentianaceae; list of genera with species numbers and general distribution

SACCIFOLIEAE
Curtia Cham. & Schltdl. (8; CA, SA)
Hockinia Gardner (1; SA)
Saccifolium Maguire & Pires (1; SA)
Tapeinostemon Benth. (7; SA)
Voyriella Miq. (1; CA, SA)
EXACEAE
Exacum L. (69; AF, AS, AU)
Exochaenium Griseb. * (22; AF)
Gentianothamnus Humbert (1; AF)
Klackenbergia Kissling * (2; AF)
Lagenias E. Mey. * (1; AF)
Ornichia Klack. (3; AF)
Sebaea Sol. ex R. Br. * (75; AF, AS, AU)
Tachiadenus Griseb. (11; AF)
CHIRONIEAE
Chironieae-Canscorinae
Canscora Lam. (9; AF, AS, AU)
Cracosna Gagnep. (3; AS)
Duplipetala Thiv * (2; AS)
Hoppea Willd. (2; AS)
Microrphium C. B. Clarke (1; AS)
Phyllocyclus Kurz (5; AS)
Schinziella Gilg (1; AF)
Chironieae-Chironiinae
Blackstonia Huds. (4; EU)
Centaurium Hill. * (20; AF, AS, AU, CA, EU, NA, SA)
Chironia L. (15; AF)
Cicendia Adans. (2; AF, AU, EU, NA, SA)
Eustoma Salisb. (3; CA, NA)
Exaculum Caruel (1; EU)
Geniostemon Engelm. & A. Gray (5; CA)
Gyrandra Griseb. * (5; CA)
Ixanthus Griseb. (1; AF)
Orphium E. Mey. (2; AF)
Sabatia Adans. (20; CA, NA)
Schenkia Griseb. * (5; AF, AS, AU, EU, PA)
Zeltnera Mansion * (25; CA, NA)
Zygostigma Griseb. (1; SA)
Chironieae-Coutoubeinae
Coutoubea Aubl. (5; CA, SA)
Deianira Cham. & Schltdl. (7; SA)
Schultesia Mart. (15; CA, SA)
Symphyllophyton Gilg (1; SA)
Xestaea Griseb. (1; CA, SA)

(continued)

Table 2.2 (continued)

GENTIANEAE
Gentianeae-Gentianinae
Crawfurdia Wall. (18; AS)
Gentiana L. (360; AF, AS, AU, CA, EU, NA, SA)
Metagentiana T. N. Ho, S. L. Chen. & S. W. Liu* (14; AS)
Tripterospermum Blume (33; AS)
Gentianeae-Swertiinae
Bartonia H. L. Mühl. ex Willd. (3; NA)
Comastoma (Wettst.) Toyok. (15; AS, EU, NA)
Frasera Walter (15; NA)
Gentianella Moench (275; AF, AS, AU, CA, EU, NA, SA)
Gentianopsis Ma (24; AS, EU, NA)
Halenia Borkh. (39; AS, CA, NA, SA)
Jaeschkea Kurz (3; AS)
Latouchea Franch. (1; AS)
Lomatogonium A. Braun (18; AS, EU, NA)
Megacodon (Hemsl.) Harry Sm. (2; AS)
Obolaria L. (1; NA)
Pterygocalyx Maxim. (1; AS)
Swertia L. (150; AF, AS, EU, NA)
Veratrilla Baill. ex Franch. (2; AS)
POTALIEAE
Potalieae-Faroinae
Congolanthus A. Raynal (1; AF)
Djaloniella P. Taylor (1; AF)
Enicostema Bl. (3; AF, AS, EU, NA)
Faroa Welw. (19; AF)
Karina Boutique (1; AF)
Neurotheca Salisb. ex Benth. (3; AF, SA)
Oreonesion A. Raynal (1; AF)
Pycnosphaera Gilg (1; AF)
Urogentias Gilg & Gilg-Ben. (1; AF)
Potalieae-Lisianthiinae
Bisgoeppertia Kuntze * (2; CA)
Lisianthius P. Browne (30; CA)
Potalieae-Potaliinae
Anthocleista R. Br. (14; AF)
Cyrtophyllum Reinw. * (5; AS)
Fagraea Thunb. * (55; AS, AU, PA)
Limahlania K.M. Wong & Sugumaran (ined.) * (1; AS)
Picrophloeus Blume * (4; AS)
Potalia Aubl. (9; CA, SA)
Utania G. Don * (15; AS)
HELIEAE
Adenolisianthus (Progel) Gilg (1; SA)
Aripuana Struwe, Maas, & V. A. Albert (1; SA)

(continued)

Table 2.2 (continued)

Calolisianthus Gilg (6; SA)
Celiantha Maguire (3; SA)
Chelonanthus Gilg (7; CA, SA)
Chorisepalum Gleason & Wodehouse (5; SA)
Helia Mart. (2; SA)
Irlbachia Mart. (9; SA)
Lagenanthus Gilg (1; SA)
Lehmanniella Gilg (2; SA)
Macrocarpaea (Griseb.) Gilg (112; CA, SA)
Neblinantha Maguire (2; SA)
Prepusa Mart. (5; SA)
Purdieanthus Gilg (1; SA)
Rogersonanthus Maguire & B. M. Boom (2; SA)
Roraimaea Struwe, Nilsson, & Albert * (2; SA)
Senaea Taub. (2; SA)
Sipapoantha Maguire & B. M. Boom (2; SA)
Symbolanthus G. Don * (37; CA, SA)
Tachia Aubl. (13; CA, SA)
Tetrapollinia Maguire & B. M. Boom (1; SA)
Yanomamua J. R. Grant, Maas & Struwe * (1; SA)
Zonanthus Griseb. (1; CA)
INCERTAE SEDIS
Voyria Aubl. (18; CA, NA, SA)

Generic changes since Struwe et al. (2002) are marked with * and discussed in the text. Estimated number of species are listed in parenthesis after each genus, as well as continental distribution (following abbreviations *AF* Africa (incl. Madagascar), *AS* Asia, *AU* Australia, *CA* Central America and Caribbean (incl. Mexico), *EU* Europe, *NA* North America (incl. Mexico), *PA* Pacific, *SA* South America. The distribution includes areas where non-native species have become naturalized

The crown-node age of the family and its major clades is not yet strongly supported, but molecular dating estimates have given some first indications. It should be remembered, however, that all dating exercises not strongly rooted in fossil records can be problematic. There are very few gentian fossils, and none from the basal clades, so calibration of internal nodes are difficult. Furthermore, fossils only provide minimum, not maximum, age estimates, so the actual ages of groups might be older, and much depends on what assumptions goes into the dating analyses.

The age of the Gentianaceae has been estimated to be about 50 million years (Yuan et al. 2003), 60 million years (Kissling 2007; Yuan et al. 2005), or in most recent estimates, slightly less than 100 million years, but with confidence intervals of 75–125 million years (Kissling, "personal communication"). In the older estimates, the family appeared to be younger than the Gondwanic continental breakup, which started about 120 million years ago, but recently the estimated family age has moved closer to a possible Gondwanic breakup of the family.

The age of the tribe Exaceae was estimated to be about 40 million years by Yuan et al. (2003) using one fossil calibration point and a nonparametric rate smoothing (NPRS) method. Two years later, Yuan et al. (2005) did a new, improved dating analysis of the whole family and its major clades/tribes, using four calibration points [minimum age of Gentianales 60 million years (Muller 1984); minimum age of *Lisianthius* 40 million years based on pollen fossil from the Eocene (Graham 1984); inferred age of Gentianeae-Swertiinae 15 million years (von Hagen and Kadereit 2001, 2002); minimum age of *Gentiana* 5 million years, from fossil seeds (Mai and Walther 1988), and a molecular clock/smoothened, penalized likelihood analysis. This gave an age estimate for Exaceae of 29–32 million years. Older and younger ages were suggested when using fewer calibration points or other analytical techniques; total age range 12–54 million years]. The most recent analysis of the age and biogeography of the family is still unpublished, but supports the fact that the Gentianaceae probably originated in the Neotropics (Merckx, "personal communication"), and that migration of tropical taxa through Laurasia during the Early Eocene may have played an important role in shaping the current global distribution patterns of the Gentianaceae.

2.5 Highlights from Each Tribe

2.5.1 Saccifolieae

No generic changes have been made in the last ten years in the classification of the neotropical tribe Saccifolieae, the most basally positioned tribe of the gentians (Fig. 2.1). It still contains five morphologically very different genera and is the most species-poor of all the tribes (Table 2.1). During the last decade, research within this group has focused on alphataxonomy, pollen morphology, and the mycoheterotrophic attributes of the achlorophyllous *Voyriella*. Current understanding and hypotheses of generic relationships in the Saccifolieae are shown in Fig. 2.2a.

Crespo and Marcondes-Ferreira (2009) revised the genus *Curtia*, a much needed treatment that clarified many species complexes, and accepted eight species. The two species *C. tenella* and *C. tenuifolia* are separated once again as different species. The heterostyly of *Curtia* (some species at least) and *Hockinia* was confirmed by Crespo and Ferreira (2006), who also investigated the variation in pollen exine morphology.

Two genera of gentians include only achlorophyllous mycoheterotrophs, *Voyria* incertae sedis and *Voyriella* (Saccifolieae), but arbuscular mycorrhiza (AM) is common in many gentians. Bidartondo et al. (2002) investigated the AM in *Voyriella* and *Voyria* and showed a high fungal/mycorrhizal species specificity. The associated fungi were shown to link with surrounding trees, indicating that these mycoheterotrophic plant species obtain carbon from surrounding trees through shared AM fungi.

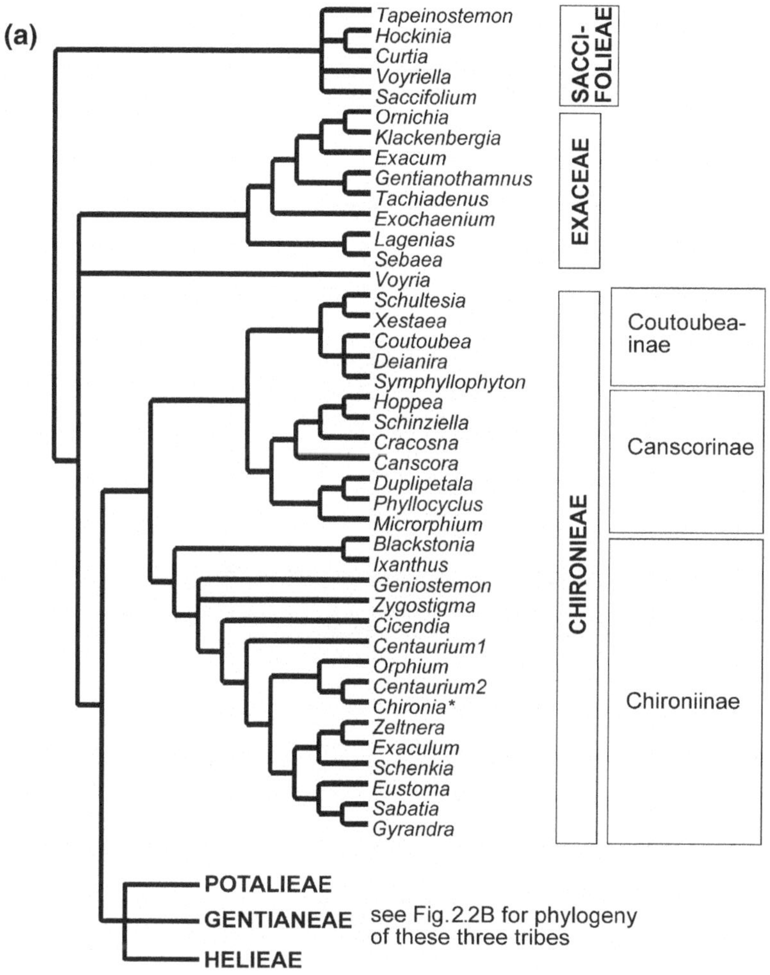

Fig. 2.2 Relationships among genera as currently understood based on studies reviewed in this chapter. This should be interpreted as a preliminary visual diagram summarizing many phylogenetic studies during the last ten years. Genera that have been shown as paraphyletic are indicated with '*' and genera placed in several clades (i.e., polyphyletic or paraphyletic) have their subclades indicated with numbers. Additional details are in the text. **a** The basal clades of Gentianaceae (Saccifolieae, Exaceae, *Voyria*, and Chironieae). **b** The sister clade to tribe Chironieae ('higher gentians'), a polytomy of tribes Helieae, Gentianeae, and Potalieae

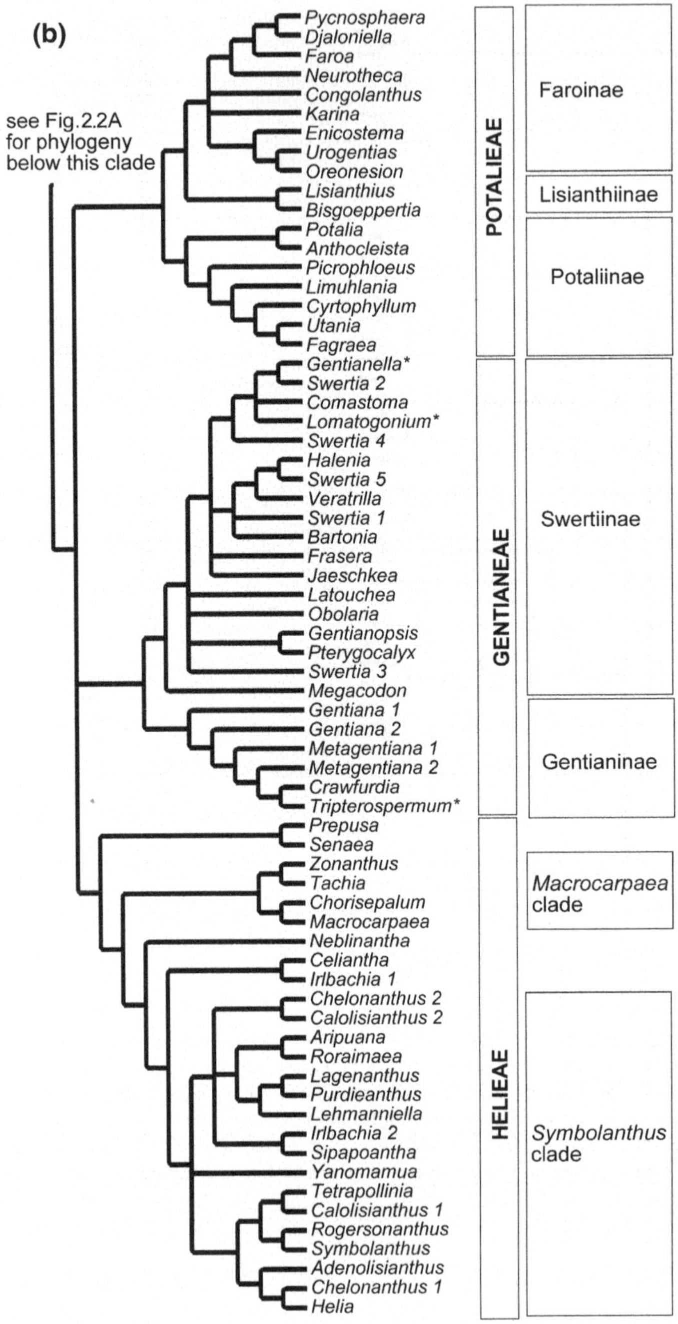

Fig. 2.2 continued

2.5.2 *Exaceae*

Extensive systematic work in the Exaceae in the last 10 years has strongly expanded taxon sampling in phylogenetic studies and clarified many species and generic relationships (Kissling et al. 2009b). The current understanding of the relationships among Exaceae genera, including the new generic circumscriptions, are shown in Fig. 2.2a, and for a generic list see Table 2.2. An updated key to all current genera of Exaceae were provided by Kissling (2012).

Sebaea turned out to be a polyphyletic genus in recent well-sampled studies (Kissling et al. 2009c; Yuan et al. 2003). This classification problem was quickly rectified through the description of a new genus, *Klackenbergia* (Kissling et al. 2009a), and the resurrection of the African genus *Exochaenium* (22 species) and the monotypic South African genus *Lagenias* (Kissling 2012). *Exochaenium* is more closely related to the *Exacum* subclade than to *Sebaea* sensu stricto, and *Exochaenium* and *Sebaea* also differ in seed testa cell anatomy. The seeds of *Exochaenium* have puzzle-shaped testa cells (as most Exaceae genera), while *Sebaea* have rectangular testa cell shapes (Kissling 2012). The rare feature of having two stigmas, one basal and one apical (diplostigmaty) in most *Sebaea* sensu stricto has also been investigated by Kissling et al. (2009c), and it was hypothesized to provide reproductive assurance.

Yuan et al. (2005) focused on the phylogeny of *Exacum*, and showed that the African *Exacum* species were nested inside a clade of Malagasy *Exacum* species, and that this African-Malagasy clade was sister to a clade formed from all Asian *Exacum* species, including one species also present in Australia. *Exacum* species from the Himalayas and Malaysia were nested inside groups present in Sri Lanka, India, Socotra, and the Arabian Peninsula, indicating a southwestern origin within Asia for this genus. It was also shown that the mycoheterotrophic genus *Cotylanthera* was a highly modified *Exacum*-clade (Yuan et al. 2005), so Klackenberg (2006) placed *Cotylanthera* and its four species into *Exacum*.

Cytological investigations of a large number of Exaceae species have shown that dysploidy and polyploidy are recurrent and common evolutionary phenomena within this tribe (Kissling et al. 2008).

2.5.3 *Chironieae*

The tribe Chironieae forms three monophyletic subclades, which are classified as subtribes Canscorinae (paleotropical), Chironiinae (world wide), and Coutoubeinae (neotropical; Fig. 2.1). Detailed generic relationships of Chironieae are shown in Fig. 2.2a, and a list of all genera is provided in Table 2.2. In order to solve the problem with the paraphyletic *Canscora* in the subtribe Canscorinae, Thiv and Kadereit (2002) segregated out the new genus *Duplipetala* from *Canscora*. A new revision of all species in all seven genera in the subtribe was published by Thiv (2003).

It was already apparent in 2002 that the subtribe Chironiinae was suffering from the "large paraphyletic-genus" problem, with *Centaurium* occurring in several different clades. Updated phylogenetic hypothesis of the subtribe Chironiinae was presented by Mansion and Struwe (2004) and Mansion et al. (2005), confirming *Blackstonia* and *Ixanthus* as the most basally positioned genera in this clade, and providing good branch support for a new generic-level classification of *Centaurium*. Accordingly, Mansion (2004) resurrected two old genus names (*Gyrandra* and *Schenkia*) and named a *Centaurium* clade as the new genus *Zeltnera*, in honor of *Centaurium* specialist and cytologist Louis Zeltner and his wife Nicole Zeltner. *Gyrandra* contains five species distributed in Mexico and Central America. *Schenkia* also has five species, but the genus is highly disjunct between endemics of Australia, Hawaii and other Pacific islands, Eurasia, and northern Africa. *Zeltnera* is strictly New World in its distribution (Mexico and the US states of Texas and California) and includes about 25 species. Left in *Centaurium* sensu stricto are about 20 species centered around the Mediterranean, but with a few widespread weedy species worldwide (Mansion 2004).

The genus *Bisgoeppertia* was placed tentatively in Chironieae-Chironiinae based only on morphological data in the 2002 classification. Sequencing of newly collected material by Thiv showed that it was clearly affiliated with *Lisianthius* in Potalieae, so it has now been moved to the subtribe Lisianthiinae in the Potalieae (Molina and Struwe 2009; Thiv 2002).

The subtribe Coutoubeinae has the unusual characteristic of having pollen in tetrads (otherwise only known from a few Helieae genera and one Potalieae genus) and, curiously, all tetrad-bearing gentians are found in Latin America. Several studies have focused on the vegetative morphology (Delgado et al. 2009) and seed morphology (Guimarães et al. 2007) of this group, and also confirmed the presence of AM. *Schultesia* has also been under further taxonomic scrutiny and several new species have been discovered (Guimarães 2004; Guimarães and Fontella-Pereira 2001; Guimarães et al. 2003). It is remarkable how many new species are being described from South America in general, indicating that the biodiversity of this continent is still very poorly understood. Whether or not *Xestaea lisianthoides* should be separated from *Schultesia* as a monophyletic genus, as argued in Struwe et al. (2002), is still under discussion and needs further study.

2.5.4 Gentianeae

The widely distributed tribe Gentianeae has received considerable attention from researchers in the last decade. Current hypotheses of generic relationships are summarized in Fig. 2.2b, and Table 2.2 includes a list of all genera.

Large taxonomic problems have been associated with the para- and polyphyly of *Gentianella* and *Swertia* (subtribe Swertiinae), nested among smaller, more well-defined genera such as *Comastoma*, *Frasera*, *Gentianopsis*, and *Halenia* (Chassot et al. 2001). Neither *Gentianella* nor *Swertia* have been reclassified into smaller monophyletic genera yet, so this is still a taxonomic problem to resolve when additional data is available on phylogenetic relationships and morphological character distributions supporting specific clades.

Kadereit and von Hagen (2003) and von Hagen and Kadereit (2003) showed that traditionally used floral characters show high homoplasy within the subtribe Swertiinae. Consequently, new characters would have to be sought as potential apomorphies for phylogenetically supported groups. The highly variable pollen morphology of *Swertia* was discussed by Chassot and von Hagen (2008) in a phylogenetic context, and showed that exine surface patterns might provide excellent characters for the definitions of several clades of *Swertia*. Seed coat morphology is also variable, as exemplified in *Gentianopsis* (Whitlock et al. 2010).

The highly diverse *Gentianella* of New Zealand, a monophyletic clade within the genus, was revised by Glenny (2004). During his revision of *Halenia* for *Flora Neotropica*, several new species were described by von Hagen (2007).

Bartonia and *Obolaria*, green but only barely so, are relatively basally positioned within Swertiinae, and their at least partially mycoheterotrophic status, long suspected and assumed, was confirmed recently by Cameron and Bolin (2010). The phylogenetic relationships of the initially four, now three, accepted *Bartonia* species were also investigated by Mathews et al. (2009), who provided a new key and species descriptions.

The subtribe Gentianinae contained the genera *Gentiana*, *Crawfurdia*, and *Tripterospermum* in 2002. No new genera or species have been moved into the subtribe, but the correct status and relationships among, and within, these genera are still unresolved (Chen et al. 2005; Favre et al. 2010; Zhang et al. 2009). *Gentiana* section *Stenogyne* was separated as the new genus *Metagentiana* by Ho and Chen (2002), and this genus is accepted by Asian contemporary botanists (Chen et al. 2008). There was new evidence that all four genera of the subtribe Gentianinae might be polyphyletic (Chen et al. 2005), which could soon led to a large reclassification of many species and redefinition of generic boundaries. However, a recent phylogenetic study with larger taxon sampling by Favre et al. (2010) showed that *Tripterospermum* and *Crawfurdia* were definitely monophyletic, but *Metagentiana* was still found to be polyphyletic and positioned in three clades. There is also uncertainty about the correct phylogenetic position of *Gentiana* section *Otophora*. No new generic combinations and circumscriptions have been proposed to date.

2.5.5 *Helieae*

The phylogeny and molecular and morphological evolution of the genera Helieae were investigated in detail by Struwe et al. (2009a), which confirmed and expanded evolutionary patterns found in the 2002 analysis. A list of the current genera is shown in Table 2.2, and our current understanding of evolutionary relationships is shown Fig. 2.2b. Southeast Brazilian *Prepusa* and *Senaea* have been placed consistently together as a clade sister to the rest of the Helieae (Struwe et al. 2009a). The remaining Helieae is informally divided into two major subclades, the *Macrocarpaea*-clade (*Chorisepalum*, *Macrocarpaea*, *Tachia*, and *Zonanthus*) and the *Symbolanthus*-clade (*Calolisianthus*, *Chelonanthus*, *Symbolanthus*). The phylogenetic positions of some genera (*Irlbachia* sensu stricto) are still unclear, and they might be positioned between the two subclades and the clade formed by *Prepusa* plus *Senaea* at the base.

Prepusa and *Senaea* were revised by Calió et al. (2008), and the biogeography and ecological vicariance patterns of *Prepusa* were investigated using the new method of spatial evolutionary and ecological vicariance analysis (SEEVA) by Struwe et al. (2011). The same method was also used to compare sister species patterns, biogeography, and speciation in Andean *Macrocarpaea* (Struwe et al. 2009b), highlighting ecological niche conservatism and dispersal. It was shown that long distance dispersal is a rather rare phenomenon in Andean *Macrocarpaea*.

No generic changes have been made in the *Macrocarpaea* subclade, but both *Tachia* and *Macrocarpaea* have been revised. Within *Tachia*, two new species were described and a revision is finished (Struwe et al. 2005; Struwe and Kinkade, 2014 (in press)). A trilingual field guide to *Tachia* was also published, and is freely available (Peters et al. 2004). Efforts with the taxonomy and phylogeny of the large genus *Macrocarpaea* continued with the publication of species, seed morphology and palynology (Grant 2004, 2005, 2007, 2009, 2011; Grant and Struwe 2003). The wood anatomy of *Macrocarpaea* showed interesting characters, suggesting a secondary derivation of woodiness within Gentianaceae (Carlquist and Grant 2005).

The *Symbolanthus* subclade has received considerable research attention, largely through the work of M. F. Calió at the University of São Paulo, and K. Lepis at Rutgers University. It was known already that a chaotic taxonomy was in use, especially related to species classified as *Calolisianthus*, *Chelonanthus*, *Helia*, *Rogersonanthus*, and *Wurdackanthus*. Through detailed alpha taxonomic and phylogenetic work, species and generic relationships have now been established (Lepis 2009; Struwe et al. 2009a). An update to the generic classification which identifies monophyletic clades as separate genera still has to be published, but is in the works (Calió et al., "personal communication"), and is incorporated into the phylogenetic relationships shown in Fig. 2.2b. There has also been new detailed anatomical work in this clade, as exemplified by studies of leaf anatomy by Delgado et al. (2011a, b) and self-pollination by Freitas and Sazima (2009).

Gould and Struwe (2004) showed that the genus *Wurdackanthus* (two species) formed a paraphyletic grade towards *Symbolanthus*, and it was therefore incorporated into *Symbolanthus* (Struwe and Gould 2004). A revision of *Symbolanthus* is ongoing, and will be published in geographically defined parts (Molina and Struwe 2008; Struwe 2003a, b). This includes the resurrection of many previously unaccepted species, as well as the description of many new morphologically well-defined species. In fact, the number of accepted species in this genus has increased from 3 to more than 30 in about 20 years.

Additionally, two new genera have been discovered in Helieae, both in the *Symbolanthus* subclade. The monotypic new genus *Yanomamua* from Brazil is a semi-woody species with very unusual lyrate leaves (Grant et al. 2006). The genus *Roraimaea* is based on a new orange-flowered Helieae species, and also includes the previously published *Roraimaea* (*Rogersonanthus*) *coccineus* (Struwe et al. 2008).

2.5.6 Potalieae

The pantropical tribe Potalieae and its three subtribes contain a fascinating diversity of habits (large timber trees to small, likely mycoheterotrophic herbs), corolla merosities (from 3 to 12-merous) and flower sizes (from a few mm to about 30 cm in diameter). A list of included, currently accepted genera is shown in Table 2.2, and a genus level estimate of phylogenetic relationships is shown in Fig. 2.2b.

Potalieae's phylogenetic relationships were investigated using internal transcriber spacer (ITS) data and the first DNA-alignment in the family based on secondary structure by Molina and Struwe (2009; for overall phylogeny see Fig. 2.2b). The study showed the conserved regions in the secondary structure of ITS and how these formed additional synapomorphies for certain clades within the Potalieae. The two subtribes, Lisianthiinae and Potaliinae, were well-sampled in this study and were supported as monophyletic, but the more poorly sampled subtribe Faroinae received no support for monophyly, possibly due to a long branch for *Neurotheca* (which would fit with it also being possibly mycoheterotrophic; Molina and Struwe 2009). Additional studies of the Faroinae, including all genera and many more species, are needed to fully understand evolution in this part of the Gentianaceae.

Potalia, the neotropical representative of the subtribe Potaliinae, was revised by Struwe and Albert (2004). This genus currently contains 11 species, of which nine were published in 2004. Its biogeographical history is very interesting, showing how lowland white sand areas are ancestral habitats within the genus, and that species on more recent, nutrient-rich and high-pH soils are derived from white sand-inhabiting ancestors (Frasier et al. 2008).

Also in the Potaliinae, the taxonomy of the species-rich and morphologically diverse genus *Fagraea* has been revised, leading to the resurrection of three genera, *Cyrtophyllum*, *Picrophloeus* and *Utania* for basally positioned clades in the genus, and the description of one new genus, *Limahlania* (Sugumaran and Wong 2012; Wong and Sugumaran 2012a, b). The revisions of *Fagraea* for peninsular Malaysia and Borneo have also led to the reacceptance of species previously considered as synonyms by Leenhouts (1962) in *Flora Malesiana*, and the discovery of many new species (Sugumaran 2010; Wong and Sugau 1996; Wong and Sugumaran 2012b). Motley (2004) also published the first treatment of the ancient and diverse ethnobotanical uses of *Fagraea* sensu lato.

As mentioned under the Chironieae, *Bisgoeppertia* was classified earlier in Chironieae-Chironiinae, but is now included in Potalieae-Lisianthiinae (Molina et al. 2009; Thiv 2002). More details on this rare taxon are found in Greuter and Rankin (2008).

2.5.7 *Incertae sedis (*Voyria*)*

The taxonomic and phylogenetic position of the enigmatic and beautiful genus *Voyria* has continued to baffle gentian taxonomists. New sequencing of nuclear and mitochondrial chloroplast genes have helped elucidate the correct evolutionary position, and recent multi-gene results suggest Voyria is a separate, monophyletic and isolated clade situated among the basal Exaceae and Saccifolieae tribes (Merckx et al. 2013). Its origin was likely in the Neotropics during the earlier part of the Eocene and it dispersed across the Atlantic to Africa at the end of the Oligocene.

Simultaneously, researchers have been investigating the mycoheterotrophic details of *Voyria* (Bidartondo et al. 2002; Courty et al. 2011; Franke 2002; Franke et al. 2006; Merckx et al. 2010). A pollen feature unique within the gentians has also been found in this genus. Pollen grains transported as one unit (through the presence of germinating pollen grains holding many grains together) was found in French Guianan *Voyria* species, as well as self-pollination and possible rodent dispersal of seeds (Hentrich et al. 2010).

2.6 Conclusions

As circumscribed here, the family Gentianaceae includes 99 genera and approximately 1,736 species distributed in seven tribes. There are still some uncertainties when it comes to some genera and their correct tribal or subtribal classification, but in general there is excellent knowledge of the overall phylogeny within the family and the tribal classification is expected to remain stable. As for genera, most appear to be monophyletic after some arranging in the last decade, but there should be reclassifications affecting poly- or paraphyletic genera such as *Chelonanthus*, *Gentiana* (in the broad sense), *Gentianella* and *Swertia*. The naming and identification of monophyletic clades within these genera will improve further the classification and make it both more practically useful in the field, as well as being more taxonomically and evolutionary correct. The next step in large-scale gentian pattern analysis will elucidate both the biogeography and the age of the family. Very little is known about this, and detailed and broad studies are needed urgently to explain not just how the different gentians are related, but where and when they evolved, why they have their current morphology, and how and when they moved to all continents. This will be an interesting challenge.

Acknowledgments Special thanks to gentian researchers Fernanda Calió, Adrian Favre, Jason R. Grant, Joseph Greimler, Bernhard von Hagen, Jonathan Kissling, Katherine Lepis, Sugumaran Manickam, Guilhem Mansion, Katherine G. Mathews, Vincent Merckx, James Pringle, and Mike Thiv, all of whom provided important data as well as useful commentary for this chapter. Grant support for this research was provided by USDA-Hatch (NJAES 17112).

References

Albert VA, Struwe L (2002) Gentianaceae in context. In: Struwe L, Albert VA (eds) Gentianaceae: systematics and natural history. Cambridge University Press, Cambridge, pp 1–20

Bidartondo MI, Redecker D, Hijri I, Wiemken A, Bruns TD, Dominguez L, Sersic A, Leake JR, Read DJ (2002) Epiparasitic plants specialized on arbuscular mycorrhizal fungi. Nature 419:389–392

Bouman F, Cobb L, Devente N, Goethals V, Maas PJM, Smets E (2002) The seeds of Gentianaceae. In: Struwe L, Albert VA (eds) Gentianaceae: systematics and natural history. Cambridge University Press, Cambridge, pp 498–572

Calió MF, Pirani JR, Struwe L (2008) Morphology-based phylogeny and revision of *Prepusa* and *Senaea* (Gentianaceae: Helieae)-rare endemics from eastern Brazil. Kew Bull 63:169–191

Cameron DD, Bolin JF (2010) Isotopic evidence of partial mycoheterotrophy in the Gentianaceae: *Bartonia virginica* and *Obolaria virginica* as case studies. Amer J Bot 97:1272–1277

Carlquist S, Grant JR (2005) Wood anatomy of Gentianaceae, tribe Helieae, in relation to ecology, habit, systematics, and sample diameter. Brittonia 57:276–291

Chassot P, von Hagen KB (2008) Pollen morphology of the Swertiinae (Gentianaceae): phylogenetic implications. Botl J Linn Soc 157:323–341

Chassot P, Nemomissa S, Yuan Y-M, Kupfer P (2001) High paraphyly of *Swertia* L. (Gentianaceae) in the *Gentianella*-lineage as revealed by nuclear and chloroplast DNA sequence variation. Plant System Evol 229:1–21

Chen SY, Xia T, Wang YJ, Liu JQ, Chen SL (2005) Molecular systematics and biogeography of *Crawfurdia*, *Metagentiana* and *Tripterospermum* (Gentianaceae) based on nuclear ribosomal and plastid DNA sequences. Ann Bot 96:413–424. doi:10.1093/aob/mci188

Chen SY, Wu GL, Zhang DJ, Gao QB, Duan YZ, Zhang FQ, Chen SL (2008) Molecular phylogeography of alpine plant *Metagentiana striata* (Gentianaceae). J Syst Evol 46:573–585. doi:10.3724/sp.j.1002.2008.07097

Courty P-E, Walder F, Boller T, Ineichen K, Wiemken A, Rousteau A, Selosse M-A (2011) Carbon and nitrogen metabolism in mycorrhizal networks and mycoheterotrophic plants of tropical forests: a stable isotope analysis. Plant Physiol 156:952–961

Crespo SRDM, Ferreira WM (2006) Morfologia polínica dos gêneros *Curtia* Cham. & Schltdl. e *Hockinia* Gardner (Gentianaceae Juss.). Acta Bot Brasil 20:273–284

Crespo SRDM, Marcondes-Ferreira W (2009) Revisão taxonômica do gênero *Curtia* (Gentianaceae). Rodriguésia 60:423–444

Delgado MN, Azevedo AA, Valente GE, Kasuya MCM (2009) Morfo-anatomia comparada de espécies da subtribo Coutoubeinae (Chironieae-Gentianaceae). Acta Bot Brasil 23:956–967. http://dx.doi.org/10.1590/S0102-33062009000400006

Delgado MN, Azevedo AA, Silva LC, Valente GE, Kasuya MCM (2011a) Comparative anatomy of *Calolisianthus* species (Gentianaceae–Helieae) from Brazil: taxonomic aspects. Edinburgh J Bot 68:139–155. doi:10.1017/S0960428610000284

Delgado MN, Da Silva LC, Báo SN, Morais HC, Azevedo AA (2011b) Distribution, structural and ecological aspects of the unusual leaf nectaries of *Calolisianthus* species (Gentianaceae). Flora 206:676–683. doi:10.1016/j.flora.2010.11.016

Favre A, Yuan Y-M, Küpfer P, Alvarez N (2010) Phylogeny of subtribe Gentianinae (Gentianaceae): biogeographic inferences despite limitations in temporal calibration points. Taxon 59:1701–1711

Franke T (2002) The myco-heterotrophic *Voyria flavescens* (Gentianaceae) and its associated fungus. Mycol Prog 1:367–376. doi:10.1007/s11557-006-0033-8

Franke T, Beenken L, Döring M, Kocyan A, Agerer R (2006) Arbuscular mycorrhizal fungi of the *Glomus*-group A lineage (Glomerales; Glomeromycota) detected in myco-heterotrophic plants from tropical Africa. Mycol Prog 5:24–31. doi:10.1007/s11557-006-0500-2

Frasier C, Albert VA, Struwe L (2008) Amazonian lowland, white sand areas as ancestral regions for South American biodiversity: biogeographic and phylogenetic patterns in *Potalia* (Angiospermae: Gentianaceae). Org Div Evol 8:44–57

Freitas L, Sazima M (2009) Floral biology and mechanisms of spontaneous self-pollination in five neotropical species of Gentianaceae. Bot J Linn Soc 160:357–368. doi:10.1111/j.1095-8339.2009.00989.x

Gilg E (1895) Gentianaceae. In: Engler A, Prantl K (eds) Die Natürlichen Pflanzenfamilien, vol 4(2). W. Engelmann, Leipzig, pp 50–108

Glenny D (2004) A revision of the genus *Gentianella* in New Zealand. PhD thesis. University of Canterbury, Christchurch, New Zealand

Gould KR, Struwe L (2004) Phylogeny and evolution of *Symbolanthus* and *Wurdackanthus* (Gentianacea-Helieae) in the Guayana Highlands and Andes, based on ribosomal 5S-NTS sequences. Ann Miss Bot Gard 91:438–446

Graham A (1984) *Lisianthius* pollen from the Eocene of Panama. Ann Miss Bot Gard 71:987–993

Grant JR (2004) De Macrocarpaeae Grisebach (ex Gentianaceis) speciebus novis V: twenty-three new species largely from Peru, and typification of all species in the genus. Harvard Pap Bot 9:11–49

Grant JR (2005) De Macrocarpaeae Grisebach (ex Gentianaceis) speciebus novis VI: seed morphology, palynology, an infrageneric classification, and another twenty-three new species largely from Colombia. Harvard Pap Bot 9:305–342

Grant JR (2007) De Macrocarpaeae Grisebach (ex Gentianaceis) speciebus novis II: typification of the Ruiz and Pavon names. Harvard Pap Bot 7:423–436

Grant JR (2009) De Macrocarpaeae Grisebach (ex Gentianaceis) speciebus novis VII: four new species and two natural hybrids. Harvard Pap Bot 11:129–139

Grant JR (2011) De Macrocarpaeae Grisebach (ex Gentianaceis) speciebus novis IX: a synopsis of the genus in Bolivia. Harvard Pap Bot 16:389–397

Grant JR, Struwe L (2003) De Macrocarpaeae Grisebach (ex Gentianaceis) Speciebus Novis III: six new species of moon-gentians (Macrocarpaea, Gentianaceae: Helieae) from Parque Nacional Podocarpus, Ecuador. Harvard Pap Bot 8:61–81

Grant JR, Maas PJM, Struwe L (2006) *Yanomamua araca* (Gentianaceae), a new genus and species from Serra do Araçá, an outlier of the Guayana region in Amazonas state, Brazil. Harvard Pap Bot 11:29–37

Greuter W, Rankin R (2008) *Bisgoeppertia* (Gentianaceae) unravelled. Account of a small genus of the Greater Antilles. Willdenowia 38:77–185

Grisebach AR (1839) Genera et Species Gentianearum Observationibus Quibusdam Phytogeographicis. J. G. Cotta, Stuttgart

Guimarães EF (2004) Novos sinônimos para espécies de *Schultesia* Mart. e *Xestea* Griseb. (Gentianaceae). Rodriguésia 55:67–72

Guimarães EF, Fontella-Pereira J (2001) Estudos em Gentianaceae II - Espécies novas para o gênero *Schultesia* Mart. Bradea 8:59–263

Guimarães EF, Mendonça CBF, Gonçalves-Esteves V, Pereira JF (2003) Palinotaxonomia de espécies de *Schultesia* Mart. Gentianaceae Juss. Arquivos Museu Nacional Rio de Janeiro 61:151–164

Guimarães EF, Saavedra MM, Costa CG (2007) Frutos e sementes em *Schultesia* Mart. e *Xestaea* Griseb. (Gentianaceae). Acta Bot Brasil 21:309–323

Hentrich H, Kaiser R, Gottsberger G (2010) The reproductive biology of *Voyria* (Gentianaceae) species in French Guiana. Taxon 59:867–880
Ho T, Chen S (2002) *Metagentiana*, a new genus of Gentianaceae. Bot Bull Academia 43:83–91
Ho T-N, Pringle JS (1995) Gentianaceae. In: Wu ZY, Raven PH (eds) Flora of China, vol 16. Science Press, Beijing, and Missouri Botanic Garden Press, St. Louis, pp 1–139
Jensen SR, Shripsema J (2002) Chemotaxonomy and pharmacology of Gentianaceae. In: Struwe L, Albert VA (eds) Gentianaceae: systematics and natural history. Cambridge University Press, Cambridge, pp 573–632
Kadereit JW, von Hagen KB (2003) The evolution of flower morphology in Gentianaceae-Swertiinae and the roles of key innovations and niche width for the diversification of *Gentianella* and *Halenia* in South America. Interl J Plant Sci 164:S441–S452
Kissling J (2007) Phylogenetics of tribe Exaceae (Gentianaceae) based on molecular, morphological and karyological data, with special emphasis on the genus *Sebaea*. Taxonomic treatment of *Exochaenium*, *Lagenias* and the new genus *Klackenbergia*. PhD thesis. Université de Neuchâtel, Neuchâtel, Switzerland
Kissling J (2012) Taxonomy of *Exochaenium* and *Lagenias*: two resurrected genera of tribe Exaceae (Gentianaceae). Syst Bot 37:238–253
Kissling J, Zeltner L, Küpfer P, Mansion G (2008) Cytogeography of Gentianaceae–Exaceae in Africa, with a special focus on *Sebaea*: the possible role of dysploidy and polyploidy in the evolution of the tribe. Bot J Linn Soc 158:556–566
Kissling J, Buerki S, Mansion G (2009a) *Klackenbergia* (Gentianaceae–Exaceae), a new endemic genus from Madagascar. Taxon 58:907–912
Kissling J, Yuan Y-M, Küpfer P, Mansion G (2009b) The polyphyletic genus *Sebaea* (Gentianaceae): a step forward in understanding the morphological and karyological evolution of the Exaceae. Mol Phylog Evol 53:734–748
Kissling J, Endress PK, Bernasconi G (2009c) Ancestral and monophyletic presence of diplostigmaty in *Sebaea* (Gentianaceae) and its potential role as a morphological mixed mating strategy. New Phytol 184:303–310
Klackenberg J (2006) *Cotylanthera* transferred to *Exacum* (Gentianaceae). Botanische Jahrbücher 126:477–481
Lepis KB (2009) Evolution and systematics of *Chelonanthus* (Gentianaceae). PhD thesis, Rutgers University, New Brunswick
Leenhouts PW (1962) Loganiaceae. In: van Steenis CGGJ (ed) Flora Malesiana, ser 1, vol 6(2). Wolters-Noordhoff, Groningen
Mai DH, Walther H (1988) Die pliozaenen floren von Thueringen, Deutsche Demokratische Republik. Quartaerpalaeontologie 7:55–297
Mansion G (2004) A new classification of the polyphyletic genus *Centaurium* Hill (Chironiinae, Gentianaceae): description of the New World endemic *Zeltnera*, and reinstatement of *Gyrandra* Griseb. and *Schenkia* Griseb. Taxon 53:719–740
Mansion G, Struwe L (2004) Generic delimitation and phylogenetic relationships within the subtribe Chironiinae (Chironieae: Gentianaceae), with special reference to *Centaurium*: evidence from nrDNA and cpDNA sequences. Mol Phyl Evol 32:951–977
Mansion G, Zeltner L, Bretagnolle F (2005) Phylogenetic patterns and polyploid evolution within the Mediterranean genus *Centaurium* (Gentianaceae–Chironieae). Taxon 54:931–950
Mathews KG, Dunne N, York E, Struwe L (2009) A phylogenetic analysis and taxonomic revision of *Bartonia* (Gentianaceae: Gentianeae), based on molecular and morphological evidence. Syst Bot 34:162–172
Merckx V, Stöckel M, Fleischmann A, Bruns TD, Gebauer G (2010) 15N and 13C natural abundance of two mycoheterotrophic and a putative partially mycoheterotrophic species associated with arbuscular mycorrhizal fungi. New Phytol 188:590–596
Merckx VS, Kissling J, Hentrich H, Janssens SB, Mennes CB, Specht CD, Smets EF (2013) Phylogenetic relationships of the mycoheterotrophic genus *Voyria* and the implications for the biogeographic history of Gentianaceae. Am J Bot. 100(4):712-21. doi:10.3732/ajb.1200330

Mészáros S, De Laet J, Goethals V, Smets E, Nilsson S (2002) Cladistics of the Gentianaceae: a morphological approach. In: Struwe L, Albert VA (eds) Gentianaceae: systematics and natural history. Cambridge University Press, Cambridge, pp 310–376
Molina J, Struwe L (2008) Revision of ring-gentians (*Symbolanthus*, Gentianaceae) from Bolivia, Ecuador and Peru, with a first assessment of conservation status. Syst Biodiv 6:477–501
Molina J, Struwe L (2009) Utility of secondary structure in phylogenetic reconstructions using nrDNA ITS sequences—an example from Potalieae (Gentianaceae: Asteridae). Syst Bot 34:414–428
Motley TJ (2004) The ethnobotany of *Fagraea* Thunb. (Gentianaceae): the timber of Malesia and the scent of Polynesia. Econ Bot 58:396–409
Muller J (1984) Significance of fossil pollen for angiosperm history. Ann Miss Bot Gard 71:419–443
Nilsson S (2002) Gentianaceae—a review of palynology. In: Struwe L, Albert VA (eds) Gentianaceae: systematics and natural history. Cambridge University Press, Cambridge, pp 377–497
Renner S (2004) Plant dispersal across the tropical Atlantic by wind and sea currents. Interl J Plant Sci 165:S23–S33
Peters W, Kinkade MP, Pohlit AM, Struwe L (2004) *Tachia* field guide. Rutgers University, New Brunswick. http://gentian.rutgers.edu/reffiles/Tachia_brochure_English.pdf. Cited 11 October 2012
Struwe L (2003a) Revision of Bolivian *Symbolanthus* (Gentianaceae–Helieae). Harvard Pap Bot 8:19–24
Struwe L (2003b) Two new winged species of *Symbolanthus* (Gentianaceae: Helieae) from Colombia. Novon 13:133–140. doi:10.2307/3393578
Struwe L, Albert VA (2002) Gentianaceae—systematics and natural history. Cambridge University Press, Cambridge
Struwe L, Albert VA (2004) A monograph of neotropical *Potalia* Aublet (Gentianaceae: Potalieae). Syst Bot 29:670–701
Struwe L, Gould KR (2004) Redefinition of *Symbolanthus* to include *Wurdackanthus* (Gentianaceae–Helieae). Novon 14:354–359
Struwe L, Kadereit JW, Klackenberg J, Nilsson S, Thiv M, von Hagen KB, Albert VA (2002) Systematics, character evolution, and biogeography of Gentianaceae, including a new tribal and subtribal classification. In: Struwe L, Albert VA (eds) Gentianaceae: systematics and natural history. Cambridge University Press, Cambridge, pp 21–309
Struwe L, Kinkade MP, Maas PJM (2005) Two new Brazilian species of *Tachia* (Gentianaceae: Helieae). Blumea 50:457–462
Struwe L, Nilsson S, Albert VA (2008) *Roraimaea* (Gentianaceae: Helieae)—a new gentian genus from white sand campinas and Cerro de la Neblina of Brazil and Venezuela. Harvard Pap Bot 13:35–45
Struwe L, Albert VA, Calió MF, Frasier C, Lepis KB, Mathews KG, Grant JR (2009a) Evolutionary patterns in neotropical Helieae (Gentianaceae): evidence from morphology, chloroplast and nuclear DNA sequences. Taxon 58:479–499
Struwe L, Haag S, Heiberg E, Grant JR (2009b) Andean speciation and vicariance in neotropical *Macrocarpaea* (Gentianaceae–Helieae). Ann Miss Bot Gard 96:450–469. doi:10.3417/2008040
Struwe L, Kinkade M (2014). Revision of Tachia (Gentianaceae: Helieae). Systematic Botany (in press)
Struwe L, Smouse PE, Heiberg E, Haag S, Lathrop RG (2011) Spatial evolutionary and ecological vicariance analysis (SEEVA), a novel approach to biogeography and speciation research, with an example from Brazilian Gentianaceae. J Biogeogra 38:1841–1854. doi:10.1111/j.1365-2699.2011.02532.x
Sugumaran M (2010) A systematic study of *Fagraea* sensu lato and a revision of the Peninsular Malaysian taxa. PhD Thesis, University of Malaya, Kuala Lumpur, Malaysia

Sugumaran M, Wong KM (2012) Studies in Malesian Gentianaceae, I. *Fagraea* sensu lato: complex genus or several genera? A molecular phylogenetic study. Gardens' Bull Singapore 64:301–332

Thiv M (2002) Gentianaceae. In: Greuter W, Rodriguez RR, Manitz H (eds) Flora de la Republica de Cuba, vol 6. Koeltz, Königstein, pp 1–40

Thiv M (2003) A taxonomic revision of *Canscora, Cracosna, Duplipetala, Hoppea, Microrphium, Phyllocyclus* and *Schinziella* (Gentianaceae-Canscorinae). Blumea 48:1–46

Thiv M, Kadereit JW (2002) A morphological cladistic analysis of Gentianaceae-Canscorinae and the evolution of anisomorphic androecia in the subtribe. Syst Bot 27:780–788

von Hagen KB (2007) Description of new taxa of *Halenia* Borkh. (Gentianaceae) from Colombia and Venezuela with significance for testing a key innovation hypothesis. Org Divers Evol 7:1–11. doi:10.1016/j.ode.2005.11.003

von Hagen KB, Kadereit JW (2001) The phylogeny of *Gentianella* (Gentianaceae) and its colonization of the southern hemisphere as revealed by nuclear and chloroplast DNA sequence variation. Org Divers Evol 1:61–79

von Hagen KB, Kadereit JW (2002) Phylogeny and flower evolution of the Swertiinae (Gentianaceae–Gentianeae): homoplasy and the principle of variable proportions. Syst Bot 27:548–572

von Hagen KB, Kadereit JW (2003) The diversification of *Halenia* (Gentianaceae): ecological opportunity versus key innovation. Evolution 57:2507–2518

Whitlock B, Silver J, Prince JS (2010) Seed coat morphology in *Gentianopsis* (Gentianaceae). Rhodora 112:58–79

Wong KM (2012) Studies in Malesian Gentianaceae II: a revision of *Picrophloeus*. Gardens' Bull Singapore 64:511–522

Wong KM, Sugau JB (1996) A revision of *Fagraea* (Loganiaceae) in Borneo, with notes on related Malesian species and 21 new species. Sandakania 8:1–93

Wong KM, Sugumaran M (2012a) Studies in Malesian Gentianaceae II: a taxonomic framework for the *Fagraea* complex, including the new genus *Limahlania*. Gardens' Bull Singapore 64:481–495

Wong KM, Sugumaran M (2012b) Studies in Malesian Gentianaceae III. *Cyrtophyllum* reapplied to the *Fagraea fragrans* alliance. Gardens' Bull Singapore 64:497–510

Yuan Y-M, Wohlhauser S, Möller M, Chassot P, Mansion G, Grant J, Küpfer P, Klackenberg J (2003) Monophyly and relationships of the tribe Exaceae (Gentianaceae) inferred from nuclear ribosomal and chloroplast DNA sequences. Mol Phylogene Evol 28:500–517. doi:10.1016/S1055-7903(03)00068-X

Yuan Y-M, Wohlhauser S, Möller M, Klackenberg J, Callmander M, Küpfer P (2005) Phylogeny and biogeography of *Exacum* (Gentianaceae): a disjunctive distribution in the Indian Ocean Basin resulted from long distance dispersal and extensive radiation. Syst Biol 54:21–34. doi:10.1080/10635150590905867

Zhang X-L, Wang Y-J, Ge X-J, Yuan Y-M, Yang H-L, Liu J-Q (2009) Molecular phylogeny and biogeography of *Gentiana* sect. *Cruciata*. Taxon 58:862–870

Chapter 3
A Monographic Revision of the Neotropical Genus *Macrocarpaea* (Gentianaceae) in Ecuador

Jason R. Grant

Abstract The genus *Macrocarpaea* is a group of woody Gentianaceae that occurs in mountainous regions of the Neotropics. The majority of species occur in the Andes, with a high percentage in the Amotape–Huancabamba zone that straddles southern Ecuador and northern Peru. Rapid diversification in several species groups in this area has led to the formation of numerous isolated and sometimes cryptic species. Here, a revision of the species that occur in Ecuador is presented, where 35 taxa (34 species, one natural hybrid) are recognized, including 28 published taxa *M. angelliae* J.R. Grant & Struwe, *M. apparata* J.R. Grant & Struwe, *M. arborescens* Gilg, *M. bubops* J.R. Grant & Struwe, *M. claireae* J.R. Grant, *M. dies-viridis* J.R. Grant, *M. elix* J.R. Grant, *M. ericii* J.R. Grant, *M. gattaca* J.R. Grant, *M. gondoloides* J.R. Grant, *M. harlingii* J.S. Pringle, *M. innarrabilis* J.R. Grant, *M. jactans* J.R. Grant, *M. jensii* J.R. Grant & Struwe, *M. lenae* J.R. Grant, *M. luna-gentiana* J.R. Grant & Struwe, *M.* × *mattii* J.R. Grant (*M. noctiluca* × *M. subsessilis*), *M. micrantha* Gilg, *M. neillii* J.R. Grant, *M. noctiluca* J.R. Grant & Struwe, *M. opulenta* J.R. Grant, *M. pringleana* J.R. Grant, *M. quizhpei* J.R. Grant, *M. sodiroana* Gilg, *M. subsessilis* Weaver & J.R. Grant, *M. voluptuosa* J.R. Grant, *M. xerantifulva* J.R. Grant, and *M. ypsilocaule* J.R. Grant. An additional seven species have yet to be described formally, but are included here for complete coverage, these being *M.* "*catherineae*," *M.* "*cortinae*," *M.* "*illuminata*," *M.* "*lilliputiana*," *M.* "*pacifica*," *M.* "*umbellata*," and *M.* "*zumbae*". Description, illustration, documented distribution, and discussion of each species is included.

An erratum to this chapter is available at 10.1007/978-3-642-54010-3_13.

J. R. Grant (✉)
Laboratoire de botanique évolutive, Institut de Biologie Université de Neuchâtel, Rue Émile-Argand 11, CH 2000 Neuchâtel, Switzerland
e-mail: jason.grant@unine.ch

J. J. Rybczyński et al. (eds.), *The Gentianaceae - Volume 1: Characterization and Ecology*, DOI: 10.1007/978-3-642-54010-3_3,

3.1 Introduction

Macrocarpaea, the most species-rich genus of the gentian tribe Helieae, is composed of shrubs or small trees with yellow to green funnel-shaped flowers that attracts diurnal visitors such as hummingbirds, ants, butterflies, and many other insects, but is perhaps most effectively pollinated nocturnally by bats and moths. The seeds are small (0.2–2.2 mm) and numerous (at least 10,000 per fruit), with taxonomically informative characters at both the species and sectional rank. It is a genus of 120 species restricted to mountainous regions of the Neotropics. *Macrocarpaea* attains its greatest diversity in the Andes (from Venezuela through Colombia, Ecuador, Peru, to Bolivia), where some 98 taxa occur. Smaller groups occur in southern Mesoamerica (five species and one natural hybrid), the Greater Antilles of the Caribbean (three species), the tepuis of the Guayana Highlands (six species), and the coastal montane Atlantic forests of Brazil (eight species).

This study is part of a series of regional taxonomic monographs within the framework of a full revision of the genus. Revisions for Cuba (Thiv and Grant 2002), Bolivia (Grant 2011), and Brazil (Grant and Trunz 2011) have already been published. In Ecuador, 35 taxa are recognized, including 1 natural hybrid and 34 species. This is a synthesis of extensive field work and herbarium studies that have led to a series of preparatory papers (Grant and Struwe 2001, 2003; Grant 2003, 2004, 2005, 2007, 2008, 2011; Grant and Trunz 2011; Grant and Weaver 2003).

3.2 Sources of Scientific Data

The aim of this project was to revise the genus *Macrocarpaea* in Ecuador through fieldwork, studies of herbarium specimens, seed micromorphology, and biogeography. In order to understand species concepts in the genus, extensive fieldwork was carried out over a period of ten years.

Eight expeditions to Ecuador took place between 2001 and 2013 with numerous colleagues, students, and friends including in 2001 Lena Struwe, February 2002 Joseph-Marie Torres, October 2002 Albert Roguenant, and Aline Raynal-Roques 2006 Meilin Cheung, Fabienne Luisier, and Neil Villard, 2008 Camille Agier, Claire Arnold, and Meilin Cheung, 2009 botanical illustrator Bobbi Angell, Wyan Grant, and Vincent Trunz, 2011 Julien Vieu, and in 2013 Jimmy Cortina.

Material for this monograph was examined during multiple visits to seven herbaria in Ecuador, namely Universidad Nacional de Loja (LOJA), Herbario del Instituto de Ciencias Naturales del Universidad Central (Q), Herbario Alfredo Paredes del Universidad Central (QAP), Pontificia Universidad Católica del Ecuador (QCA), Museo Ecuatoriano de Ciencias Naturales (QCNE), Herbario Luis Sodiro, Biblioteca Ecuatoriana Aurelio Espinosa Pólit (QPLS), and the Universidad San Francisco de Quito (QUSF). The herbaria LOJA, QCA, and QCNE were visited on numerous occasions to examine their rich collections.

Additionally, the most important neotropical collections of Gentianaceae are located at the New York Botanical Garden (NY) and the Missouri Botanical Garden (MO). Numerous visits and exchange of specimens with these institutions have been invaluable for this research. Specimens have been examined in situ or on loan from the following herbaria (acronyms according to Index Herbariorum, Thiers 2008), these being AAU, AFP, ALA, B, BM, BP, BR, BRIT, BSB, C, CAS, CAUP, CEPEC, CESJ, CHOCO, CHRB, COAH, COL, CONN, CR, CUVC, CUZ, DAV, DUKE, E, EHH, F, FAUC, FI, FLAS, FMB, FR, FTG, G, GB, GH, GOET, HAC, HAL, HAM, HAO, HB, HUA, HUCP, HUEFS, HUQ, HUT, IAN, INB, INPA, JAUM, JBSD, JE, K, L, LD, LINN, LPB, LS, m, MA, MANCH, MARY, MBM, MBML, MEDEL, MER, MG, MICH, MIN, MOL, MSB, MU, MY, NA, NEU, NO, NSW, OXF, P, PH, PORT, PR, PRC, R, RB, RNG, S, SBBG, SEL, SP, SPF, TEX, U, UC, UCWI, UDBC, UPCB, UPS, UPTC, US, USM, VALLE, VEN, W, WIS, WU, YU, and Z.

3.3 Results and Discussion

3.3.1 Morphology

3.3.1.1 Habit

Species of *Macrocarpaea* in Ecuador range from nearly herbaceous/woody herbs from 1 to 3 m (e.g., *M. claireae, M. xerantifulva*), to 1–4 m woody bushes with thickened leaves (e.g., *M. arborescens, M. harlingii, M. subsessilis,*), to standard 2–4 m tall bushes or small trees (e.g., *M. dies-viridis, M. lenae, M. micrantha*), to trees growing to 10 m in height (e.g., *M. apparata, M. bubops, M. elix, M. noctiluca*), to high elevation paramo shrubs (*M. angelliae, M. luna-gentiana*).

3.3.1.2 Leaves

The leaves of *Macrocarpaea* have a broad range of morphological variation (Fig. 3.1). They are always entire, yet vary from sessile to petiolate, glabrous to pubescent, thick and coriaceous to thin and papery, and in shape from linear-lanceolate, lanceolate, oblong, ovate, oval, elliptic, oblanceolate, to obovate. The leaf base is aequilateral, oblique, attenuate, cuneate, to rounded; the leaf apex is acuminate, acute, mucronulate, obtuse to rounded. The leaves are generally glabrous, but hairs have evolved independently in a few species, although none of these occur in Ecuador. The interpetiolar ridge has taxonomically useful characters, and is defined as the area between paired opposite leaves where the petioles are fused. Petioles range from unvaginated to having a long scooped out groove or vagination on the upper surface.

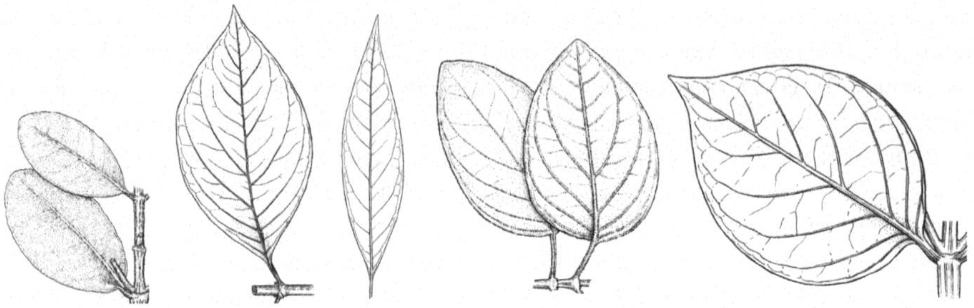

Fig. 3.1 Morphological variation in the leaves of *Macrocarpaea*. Illustrations by Bobbi Angell

3.3.1.3 Flowers and Pollen

Flowers have many important taxonomically rich characters in *Macrocarpaea* (Figs. 3.2, 3.3). A flower consists of a pedicel, bracteoles, calyx, and corolla, as well as the androecium/stamens (filaments and anthers), and gynoecium (ovary, style, and stigma). Comparison of these discrete characters provides a suite from which taxonomic decisions are based. The flowers of *Macrocarapea* display protandry, where the anthers mature first. Only once they have finished does the stigma open and become receptive to pollen (Fig. 3.3).

Calyx. The single-most important morphological feature in terms of taxonomic utility that is readily visible on all specimens regardless of phenological state (flowering to fruiting) is the calyx formed of five sepals. Calyces vary considerably in length and width, in shape from narrowly to broadly campanulate, ecarinate to strongly carinate, and glabrous, spiculate to pubescent in vestiture. Calyx lobes vary from scarcely present to dividing the calyx to nearly the base. The lobes range in shape from ovate, oblong, rotund, elliptic, lunate, to small and triangular, with apices that vary from acuminate, acute, obtuse, to rounded.

Corolla. The corollas of *Macrocarpaea* also display a wide range of morphological diversity. In one species, perhaps the basal-most species in the genus, *M. rubra* of Brazil, the corolla is urceolate in shape. All remaining species of the genus including all species in Ecuador have variations on a funnel-shaped corolla. Useful characters include length and width, and shape of the corolla lobes. Lobes vary from ovate, elliptic, lanceolate, to lunate, with apices that vary from acute, obtuse, retuse, to rounded. Corolla lobes range in shape from ovate, elliptic, lanceolate, large and lunate, to small and triangular, with apices that vary from acute, obtuse, retuse to rounded.

Pollen. Two morphological types of pollen grains occur in *Macrocarpaea* (Nilsson 1968; Grant 2006). The *Glabra*-type pollen (Fig. 3.4a–d) is present in most species of the genus and has be described as "3-colporate, rarely porate, 23–44 × 26–42 μm, spheroidal to subspheroidal, with a reticulate to verrucose-gemmate exine, and usually relatively wide muri" (Nilsson 1968, 2002).

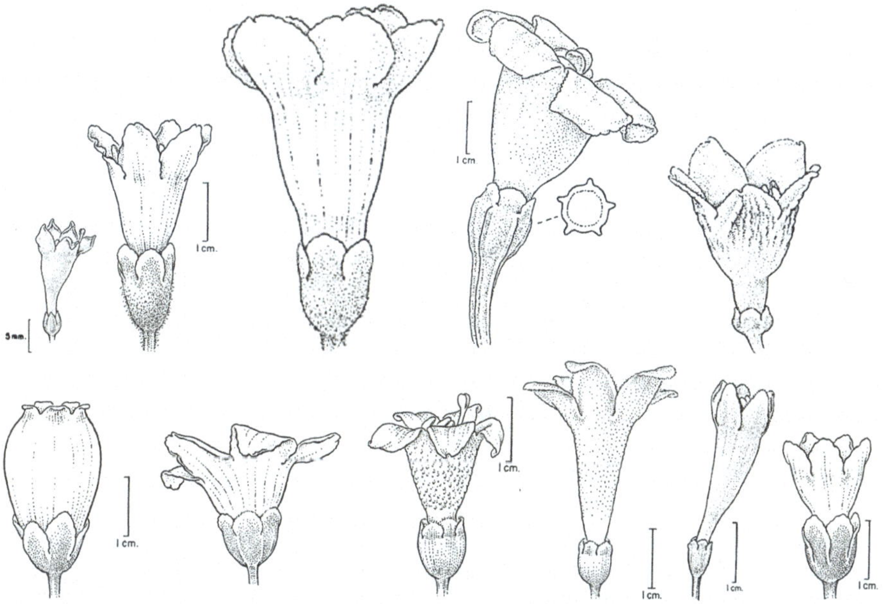

Fig. 3.2 Morphological variation in the calyx and corolla or *Macrocarpaea*. Illustrations by Bobbi Angell

Fig. 3.3 Protandry in *Macrocarpaea*. In the flower on the *right* the anthers are open, but the stigma is closed. In the flower on the *left*, the anthers are finished and recoiled backward, and the stigma has opened. *Macrocarpaea apparata* J.R. Grant & Struwe. *Photo* Neil Villard

Corymbosa-type pollen (Fig. 3.4e–h) is "3-colporate to colporate, 28–35 × 28–39 μm, subspheroidal to spheroidal, and with warty exine (verrucae, gemmae, pila, and clavae are present)" (Nilsson 1968, 2002). The *Corymbosa*-type conforms to section *Magnolifoliae*, and provides an important character to define the section. It has been identified in 13 species, with two species in Ecuador, namely *M. harlingii* and *M. jactans*.

3.3.1.4 Fruit and Seeds

The capsules of *Macrocarpaea* are bilocular and filled with upwards of 10,000 wind-dispersed seeds (Grant 2006). The seeds are minute in size, ranging from 0.2–2.2 × 0.2–1.0 (–2.2) mm, yet have a great amount of morphological diversity useful for taxonomic purposes. The four seed types recognized in Grant (2005) are:

1. *Flattened-type seeds* (Fig. 3.5a–b) correspond to section *Tabacifoliae*. The seeds are generally triangular in outline, flattened to polygonal, and lack wings. Some species have extended chalazal and micropylar ends while it either falls off before seed maturity or is lost during seed development in other species.
2. *Perimetrically winged-type seeds* (Fig. 3.5c–h) correspond to the section *Choriophylla*. The seeds are flattened with wings surrounding the perimeter of the seed. They are typically very light in weight, likely an adaptation for wind dispersal.
3. *Rimmed-type seeds* (Fig. 3.6a–f) correspond to the section *Macrocarpaea*. The seeds are polygonal to slightly flattened, and rimmed (typically without, yet sometimes with, very short wings).
4. *Winged-type seeds* (Fig. 3.6g–h) correspond to the section *Magnolifoliae*. The seeds are flattened and prominently winged at their chalazal and micropylar ends, or rarely all around. This seed type is an important character used to define the section *Magnolifoliae*.

3.3.1.5 Wood Anatomy

The Gentianaceae is a largely herbaceous plant family, yet several genera are composed almost entirely of shrubs and trees, including *Anthocleista*, *Fagraea*, *Potalia*, *Symbolanthus*, and *Macrocarpaea*. The general rarity of gentian trees has led to a broad characterization of the family as composed of herbs and shrubs, and few dendrologists have ever studied the gentian wood. Therefore, a collection of *Macrocarpaea* wood from 17 species was made for the purpose of a study of its anatomy, including *M. angelliae, M. apparata, M. arborescens, M. bubops, M. gattaca, M. harlingii, M. jensii, M. lenae, M. luna-gentiana, M. micrantha, M. noctiluca, M. pachystyla, M. rubra, M. sodiroana, M. subsessilis, M. wallnoeferi,* and *M. wurdackii*. This study resulted in a paper entitled "Wood anatomy of Gentianaceae, tribe Helieae, in relation to ecology, habit, systematics, and sample diameter" (Carlquist and Grant 2005). Perhaps the most notable result was the identification of juvenile characteristics in the rays and pseudoscalariform lateral wall pitting of vessels that suggest paedomorphosis (retention of juvenile characters in adults) (Figs. 3.7, 3.8). This suggests that the evolution of *Macrocarpaea* into trees, and perhaps the entire family, is derived from perennials that have at least a somewhat woody base.

Fig. 3.4 Pollen grains of *Macrocarpaea*. **a–d** Glabra type; **e–h** Corymbosa type. **a** equatorial view of 3-colporate reticulate pollen of *M. subcaudata* from Luteyn 3918 (NY); **b** detail of reticulate exine in pollen of *M. kuepferiana*, from Vargas 18928 (CUZ); **c** polar view of 3-colporate reticulate pollen of *M. wurdackii*, from Smith 4793 (NEU); **d** detail of reticulate exine in pollen of *M. wurdackii*, from Smith 4793 (NEU); **e** polar view of 3-colporate verrucose pollen of *M. revoluta*, from Smith 2748 (NY); **f** detail of intergrading reticulate to verrucose warty exine in pollen of *M. revoluta*, from Smith 2748 (NY); **g** polar view of 3-colporate verrucose pollen of *M. canoëfolia*, from Woytkowski 35417 F **h** detail of verrucose warty exine in pollen of *M. canoëfolia*, from Woytkowski 35417 F. Reprinted from Grant (2005)

Fig. 3.5 Seed morphology in *Macrocarpaea*. **a–b** "flattened seeds," *M. illecebrosa*, from Grillo & Conceição 217 (SPF); **c–d** "perimetrically winged seeds," *M. xerantifulva*, from Campos & Garcia 3960 (MO); **e–f** "perimetrically winged seeds," *M. gulosa*, from Ramos 1477A (MO); **g–h** "perimetrically winged reversal to rimmed seeds," *M. umbellata* , from Rubio 1636 (MO). Reprinted from Grant (2005)

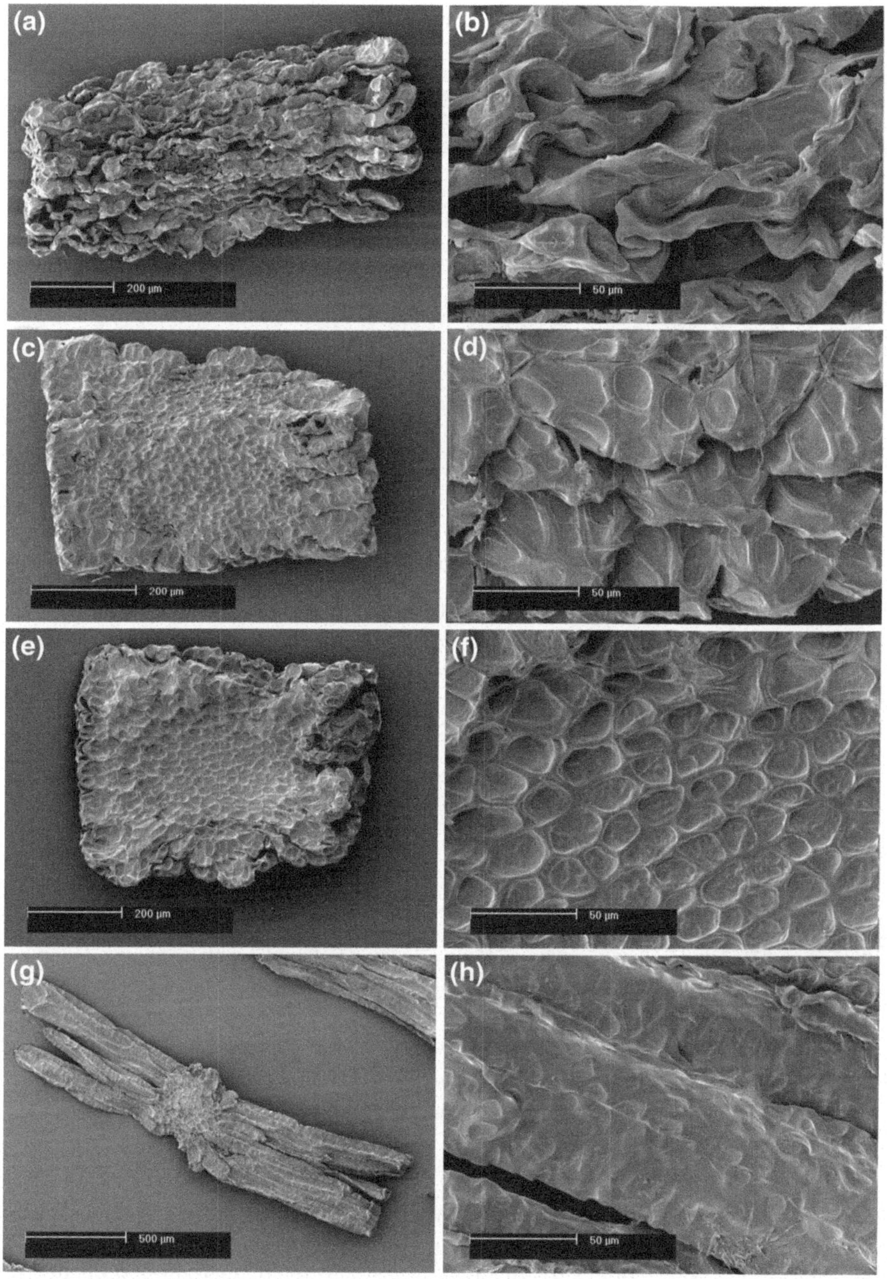

Fig. 3.6 **a–b** "rimmed seeds," *Macrocarpaea laudabilis*, from Garcia Barriga & Hawkes 12884 (COL); **c–d** "rimmed seeds," *M. gaudialis*, from Schultes 10415 (NY); **e–f** "winged seeds," *M. cinchonifolia*, from Solomon 7492 (U); **g–h** "winged seeds," *M. jactans*, from Vasquez 20219 (MO). Reprinted from Grant (2005)

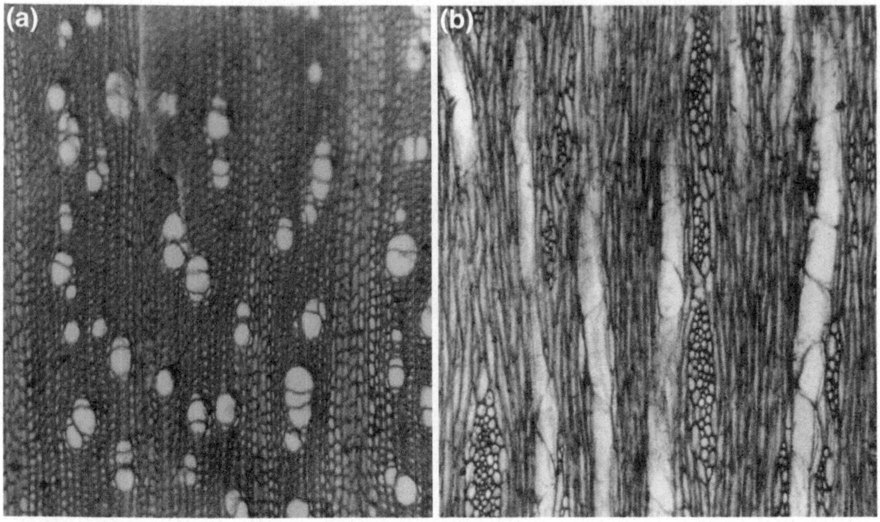

Fig. 3.7 Wood sections of *Macrocarpaea bubops*. **a** Transection, vessels are mostly grouped. **b** Tangential section, several wide multiseriate rays with procumbent cells in wider portions present. Reprinted from Carlquist and Grant (2005)

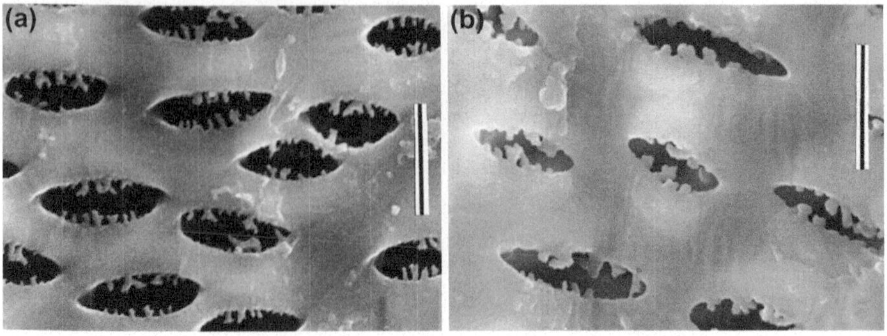

Fig. 3.8 SEM photographs of tangential wood sections of *Macrocarpaea bubops* (**a**) and *M. gattaca* (**b**) showing vestured pits. Reprinted from Carlquist and Grant (2005)

3.3.2 Phylogeny and Evolution

A recent molecular phylogeny of the gentian tribe Helieae indicates that *Macrocarpaea* has the greatest number of species in the tribe (Struwe et al. 2009a, b). The first phylogeny (Fig. 3.9) of *Macrocarpaea* itself was prepared in the framework of a newly designed software program entitled *Spatial Evolutionary and Ecological Vicariance Analysis* (SEEVA) to determine environmental characteristics between clades and sister species (Struwe et al. 2009a, b). A new phylogeny is currently being prepared with multiple accessions per species with a

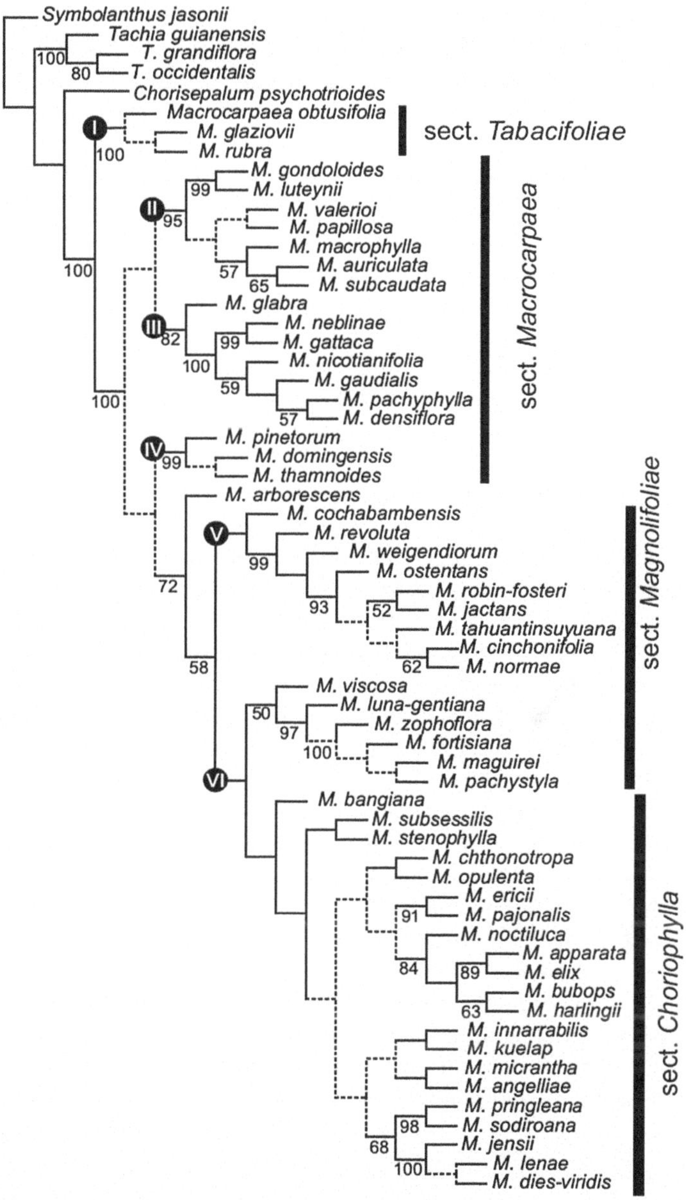

Fig. 3.9 Phylogeny of *Macrocarpaea* from based on combined nuclear ITS and 5 s-TS sequences with outgroups of *Symbolanthus*, *Tachia*, and *Chorisepalum*. Reprinted from Struwe et al. (2009a)

high percentage of representatively in species (Grant, Vieu et al. in preparation) *Macrocarpaea* has been divided into four taxonomic sections based on pollen and seed characters by Grant (2005). The phylogenies indicate that as currently defined, section *Tabacifoliae* of Brazil (Grant and Trunz 2011), section *Macrocarpaea* largely of the northern Andes, and section *Choriophylla* largely of the Amotape–Huancabamba zone of Ecudor and Peru, are monophyletic. However, the section *Magnolifoliae* is paraphyletic and requires further investigation.

Section *Macrocarpaea*. Four species in Ecuador belong to this section, these being *M. gattaca* (endemic), *M. gondoloides, M. voluptuosa* (endemic), and *M. ypsilocaule*. Two of these species are endemic to Ecuador, while the remaining two also occur in adjacent regions of Colombia. All of these species occur generally north of Quito, which make sense since the greatest diversity of this section is in Colombia. These species all have pollen with reticulate exine, and rimmed type seeds (except *M. gondoloides* which has a reversal to perimetrically winged seeds).

Section *Magnolifoliae*. This section has its greatest diversity in Peru, yet two species are found in Ecuador, namely *M. jactans* and *M. luna-gentiana*. That belong to different clades in the paraphyletic section. While *M. jactans* has pollen grains with a warty exine pattern, *M. luna-gentiana* has a reticulate exine more characteristic of section *Choriophylla*.

Section *Choriophylla* attains its greatest diversity in the Amotape–Huancabamba zone that straddles the border of Ecuador and Peru. In Ecuador, 28 species occur, 20 of which are endemic, 7 that range into Peru, and one that ranges into Colombia. These species include *M. angelliae* (endemic), *M. apparata* (endemic), *M. bubops* (with Peru), *M. catherineae* (endemic), *M. claireae* (endemic), *M. cortinae* (endemic), *M. dies-viridis* (endemic), *M. elix* (endemic), *M. ericii* (with Peru), *M. harlingii* (with Peru), *M. illuminata* (endemic), *M. innarrabilis* (with Peru), *M. jensii* (endemic), *M. lenae* (endemic), *M. lilliputiana* (endemic), *M.* × *mattii* (endemic), *M. micrantha* (with Peru), *M. neillii* (endemic), *M. noctiluca* (with Peru), *M. opulenta* (endemic), *M. pacifica* (endemic), *Macrocarpaea pringleana* (endemic), *M. quizhpei* (endemic), *M. sodiroana* (endemic), *M. subsessilis* (endemic), *M. umbellata* (with Colombia), *M. xerantifulva* (with Peru), and *M. zumbae* (endemic). The species of this section have perimetrically winged seeds and pollen with reticulate exine. *Macrocarpaea arborescens* has a unique morphology and is sister to section *Choriophylla*.

3.3.3 Distribution

Macrocarpaea is a genus of 120 species restricted to mountainous regions of the Neotropics. It attains its greatest diversity in the Andes (from Venezuela through Colombia, Ecuador, Peru, to Bolivia) where some 98 taxa (98 species and 1 natural hybrid) occur. Smaller groups occur in southern Mesoamerica (five species and one natural hybrid), the Greater Antilles of the Caribbean (three species), the

Tepuis of the Guayana Highlands (6 species), and the coastal montane Atlantic forests of Brazil (eight species). Three countries harbor the greatest number of species and diversity, these being Ecuador, Peru, and Colombia.

In Ecuador, 35 taxa are recognized including one natural hybrid and 34 species (Figs. 3.10, 3.11, 3.12, 3.13). Twenty-five of these species are endemic to Ecuador, while 11 others range into adjacent countries. Three species of section *Macrocarpaea* occur in both southern Colombia and northern Ecuador (*M. gondoloides*, *M. umbellata*, and *M. ypsilocaule*), while eight species of section *Choriophylla* occur in the Amotape–Huancabamba zone of both southern Ecuador and northern Peru (*M. bubops*, *M. ericii*, *M. harlingii*, *M. innarrabilis*, *M. jactans*, *M. micrantha*, *M. noctiluca*, and *M. xerantifulva*.

Three distinct mountainous biogeographical zones are found in Ecuador; the Cordillera Occidental and Choco have six species of *Macrocarpaea* (Fig. 3.13, region F) [*M. gattaca*, *M. gondoloides*, *M. pacifica, M. sodiroana*, *M. umbellata*, and *M. voluptuosa*], The Cordillera Central with two species (Fig. 3.10, region E) [*M. cortinae* and *M. ypsilocaule*], and the Amotape–Huancabamba region (Fig. 3.10, region h) where the remaining 27 taxa occur (Struwe et al. 2009a, b). The Amotape–Huancabamba region of southern Ecuador and northern Peru (Weigend 2002, 2004; Young and Reynel 1997) has the largest number of species and greatest morphological diversity in *Macrocarpaea*. Intense fieldwork in both Ecuador and Peru in recent years has uncovered many species previously unknown to science, and dramatically changed the previously broad circumscriptions of some species.

3.3.4 Pollination and Phylogeny

Macrocarpaea appears to be geared toward nocturnal pollination, yet perhaps functions as a generalist flower. Observations on effective pollination still need to be conducted. The flowers are yellow to green, and begin to blossom in the evening and emit a fragrance. The flowers are funnel-shaped or bell-shaped and produce quantities of nectar, perfect for nocturnal visits by bats and large moths. However, they remain open for several days also attracting diurnal visitors. Many other small insects, including beetles and parasitoids, are found inside some flowers during the night. During the day, ants, butterflies, and hummingbirds, all attracted by the nectar or pollen, visit the flowers.

Joseph Marie Torres of the University of Neuchâtel conducted a study of the pollinators of *Macrocarapea sodiroana* in the Reserva Maquipucuna in northern Ecuador and identified many nocturnal and diurnal pollinators including at least 11 different species of nocturnal moths and sphinx moths (Fig. 3.14). Matt (2001) described the behavior of bats in southern Ecuador and reported on the visits of bats to a series of species of *Macrocarpaea*. Recently, Nathan Muchhala has photographed the bat *Anoura geoffreyi* pollinating *Macrocarpaea sodiroana* in southern Ecuador (Fig. 3.15).

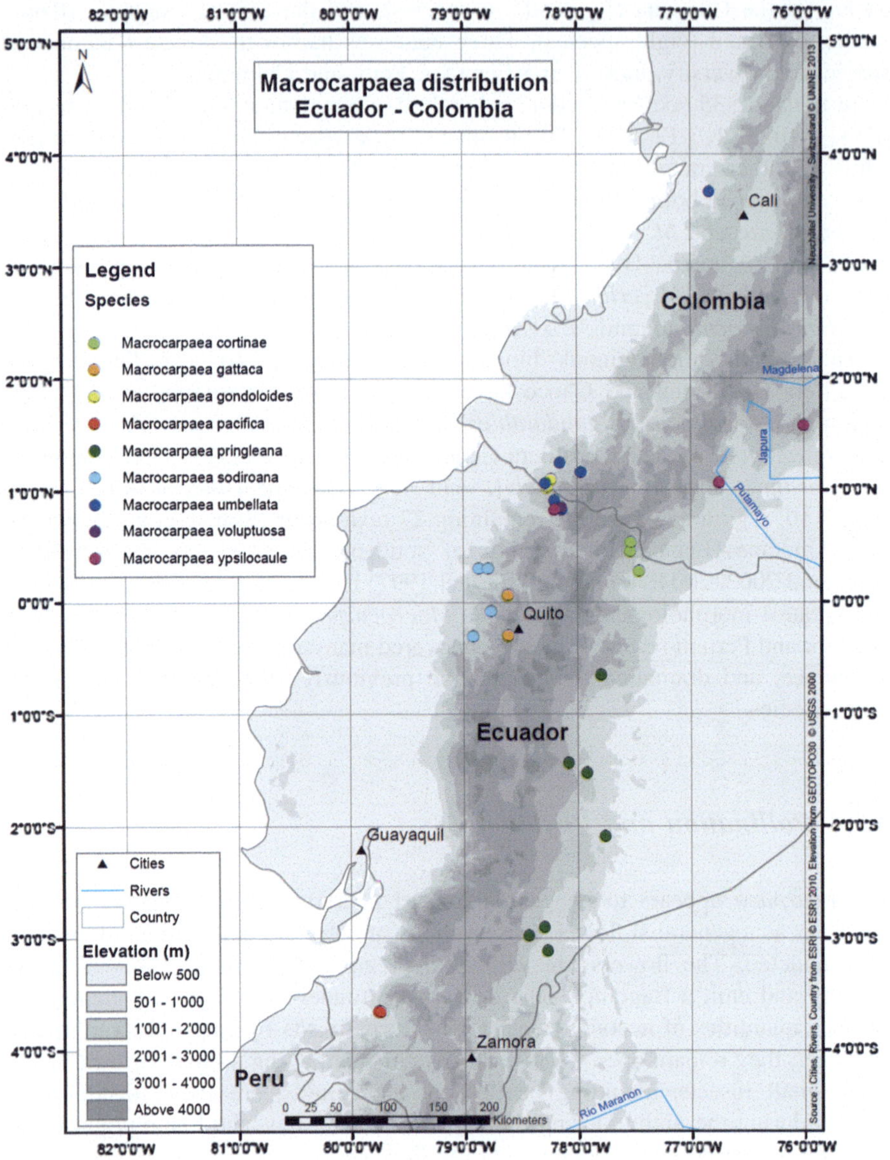

Fig. 3.10 Map of species of *Macrocarpaea* in Ecuador

3.3.5 Taxonomic History

The only revision of *Macrocarpaea* is that of Ewan (1948) who recognized 30 species based on 114 specimens. The specimens used in that revision largely came

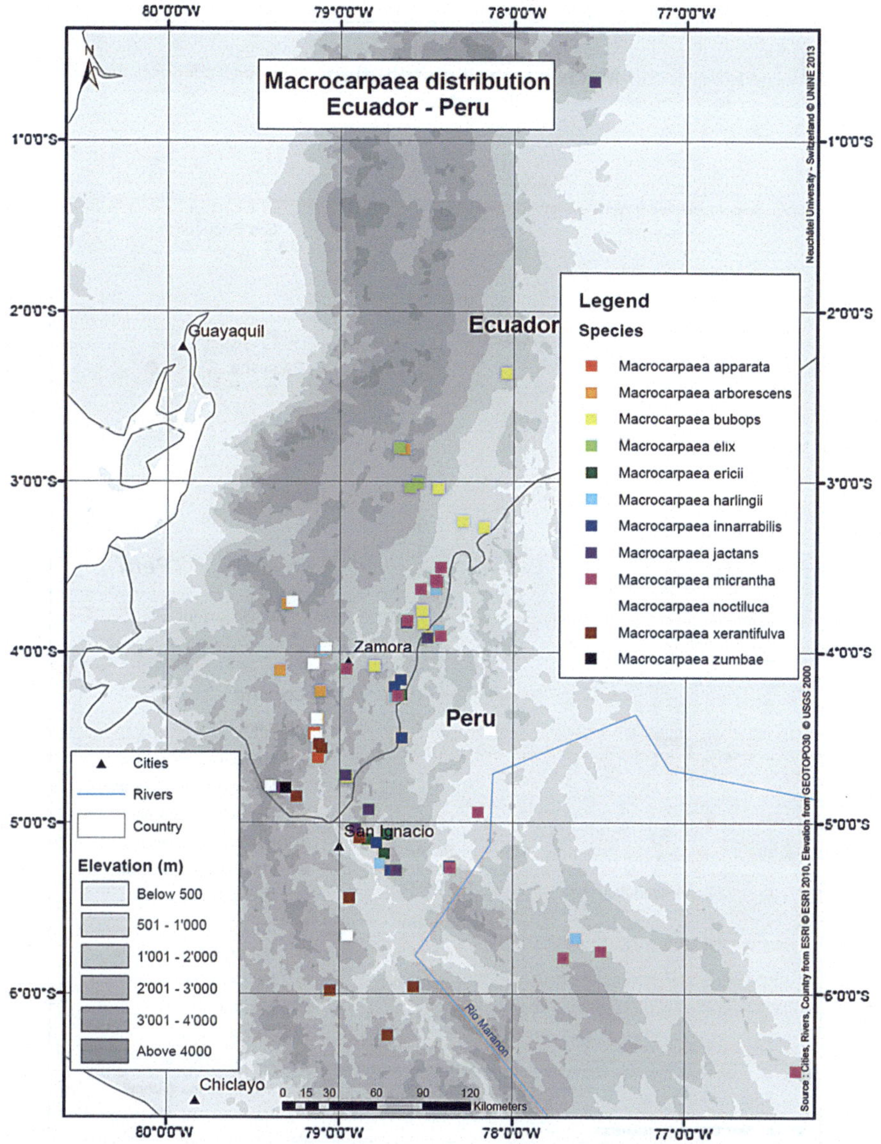

Fig. 3.11 Map of wide-ranging species of *Macrocarpaea* in the Amotape–Huancabamba zone of southern Ecuador and northern Peru

from areas near to major cities such as Bogotá, Quito, or Rio de Janeiro. Since then, vast unexplored regions of South America have opened up that botanists have explored in depth. Researchers from throughout the world, as well as those from their own countries have conducted detailed inventories of previously

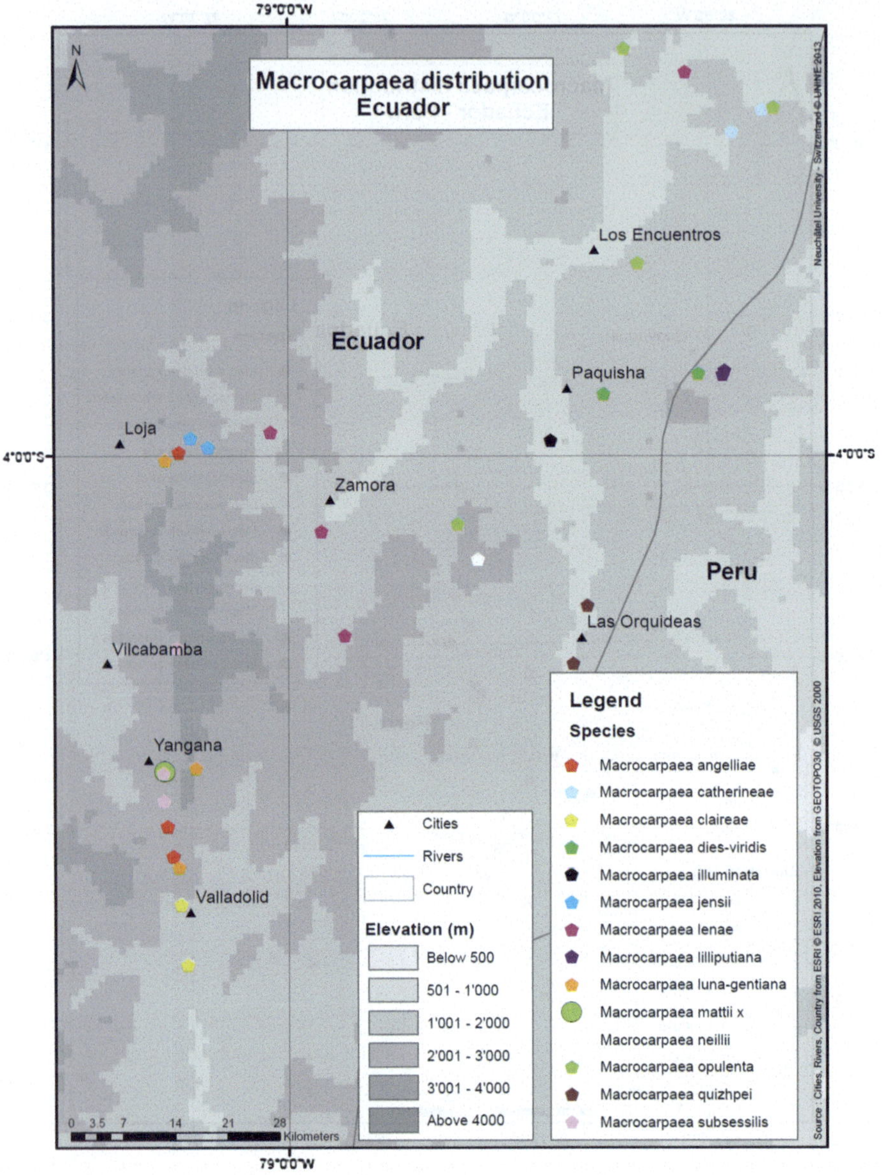

Fig. 3.12 Map of endemic species of *Macrocarpaea* in the Amotape–Huancabamba zone of southern Ecuador and northern Peru

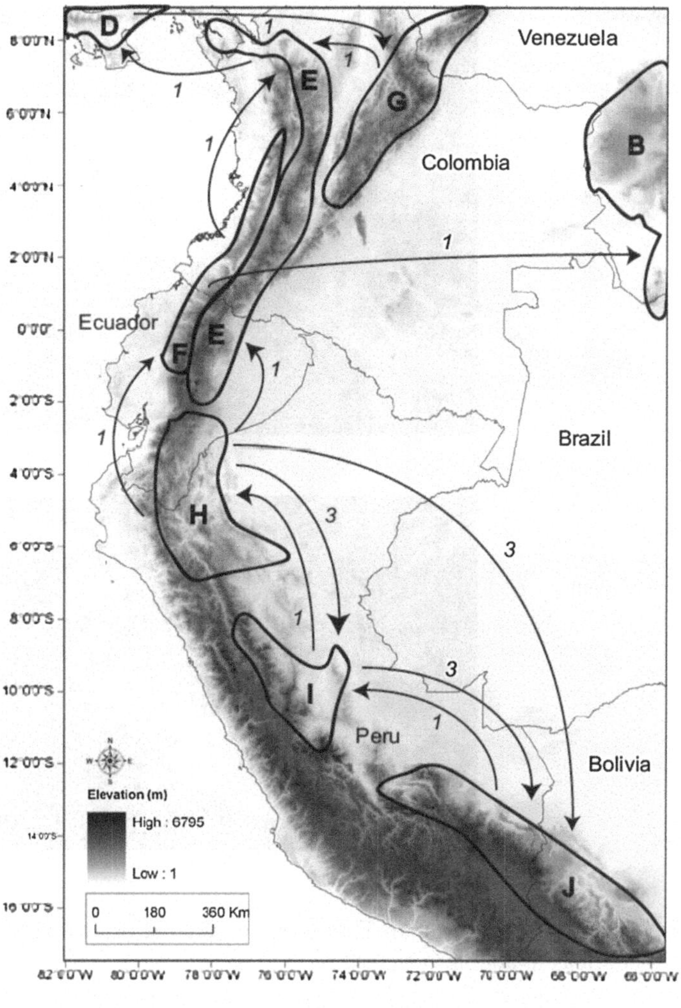

Fig. 3.13 Map of northwestern South America and southern Mesoamerica with the areas used for biogeographic analysis indicated with letters and solid lines indicating *Macrocarpaea* species distributions. *Arrows* with numbers indicate the number of dispersal events as hypothesized from the DIVA analysis. Biogeographic areas are: **A** Southeastern Brazil; **B** Pantepui of the Guayana Shield; **C** Greater Antilles of the Caribbean; **D** Mesoamerica; **E** Cordillera Central in Colombia and Cordillera Oriental in Ecuador; **F** Chocó and Cordillera Occidental in Colombia and Ecuador; **G** Cordillera Oriental and Merida in Colombia and Venezuela; **H** Amotape–Huancabamba region in Ecuador and Peru; **I** Cordillera Central in central Peru; **J** Bolivia and Cordillera Central in southern Peru; **K** Amazon Basin in Brazil, Colombia, Peru, and Venezuela. Not indicated on are areas **A**, **C**, and **K**. Reprinted from Struwe et al. 2009a

Fig.3.14 Sphinx moth visiting *Macrocarpaea sodiroana* in Ecuador. *Photo* Joseph Marie Torres

Fig. 3.15 The bat species *Anoura geoffroyi* visiting *Macrocarpaea harlingii* in Ecuador. *Photo* Nathan Muchhala

inaccessible regions of the continent. This broad collecting has resulted in a huge backlog in specimens available for study.

Specifically for Ecuador, Ewan (1948) recognized three species to occur there based on 11 specimens made around the cities of Quito and Loja (*M. ovalis* [*Lobb s.n, André 4549, Steyermark, J. 54805, Steyermark, J. 54436, Steyermark, J. 54758*] *M. pachyphylla* [*Andre 4513*], and *M. sodiroana* [*Espinosa 198, Mexia 6977, Sodiro 101/1, Steyermark, J. 53548, Steyermark, J. 54812*]). The only species that Ewan recognized based on a type from Ecuador is *M. sodiroana*, *Sodiro 101/1*. However, he also placed here four specimens that actually represent three other species: *Mexia 6977* = *M. pringleana*; *Steyermark, J. 53548* = *M. elix*; *Steyermark, J. 54812* = *M. noctiluca*; and *Espinosa 198* = *M. noctiluca*. Additional new determinations among these 11 specimens are: *Andre 4513* = *M. luna-gentiana*; *André 4549* = *M. arborescens*; and four specimens *Lobb s.n.*, *Steyermark, J. 54805*, *Steyermark, J. 54436*, and *Steyermark, J. 54758* = *M. subsessilis*.

In the only regional revision of Ecaudorian Gentianaceae, Pringle (1995) provided a revision of *Macrocarpaea* recognizing eight species, *M. cinchonifolia*,

M. harlingii, *M. micrantha*, *M. ovalis*, *M. pachyphylla*, *M. revoluta*, *M. sodiroana*, and *M. stenophylla*. This revision increased both the number of species recognized and specimens on which the species were characterized. A single new species was described, *M. harlingii* J.S. Pringle. Here, *M. cinchonifolia* is recognized to be endemic to Bolivia, *M. pachyphylla* endemic to Colombia, and both *M. revoluta* and *M. stenophylla* as endemic to Peru. The name *M. ovalis* has been suppressed (Grant 2004). Therefore, the only names used the same as in Pringle (1995) are *M. harlingii*, *M. micrantha*, and *M. sodiroana*, yet each has a much narrower circumscription here than that of Pringle.

From 1998 to 2003, this author revised the genus *Macrocarpaea* for his Ph.D. dissertation. What seemed like an easy task at the beginning resulted in a daunting realization that numerous new species needed to be described, and previously broad species concepts neither made taxonomic nor biogeographical sense. Thousands of specimens were studied on loan from abroad, and numerous institutions in Europe, North America, and South America were visited. I have databased some 4,000 specimens, an exponential increase in material from the 114 specimens of Ewan. The 35 taxa currently recognized in Ecuador are based on a sizeable of these specimens, especially due to fieldwork from 2001 to 2013.

3.3.6 Species Concept

Species concepts vary considerably between author and taxonomic group. In *Macrocarpaea*, I define a species as a group/population of interbreeding individuals that are morphologically identifiable by discrete characteristics. These species must originate from the same biogeographical area and habitat and have also been identified as distinct by DNA sequences (DNA barcoding). A suite of characters has been used to circumscribe species, species groups, and to formulate an infrageneric classification.

The most important characters are those of gross plant morphology [especially leaf morphology, the architecture of the inflorescence (a thyrse, where the primary branching is racemose, and the secondary is cymose), the position of the flowers before and after anthesis (erect, spreading, horizontal, nodding), calyx size, shape (lobes acute, acuminate, cuspidate, obtuse, rounded), and vestiture (glabrous, scabrous, hairy), and leaf size and shape (linear, lanceolate, ovate, elliptical, obovate)]. Additional sources of taxonomically useful information come from seed anatomy, pollen morphology, and molecular data.

It is true that I have described most of the species currently recognized in Ecuador. This is due to the great number of specimens available for study to which previous authors did not have access. Ecuador is also one of the most biodiversity rich countries in the world and, therefore, the discovery of numerous new species has not been surprising. A few additional new species will probably be found in remote unexplored areas.

3.3.7 Key to the Species of Macrocarpaea *in Ecuador*

To simplify identification, this key is based on dried herbarium specimens alone. However, most species can also be identified from fresh material with this key.

1a. Secondary leaf veins absent or scarcely visible on a smooth upper leaf surface ...2

2a. Leaves lanceolate, apex acute to acuminate ...3

3a. Plants generally single-stemmed or sparsely branched; calyx rugose, calyx lobes drying light brown, and unlobed portion dark brown...***M. harlingii***

3b. Plants much branched; calyx strongly rugose, drying dark brown throughout ...***M. zumbae***

2b. Leaves obovate, orbicular to elliptical, apex rounded to obtuse ...4

4a. Bracteoles small and inconspicuous (2–5 mm long), positioned remotely about halfway along the pedicel; flowers spreading to pendent ...***M. arborescens***

4b. Bracteoles conspicuous (10–20 mm long), positioned just below the calyx, and exceeding the calyx in length; flowers erect to rarely spreading, but never pendent ...5

5a. Inflorescences paniculate; bracteoles linear to lanceolate ...***M. ericii***

5b. Inflorescences short to nearly sessile within the upper leaves; bracteoles spathulate ...6

6a. Plants robust; leaves 2.5–9 cm long; capsules 12–20 mm long ...***M. subsessilis***

6b. Plants slender; leaves 2.0–3.5 cm long; capsules 10–12 mm long ...***M. lilliputiana***

1b. Secondary (and sometimes tertiary) leaf veins prominent, usually several pairs of secondary veins arch toward the leaf apex (readily visible on the upper leaf surface) ...7

7a. Leaves brittle coriaceous, plants of high elevation (2,700–3,100 m)

8a. Corolla green, 41–62 mm long; leaves with strong prominent rugose venation, veins strongly impressed on upper surface ...***M. luna-gentiana***

8b. Corolla white to yellow–green, 35–55 mm long; leaves with prominent venation, but not so much impressed on the upper surface ...9

9a. Leaves long-petiolate (petioles 10–35 mm long) ...***M. bubops***

9b. Leaves sessile to short-petiolate (petioles 0–10 mm long) ...10

10a. Leaves lanceolate to elliptic, apex acute to acuminate …11

11a. Bracts large and broadly ovate, exceeding the length of the cymes; cymes nodding; calyx 6–8 mm long …***M. angelliae***
11b. Bracts smaller and not exceeding the length of the cymes; cymes not nodding; calyx 15–22 mm long …***M. ypsilocaule***

10b. Leaves broadly ovate to oval, apex obtuse to acute …12

12a. Calyces 8–11 × 5–6 mm; corolla lobes undulate …***M. gattaca***
12b. Calyces 9–13 × 8–10 mm; corolla lobes not undulate …***M. voluptuosa***

7b. Leaves tender smooth, plants of generally lower clevations (less than 2,700 m) …13

13a. Secondary veins pinnate and nearly parallel to another; corolla broadly flared open, funnelform …***M. jactans***
13b. Secondary veins arching, often with several prominent pairs that arch toward the leaf apex …14

14a. Calyces puberulent hairy …15

15a. Shrubs 2–3 m; veins on lower leaf surface glabrous; calyx scabrous …***M. dies-viridis***
15b. Robust trees 2–5 m; veins on lower leaf surface hirsute to hispid; calyx scabrous, hirsute to hispid …16

16a. Calyx lobes rounded to obtuse …***M. apparata***
16b. Calyx lobes ovate …***M. elix***

14b. Calyces glabrous and smooth …17

17a. Corolla broadly flared open, funnelform 40–45 mm wide at broadest; corolla lobes distinctly reflexed; calyx 13–15 × 11–17 mm …***M. opulenta***
17b. Corolla tubular funnelform (10–25 mm wide at broadest); corolla lobes generally not reflexed …18

18a. Leaves obovate, generally widest above the middle, leaf apex rounded to obtuse …19

19a. Pedicels 12–45 mm long …***M. innarrabilis***
19b. Pedicels 2–15 mm long …20

20a. Petioles 10–12 mm; bracteoles 1–3 mm …***M. neillii***
20b. Petioles 0–10 mm; bracteoles 5–18 mm … ***M. × mattii (M. noctiluca × M. subsessilis)***

18b. Leaves ovate, oval, elliptic to obovate, if obovate leaf apex acuminate …21

21a. Cymes simple to compound, rarely subumbelliform; leaves firm, with a pair of "dots" at the base of the lower side of each leaf adjacent to the petiole (easily visualized on living plants, but more difficult on herbarium material) …22.

22a. Leaves coriaceous …***M. bubops***
22b. Leaves smooth …23

23a. Leaves ovate, oblong to elliptic, matte green; bracteoles linear to lanceolate to obovate; common species of southern Ecuador…***M. noctiluca***
23b. Leaves elliptic to obovate; shiny green; bracteoles spathulate to oblanceolate; rare hybrid …***M.*** × ***mattii (M. noctiluca*** × ***M. subsessilis)***

21b. Cymes umbelliform; leaves smooth, lacking pair of "dots" at the base of the lower side of each leaf adjacent to the petiole …24
24a. Calyx lingulate to obling, 11–13 mm long; plants of section *Macrocarpaea* …***M. gondoloides***
24b. Calyx ovate, 5–10 mm long; plants of section *Choriophylla* …25
25a. Stem leaves short to long-petiolate; leaves long-decurrent on petiole, narrowing and tapering to the base of the petiole; plants from north of the Amotape–Huancabamba zone (except *M. jensii*), occurring on both Pacific-facing (western) and Amazon-facing (western) slopes of the Andes …26

26a. Leaves obovate to elliptic; calyx urceolate to campanulate; corolla fleshy …***M. jensii***
26b. Leaves lanceolate, ovate, oval to broadly elliptic; calyx campanulate; corolla not fleshy …27

27a. Umbelliform cymes diffuse; plants of Pacific-facing (western) slopes of the Andes …28

28a. Stem leaf petioles 3–7 cm long; plants of El Oro province …***M. pacifica***
28b. Stem leaf petioles 0.5–3.0 cm long; plants largely north of Quito …29

29a. Calyx 5–10 mm long; plants of Pichincha and Imbabura provinces ……***M. sodiroana***
29b. Calyx 8–10 mm long Plants of Carchi province …***M. umbellata***

27b. Umbelliform cymes compact; plants of Amazon-facing (eastern) slopes of the Andes ...30

30a. Calyx lobes green with yellow/orange on dorsal surface; bracts subtending branches elliptic; plants of Sucumbios and Napo ...***M. cortinae***

30b. Calyx lobes green; bracts subtending branches broadly ovate; plants widespread, ranging from about Baeza (Napo) to Limón (Morona-Santiago) ...***M. pringleana***

25b. Stem leaves long-petiolate, petiole slender; leaves either not at all decurrent on petiole, or only very shortly decurrent; plants of the Amotape–Huancabamba zone and Cordillera del Condor of southern Ecuador (almost exclusively in Zamora-Chinchipe province) ...31

31a. Calyx scabrous puberulent ...***M. dies-viridis***

31b. Calyx glabrous ...32

32a. Inflorescences large, diffuse; pedicels long (20–52 mm) ...***M. quizhpei***

32b. Inflorescences paniculate, but not diffuse; pedicels shorter (15–25 mm) ...33

33a. Leaves drying thin membranaceous; plants typically much-branched ...***M. micrantha***

33b. Leaves drying papery; plants single stemmed to branched ...34

34a. Leaves smooth coriacous; plants of the Cordillera del Cóndor area ...***M. catherineae***

34b. Leavs smooth but not coriaceous; plants of the Andes, but not occurring on the Cordillera del Cóndor ...35

35a. Calyx drying brownish; stems drying yellowish gold in color ...***M. xerantifulva***

35b. Calyx drying black; stems drying green ...36

36a. Calyx not glaucous ...***M. illuminata***

36b. Calyx glaucous ...37

37a. Leaf veins strongly impressed giving a quilted appearance to the leaves; calyx faintly scabrous; plants from the Valladolid region...***M. claireae***

37b. Leaf surface smooth; calyx smooth glabrous; plants from the Zamora region...***M. lenae***

3.3.8 Taxonomic Treatment

3.3.8.1 Genus Description

Macrocarpaea (Griseb.) Gilg in Engl. & Prantl, Nat. Pflanzenfam. 4(2): 94. 1895. Based on: *Lisianthus* section *Macrocarpaea* Griseb., Gen. Sp. Gent. 173. 1839. T.: *Lisianthus glaber* L.f., Suppl.: 134. 1781. *Macrocarpaea glabra* (L.f.) Gilg in Engl. & Prantl, Nat. Pflanzenfam. 4(2): 94. 1895.

= *Rusbyanthus* Gilg in Engl. & Prantl, Nat. Pflanzenfam. 4(2): 95. 1895. T.: *Rusbyanthus cinchonifolius* Gilg, in Engl. & Prantl, Nat. Pflanzenfam. 4(2): 95. 1895. *Macrocarpaea cinchonifolia* (Gilg) Weaver, J. Arnold Arbor. 55(2): 300. 1974.

Plants branched to unbranched; 1 m shrubs, treelets, to robust trees to 10 m, facultative epiphytes (*M. browallioides*, *M. luteynii*, and *M. subcaudata*), and upper páramo shrubs; glabrous to pubescent. **Trunk** 0.5–9.5 cm in diameter, wood hollow (typically narrower in trunk and wider in younger branches) to solid, growth rings faint to prominent, bark papery thin, to thick and pithy to 4 mm wide. **Stems** terete to quadrangular, solid to hollow. **Leaves** entire, petiolate to sessile; glabrous to pubescent; small thick and coriaceous, to large thin and papery; linear-lanceolate, lanceolate, oblong, ovate, oval, elliptic, oblanceolate, to obovate in shape; **leaf base** aequilateral, oblique, attenuate, cuneate, to rounded; **leaf apex** acuminate, acute, mucronulate, obtuse to rounded. **Petioles** short, to long and slender; unvaginated to having a long scooped groove or vagination on the upper surface; **interpetiolar ridge** prominent, exstipulate to stipulate (*M. zophoflora*). **Inflorescence** a thyrse, composed of branches of cymes in a racemose branching pattern. **Bracts** leaf-like and found throughout the inflorescence. **Bracteoles** scarious to bract-like. **Flowers** pedicellate, remaining erect or becoming nodding in maturity. **Calyx** narrowly to broadly campanulate; ecarinate to strongly carinate; glabrous, spiculate to pubescent in vestiture; **calyx lobes** ovate, oblong, rotund, elliptic, to lunate in shape, with apices that vary from acuminate, acute, obtuse, to rounded; persistent in fruit. **Corolla** 5–merous, funnel-shaped and actinomorphic, rarely urceolate (*M. rubra*), yet when androecium and gynoecium are included, the overall flower is zygomorphic; leathery to fleshy; white, cream, yellow to light green in color; **corolla lobes** ovate, elliptic, lanceolate, to lunate, with apices that vary from acute, obtuse, retuse to rounded. **Stamens** of unequal length, originating from about halfway up the corolla tube, displaced such as to aggregate together on the *lower* part of the corolla mouth; **anthers** linear to elliptical, versatile, sagittate; **pollen** in monads with either *glabra*-type (3-colporate, rarely porate, 23–44 × 26–42 μm, spheroidal to subspheroidal, with reticulate to verrucose-gemmate exine, and usually relatively wide muri), or *corymbosa*-type (3-colporate to colporate, 28–35 × 28–39 μm, subspheroidal to spheroidal, and with warty exine [verrucae, gemmae, pila, and clavae are present]). **Ovary** sessile, bilocular, placentation axile or parietal with strongly inrolled placentas; **style** persistent in fruit; **stigma** bilamellate with broad lobes. **Capsule** dry,

Fig. 3.16 *Macrocarpaea angelliae* J.R. Grant. *Photo* J.R. Grant

woody, 2–carpelled, bilocular and dehiscing medially; **seeds** flattened, polygonal, rimmed, winged or winged perimetrically, miniscule, ca. 0.2–1.6 × 0.2–0.6 mm.

3.3.8.2 Species Descriptions

1. ***Macrocarpaea angelliae*** J.R. Grant & Struwe

Harvard Pap. Bot. 5(2): 490. 2001. TYPE: ECUADOR. Zamora-Chinchipe: Carretera Loja-Zamora, cerca al Paso, 2,800–2,900 m, 15 March 1989, *Romoleroux 804* (Holotype NY; Isotypes AAU, QCA) (Figs. 3.16, 3.17).

Shrubs 1.0–1.5 m tall; stems terete, solid or hollow, up to 8 mm in diameter, glabrous throughout except lower leaf surfaces. Leaves short-petiolate, ovate, elliptical to lanceolate, 55–95 × 25–40 mm, glabrous, edges strongly revolute (at least when dried), the base rounded to slightly attenuate, the apex acute to acuminate, leathery-coriaceous, the upper side dark green with slightly impressed veins, the lower contrasting light green with raised veins and short brown sessile stellate hairs; ochrea 5–6 mm high; petiole 8–11 mm long. Inflorescences 150–250 mm long, 6–12-flowered; bracts leaf-like, ovate, apex obtuse, 25–41 × 19–29 mm. Flowers pedicellate, erect, not spreading; pedicel 8–11 mm. Calyx campanulate, 6–8 mm long, yellowish-green, the lobes rounded, 1–2 mm, with hyaline margins, remaining erect in fruit. Corolla yellowish-green, 33–36 mm, funnelform, the lobes ovate to elliptical, 8–12 × 8–11 mm, lobe apex acute. Stamens reaching the base of the corolla lobes; filaments of unequal length, filiform, straight; anthers linear, 4.5–5.0 mm long, sagittate, versatile. Gynoecium 30 mm long, ovary 5–6 × 3–4 mm, style 24–25 × 1.0 mm, stigma 2-lobed, each lobe circular, 2–3 × 2 mm. Capsules erect to slightly spreading from another, slightly woody, 25–30 mm (excluding the style base), rugose, shiny, with a persistent, elongate style base to 15 mm long. Seeds unknown.

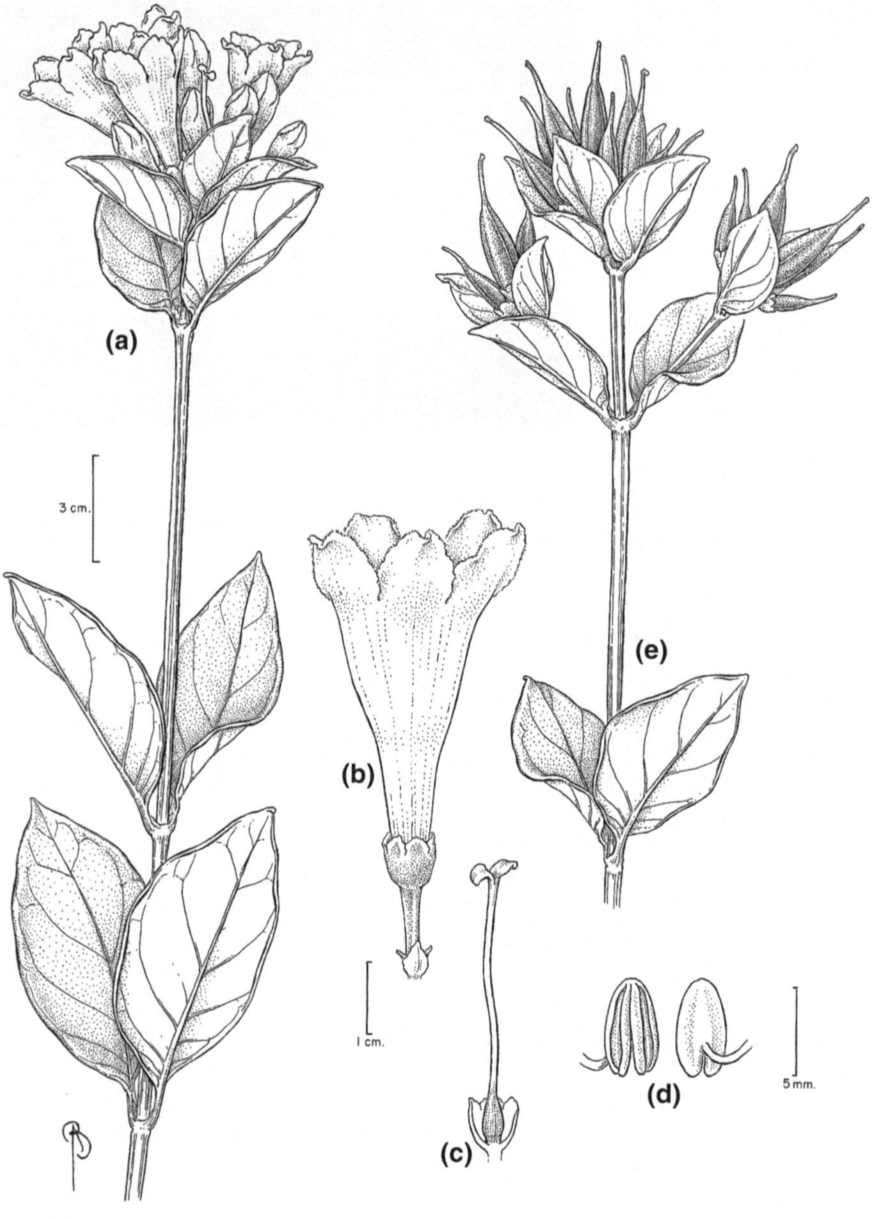

Fig. 3.17 *Macrocarpaea angelliae* J.R. Grant. **a** habit of flowering stem; **b** flower; **c** gynoecium; **d** anthers; **e** fruiting stem. From Øllgaard 91065 (AAU)

Macrocarpaea angelliae belongs to *Macrocarpaea* section *Choriophylla* and occurs from 2,500 to 3,000 m in the Amotape–Huancabamba region of southern Ecuador. It is characterized by large bracts ½ to ¾ the length of the cymes they subtend. The inflorescences are generally nodding. The leaves are also rather leathery, typical of high elevation species.

Etymology. Named for Bobbi Angell (1955–), Marlboro, Vermont, USA, who has prepared the scientific illustrations for all the new species of *Macrocarpaea.*

Specimens examined: ECUADOR. **Loja:** Carr. Yangana-Valladolid, antes del refugio de Quebrada Honda, 04° 28′ 59″ S, 079° 8′ 20″ W, 2,720 m, 8 October 1995, *Garmendia, A. & G. Paredes 678* (QCNE); *Garmendia, A. & G. Paredes 684* (LOJA, QCA, QCNE); 21.3 km past Yangana toward Valladolid; 5 November 2002, *Grant, J.R., A. Roguenant & A. Raynal-Roques 02-4289* (MO, NEU, NY, US); Carr. Loja-Zamora, collecciones en el km 15, 2500 m, 16 Aug 1983, *Jaramillo & Winnerskjold 5806* (QCA); P.N. Podocarpus, ca. 1 km along trail toward Numbala at pass on the Yangana-Valladolid road, 4° 28′ S, 79° 10′ W, 3,000 m, 16 January, 1990, *Knudsen 8* (LOJA, QCA, S); El Firo Oriental; 03° 59′ 43″ S, 079° 07′ 53″ W, 2,800 m, 16 January 2002, *Lozano, P.* et al. *E-290* (LOJA); Nudo de Sabanilla, Sendero a Achupallas, 2,900 m, 21 February 2002, *Lozano, P.* et al. *E-933* (LOJA); Yangana-Valladolid Rd., 2,500–3,000 m, 10 December 1989, *Madsen 86683* (AAU, LOJA, QCNE); **Zamora-Chinchipe:** Summit of Nudo de Cajanuma, 04° 27′ 019″ S, 079° 08′ 904″ W, 2,745 m, 29 January 2009, *Grant, J.R., B. Angell, W. Grant & V. Trunz 09-4568* (NY); Summit of Nudo de Sabanilla, 04° 26′ 495″ S, 079° 08′ 44.8″ W, 2,877 m, 4 February 2011, *Grant, J.R. & J. Vieu 11-4664* (NY); Palanda, Podocarpus National Park, crest of the Nudo de Sabanilla at the continental divide, east of the road pass between Yangana and Valladolid, above the microwave antenna; 04° 26′ 10″ S, 079° 08′ 33″ W, 2,870 m, 3 December 2006, *Neill, D. & L. Jost 15345* (LOJA, MO, NY, QCNE); P.N. Podocarpus, road Yangana-Valladolid, just S and E of the pass (Nudo de Sabanilla), muletrack from pass toward Quebrada Honda, 4° 27′ S, 78° 08′ W, 2,750–2,950 m, 18 February 1989, *Øllgaard 90630* (AAU, QCA); P.N. Podocarpus, 2,750–2,950 m, 14 March 1989, *Øllgaard 91065* (AAU, QCA).

2. *Macrocarpaea apparata* J.R. Grant & Struwe

Harvard Pap. Bot. 8(1): 63. 2003. TYPE: ECUADOR. Loja: 28.8 km S of Yangana, 04° 27′ 59″ S, 079° 08′ 44″ W, 2,450 m, 11 February 2001, *Grant, J.R & L. Struwe 01-4002* [Holotype: US (3 sheets); Isotypes: G (2 sheets), LOJA, MO, NEU (2 sheets), NY, QCA, QCNE, S, SBBG] (Figs. 3.3, 3.18, 3.19).

Dichotomously branched tree, overall branching pattern triangular in outline, 4–5 m tall, hyaline hispid with short simple hairs on stems, petioles, leaves, inflorescences, calyces, and corolla lobes. Trunk to 6.5 cm in diameter, wood hollow (pith 3–19 mm, narrower in trunk, and wider in younger branches), growth rings detectable but not prominent; bark papery thin to scarcely measurable (to 0.05 mm), outer surface smooth to rugose, tan-brown. Stems terete to slightly quadrangular, hollow, 8–16 mm in diameter just below inflorescence. Leaves broadly to narrowly

Fig. 3.18 *Macrocarpaea apparata* J.R. Grant & Struwe. *Photo* J.R. Grant

ovate to elliptic, petiolate, 27–60 cm; blades 24–50 × 14–31 cm, base aequilateral to slightly oblique, apex acute, entire, not revolute, papery thin in texture, dark green, with slightly impressed veins above and slightly raised veins below, glabrous above, hyaline to tan pubescent on lower surfaces and especially veins; interpetiolar ridge 3–10 mm high; petioles robust with strong open vagination nearly equaling the length of the petiole, (30–) 60–100 mm. Branches of the inflorescence 23–40 cm, 14–30 flowered per branch; bracts ovate to lanceolate to linear, 20–165 × 4–100 mm, base aequilateral to oblique, apex acute to acuminate, petiolate; bract petioles 1–20 mm. Flowers pedicellate, erect to horizontal to oriented in all directions in the inflorescence; pedicels 2–26 mm; bracteoles linear to lanceolate to ovate, 3–25 × 1–8 mm. Calyx narrowly campanulate, 7–10 × 6–8 mm, hyaline puberulent, smooth, medium to dark green, ecarinate, no ridges extend down from calyx lobes; calyx lobes dividing calyx to one-third, rounded to obtuse, 2–3 × 3–4 mm. Corolla funnel-shaped, 39–53 mm, 18–22 mm wide at corolla lobe sinuses, whitish-green, rugose to smooth; corolla lobes ovate, apex obtuse to

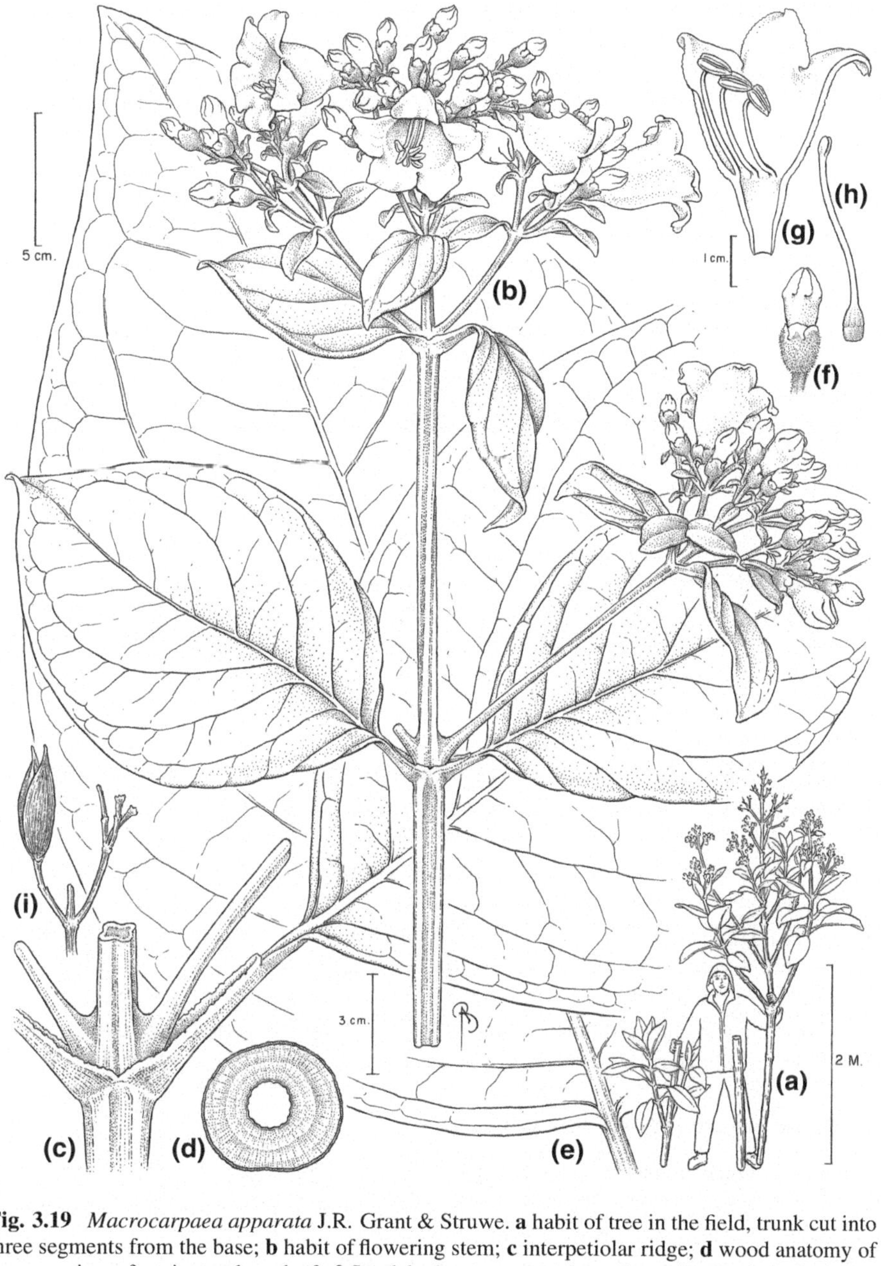

Fig. 3.19 *Macrocarpaea apparata* J.R. Grant & Struwe. **a** habit of tree in the field, trunk cut into three segments from the base; **b** habit of flowering stem; **c** interpetiolar ridge; **d** wood anatomy of cross-section of main trunk; **e** leaf; **f** floral bud; **g** cross-section of corolla; **h** pistil; **i** mature dehisced capsule. **a**, **d**, **f**, **h** from pickles, *photos* and pressed specimens of Grant, J.R & L. Struwe 01-4002; **b**, **c** from Grant, J.R & L. Struwe 01-3999 and *photos* of Grant, J.R & L. Struwe 01-4002; **e** from Grant, J.R & L. Struwe 3998; I from Grant, J.R & L. Struwe 4001

rounded to retuse, 7–14 × 8–12 mm. Stamens 27–33 mm; filaments 20–26 mm, filiform, terete; anthers linear to linear-elliptic, 5.5–7.0 × 2.0–2.5 mm, sagittate, versatile. Pistil 35–39 mm; ovary 7–8 × 3.0–3.5 mm; style 22–26 × 1.0–1.5 mm; stigma 2-lobed, lobes 4–5 × 2.0–2.5 mm, rounded to spathulate. Capsules and seeds unknown.

Macrocarpaea apparata belongs to *Macrocarpaea* section *Choriophylla* and occurs from 2,400 to 1,600 m in the Amotape–Huancabamba region of southern Ecuador. It is a robust treelet to tree to 4–5 m tall with some of the largest leaves known to date within the genus. It is mostly closely related to *M. elix* with which it shares hairy leaves and calyces. This species is part of a distinctive monophyletic group including *M. apparata*, *M. bubops*, *M. elix*, *M. jensii*, *M.* × *mattii*, and *M. noctiluca*.

Etymology. From the English verb "to apparate."

Specimens examined: ECUADOR. **Zamora-Chinchipe:** Nudo de Sabanilla, ca. 24 km S of Yangana, at border of P.N. Podocarpus, roadside right after stream that crosses the road, 2,550 m, 11 February 2001, *Grant, J.R & L. Struwe 01-3995* [LOJA, QCNE, NEU, NY, US (2 sheets)]; 28.8 km S of Yangana, 04° 27′ 59″ S, 079° 08′ 44″ W, 2,560 m, 11 February 2001, *Grant, J.R & L. Struwe 01-3997* [LOJA, QCNE, US (2 sheets)]; 29.6 km S of Yangana, 04° 27′ 59″ S, 079° 08′ 44″ W, 2,450 m, 11 February 2001, *Grant, J.R & L. Struwe 01-3998* [LOJA, QCNE, NEU, NY, US (2 sheets)]; 29.6 km S of Yangana, 04° 27′ 59″ S, 079° 08′ 44″ W, 2,450 m, 11 February 2001, *Grant, J.R & L. Struwe 01-3999* [LOJA, QCNE, NEU, NY, US (2 sheets)]; 30.9 km S of Yangana, 04° 27′ 59″ S, 079° 08′ 44″ W, 2,450 m, 11 February 2001, *Grant, J.R & L. Struwe 01-4001* [LOJA, QCNE, NEU (4 sheets), NY, US (4 sheets)]; 30.1 km past Yangana toward Valladolid, 5 November 2002, *Grant, J.R., A. Roguenant & A. Raynal-Roques 02-4292* (NEU, NY, US). Nudo de Sabanilla, ca. 24 km S of Yangana toward Valladolid, at border of P.N. Podocarpus, Tapichalaca Reserve, 2,450 m, 17 February 2006, *Grant, J.R., M.L. Cheung, F. F. Luisier, & N. Villard 06-4340* (CHRB, G, LOJA, MO, NY, QCA, QCNE); Around 30 km south of Yangana toward Valladolid, south of the continental divide at Nudo de Sabanillas, 04° 28′ 26.4″ S, 079° 08′ 50.4″ W, 2,601 m, 4 February 2011, *Grant, J.R. & J. Vieu 11-4665* (NY); Región del Río Chinchipe, por la carretera entre Palanda y Valladolid, 04° 36′ 47″ S, 079° 07′ 20″ W, 1,300 m, 11 March 2007, *Quizhpe, W. & A. Wisum 2653* (LOJA, MO, NY).

3. ***Macrocarpaea arborescens* Gilg**

Bot. Jahrb. Syst. Beibl. 111: 50. 1913. TYPE: Colombia. Cauca: W Andes of Popayan, 1,800–2,500 m, 1 January 1882, *Lehmann 5450* [lectotype GH; isolectotypes F, K, US, selected by Grant (2003)] (Fig. 3.20).

Shrub or small tree, 0.5–6.0 m, woody throughout, often much branched, glabrous. Stems quadrangular, sometimes less sharply angles or terete distally; internodes much shorter than to about equaling leaves, 0.5–12.0 cm long. Leaves coriaceous; secondary veins slender, branching from midrib mostly at 50–80 degrees, nearly straight, becoming indistinct toward margin, not or imperfectly

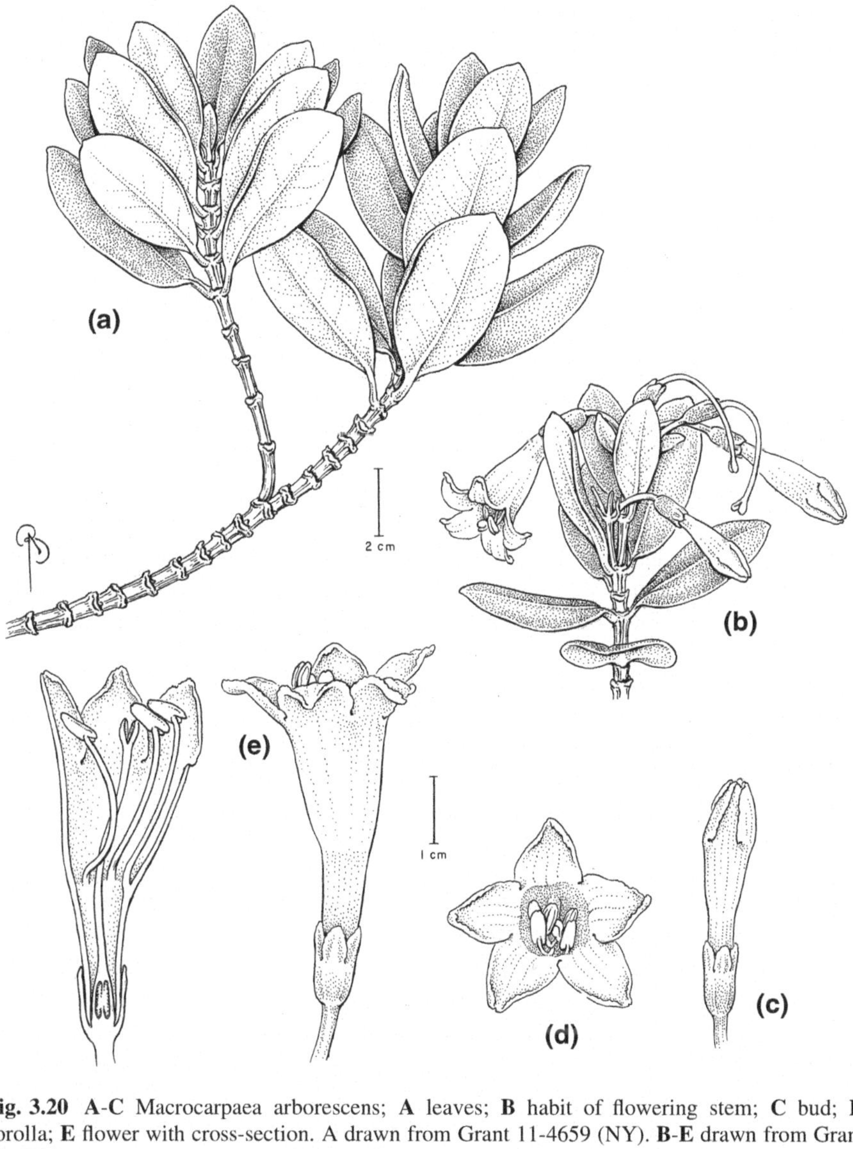

Fig. 3.20 **A**-**C** Macrocarpaea arborescens; **A** leaves; **B** habit of flowering stem; **C** bud; **D** corolla; **E** flower with cross-section. A drawn from Grant 11-4659 (NY). **B**-**E** drawn from Grant 01-4084

brochidodromous. Blades of lower and middle cauline leaves elliptic to obovate, 3–12 cm long, 2–7.5 cm wide; base rounded or abruptly tapering to a clasping petiole 2–10 mm apex rounded to obtuse, more or less mucronulate; margin narrowly cartilaginous, revolute. Upper carlines leaves similar or gradually smaller, subsessile or sessile. Inflorescence a more or less compact cyme, 1–5 times divided, leafy throughout or on larger plants with lower divisions subtended by reduced leaves, upper by minute bractlets; flowers 3–35. Pedicels erect or arcuate, 2–20 mm, sometimes appearing longer because uppermost bractlets often subtend a single flower. Calyx 7–14 mm; lobes semicircular to ovate-triangular, 0.5–1 times as long as tube, obtuse to subacute, margins narrowly hyaline distally, erose-fimbriate. Corolla 25–75 mm, pale greenish yellow, funnel-form to campanulate; tube nearly straight or moderately curved, usually scarcely widening for one-third to nearly half its length, then flaring widely; lobes 7–18 mm, ovate-triangular, obtuse to subacute. Capsule ovoid, 15–25 mm. This description is largely from Pringle (1995).

M. arborescens occurs from 2,400 to 3,300 m in the Amotape–Huancabamba region of southern Ecuador and has no known close relatives. It is a characteristic species in being a much-branched shrub with distinctive flowers that are always spreading to pendent. This species is part of a group of species with smooth leathery leaves where the secondary leaf veins absent or scarcely visible including *M. arborescens, M. ericii, M. harlingii, M. lilliputiana, M. subsessilis*, and *M. zumbae*.

Specimens examined: ECUADOR. 1 January 1840, *Lobb 22* (F, K, W); **Azuay**: Páramo de Castillo, 2,745–3,355 m, 18 August 1945, *Camp 4795* (NO, NY, S); Sevilla de Oro, en el Camino Sevilla de Oro-Mendez, llegando al Páramo del Castillo, 3,300 m, 29 August 1996, *Garmendia & Cisneros 1113* (LOJA, QCNE); Sevilla de Oro, en el Camino Sevilla de Oro-Mendez, llegando al Páramo del Castillo, 3,300 m, 31 August 1996, *Garmendia & Cisneros 1171* (LOJA, QCNE); Road in construction Sigsig-Gualaquiza, Campamento Molón, ca. 1 km below Molón, 2,800–3,000 m, 11 December 1968, *Harling, G.* et al. *8236* (GB, MO, QCA); Canton Oña, Vía Urdaneta Yacuambi km 15, 3136 m, 7 December 2008, *Jadán, P. & P. Pesantes 55* (LOJA); Pica Sevilla de Oro-Méndez "collecctiones entre Sevilla de Oro-Páramo del Castillo," 2,800–3,250 m, 7 August 1983, *Jaramillo 5588* (MO, QCA); Carr. Zigzig-Molón-Gualquiza, 2,810 m, 6 August 1986, *Jaramillo* et al. *8856* (AAU, HUA, NY, QCA); Km 74 de Cuenca, Carr. Zigzig-Molón-Gualaquiza, 2,790 m, 6 August 1986, *Jaramillo* et al. *8866A* (QCA). **Cañar**: At pass on road Pindilig-Rivera ("La Virgen"), 3,200 m, 9 March 1985, *Harling, G. & Andersson 22989* (GB, MO, QCA). **Loja**: Loja, 3,000 m, 1 Dec 1876, *André 4549* (F, GH, K, NY); Loma de Oro, 10 km S of Saraguro, 3,300 m, 2 January 1981, *Balslev 1385* (AAU, NY); Road Loja-Saraguro, 3,000 m, 15 February 1987, *Bohlin* et al. *1357* (GB, QCA); Nudo de Cajanuma, 3,000 m, 2 July 2000, *Chassot 00-15* (NEU); 5 km S of Saraguro, 2,600–2,700 m, 7 October 1988, *Ellemann 66594* (AAU, LOJA, QCA, QCNE); Hac. Horta-Naque, 3,100 m, 7 November 1946, *Espinosa, R. 897* (K, LOJA); Cajanuma Field Station, and trail to Mirador, 2,950 m, 17 February 2001, *Grant, J.R. & L. Struwe 01-4066* (G, LOJA, NEU, NY, QCA, QCNE, US); From Loja-Saraguro road, then 2.2 km on road toward Fierro Urco, 2,900 m, 18 February 2001, *Grant, J.R. & L. Struwe 01-4075* (G, LOJA, NEU, NY, QCA, QCNE); From Loja-Saraguro road, then 3 km on road toward Fierro Urco, 2,900 m, 18 February 2001, *Grant, J.R. & L. Struwe 01-4084* (G, LOJA, NEU, NY, QCNE); 21.3 km past Yangana toward Valladolid, 5 November 2002, *Grant, J.R., A. Roguenant & A. Raynal-Roques 02-4290* (NEU, NY); Km 8.9 on road from Yangana-Cerro Toledo, 2,450 m, 29

December 2007, *Grant, J.R., C. Agier, C. Arnold & M.L. Cheung 07-4506* (LOJA, MO, NY); Between km 10–20 on road from Yangana up Cerro Toledo, 04° 23′ 05.2″ S, 079° 07′ 29.4″ W, 2,892 m, 3 February 2011, *Grant, J.R. & J. Vieu 11-4659* (NY); Loma de Loro, 3,200 m, 11 February 1985, *Harling, G. & Andersson 21918* (GB); Parque Nacional Podocarpus, Cajanuma entrance, upper montane forest and elfin forest, 04° 07′ W, 079° 10′ W, 2,980 m, 25 September 2007 *Homeier, J. 2637* (LOJA, QCNE); Saraguro, area de Inka Pirca, colectado en la parcela 1, 2,800 m, 26 November 2003, *Jadán & Veintimilla 15* (LOJA); Mountains below Saraguro and San Lucas, 8 Sep 1865, *Jameson 67* (K); Loma del Oro, 2,800–3,200 m, 4 August 1986, *Jaramillo* et al. *8816* (MO, QCA); Saraguro-Loja, km 12.4, turnoff toward Fierro Urco, 3,120–3,390 m, 7 December 1994, *Jørgensen* et al. *1314* (LOJA, MO, NY, QCA, QCNE); P.N. Podocarpus, Cajanuma, 3,000 m, 29 October 1977, *Lewis 3674* (AAU, K, LOJA, MO QCNE); Parque Nacional Podocarpus, "Lagunas de Banderillas," 04° 13′ 26″ S, 079° 6′ 38″ W, 15 December 1994, *Lozano, P.* et al. *55* (LOJA); Cajanuma Field Station, 3,000 m, 29 January 2002, *Lozano, P.* et al. *E-487* (LOJA); Yangana-Cerro Toledo, 04° 23′ 04″ S, 079° 07′ 44″ W, 2,900 m, 20 February 2002, *Lozano, P.* et al. *E-760* (LOJA); Páramos de Saraguro, páramo ca. 7 km S of Saraguro, 3,000 m, 2 January 1979, *Luteyn* et al. *6669* (NY, QCA); P.N. Podocarpus, Nudo de Cajanuma, 3,000–3,150 m, 6 September 1988, *Madsen & Elleman* 75280 (AAU, LOJA, MO, NY, QCA, QCNE); P.N. Podocarpus, above the refuge of Cajanuma, at the point 'Mirador', 2,950 m, 1 November 1997, *Matt, F. 2* (ER); P.N. Podocarpus, above the refuge of Cajanuma, at the point 'Mirador', 2,950 m, 1 November 1997, *Matt, F. 3* (ER); Ca. Km 5 rd Fierro Ucu from Pichig, 3,100–3,200 m, 12 February 1989, *Øllgaard & Madsen 90494* (AAU, LOJA, MO, QCA, QCNE); P.N. Podocarpus, S of Loja, 2,950–3,000, 22 February 1985, *Øllgaard* et al. *57953* (AAU, LOJA, MO, QCA, QCNE); P.N. Podocarpus, S of Loja, 3,000–3,200 m, 22 February 1985, *Øllgaard* et al. *57983* (AAU, LOJA, MO, NY, QCA, QCNE); Páramo de Saraguro, 17 March 1983, *Pipoly 6393* (GB, NY); Hualaseo, 1 January 1805, *Tafalla 540* (MA-RUIZPAVON); Loja, 1 January 1805, *Tafalla 572* (MA-RUIZ PAVON); Hualaseo or Loja, 1 January 1805, *Tafalla 540 or 572* (BM, F, G); Cajanuma, 3,000 m, 23 January 2001, *Wolff, D. & F. Matt 56* (UTB). **Morona-Santiago**: Sigsig-Gualaquiza, 9 April 1968, *Harling, G.* et al. *8117* (GB). **Zamora-Chinchipe**: Río Yacuambi, region de Rumetranca, Cooral-Huaico, 2,300–3,000 m, 1 January 1944, *Goetschel, J. 94945* (VEN).

4. *Macrocarpaea bubops* J.R. Grant & Struwe

Harvard Pap. Bot. 8(1): 66. 2003. TYPE: ECUADOR. Zamora-Chinchipe: Along new road Loja-Zamora, 2,500 m, 25 April 1987, *van der Werff, H. & W. Palacios 8986* (Holotype: NY; Isotypes: AAU, G, GB, MO, QCNE) (Figs. 3.21, 3.22. 3.28 and 3.35).

=*Macrocarpaea berryi* J.R. Grant, Harvard Pap. Bot. 9(2): 332. 2005. TYPE: ECUADOR. Zamora-Chinchipe: *Berry, P. & D. Neill 7665* (Holotype QCNE; Isotypes, MO, NY).

Fig. 3.21 *Macrocarpaea bubops* J.R. Grant & Struwe. *Photo* J.R. Grant

Dichotomously branched tree, much branched from about 2 m above the ground; overall branching pattern conical in outline, 1–10 m tall, glabrous throughout except lower leaf surfaces with short papillae. Trunk to 9.5 cm diameter wood always solid without any hollow cavities, growth rings prominent; bark thick, pithy to 4 mm thick, outer surface rugose, brown. Stems terete, solid, 5–12 mm in diameter just below inflorescence. Leaves narrowly ovate to elliptic, petiolate, 8.5–22.0 cm; blades 7.5–19.5 × 4.0–11.5, cuneate to rounded, margin entire, slightly revolute, base aequilateral to slightly oblique, apex acute, thick, leathery-coriaceous, dark green, with no or few impressed veins above, and slightly raised veins below, glabrous above and typically below, yet often with short tuberculate hairs on lower veins, and on most herbarium specimens (and especially visible on living material) on the underside of most leaves, at the base, on either side of the midrib just above the petiole, there are two orange dots; interpetiolar ridge 1–5 mm high; petioles robust with strong open vagination nearly equaling the length of the petiole, 10–35 mm. Branches of the inflorescence 14–36 cm, 5–20 flowered per branch. Bracts narrowly ovate to elliptic to obovate to rounded, 22–60(−170) × 6–32(–80) mm, base aequilateral, apex acute, petiolate; bract petioles 2–14(–30) mm. Flowers pedicellate, erect to horizontal to oriented in all directions in the inflorescence; pedicels 4–20 mm; bracteoles linear to ovate to obovate, 2–20 × 2–9 mm. Calyx narrowly to broadly campanulate, 7–11 × 7–10 mm, glabrous, smooth, green, ecarinate, often with pronounced ridges between calyx lobes extending to the base of the calyx; calyx lobes dividing calyx one-third to one half, rounded to obtuse, 3–4 × 3–5 mm, often with a slightly raised orange keel. Corolla funnel-shaped, 32–51 mm, 13–22 mm wide at corolla lobe sinuses, white to greenish yellow, smooth; corolla lobes ovate to elliptic, apex obtuse to rounded to retuse, 5–10 × 7–11 mm. Stamens 17–23 mm; filaments 12–19 mm, filiform, terete; anthers elliptic to oblong, 4–5 × 2.0–2.5 mm, sagittate, versatile. Pistil 22–29 mm; ovary 5–6 × 2.5–4.0 mm; style 14–19 × 0.5–1.0 mm; stigma 2-lobed, lobes 3–4 × 1–2 mm. Capsules and seeds unknown.

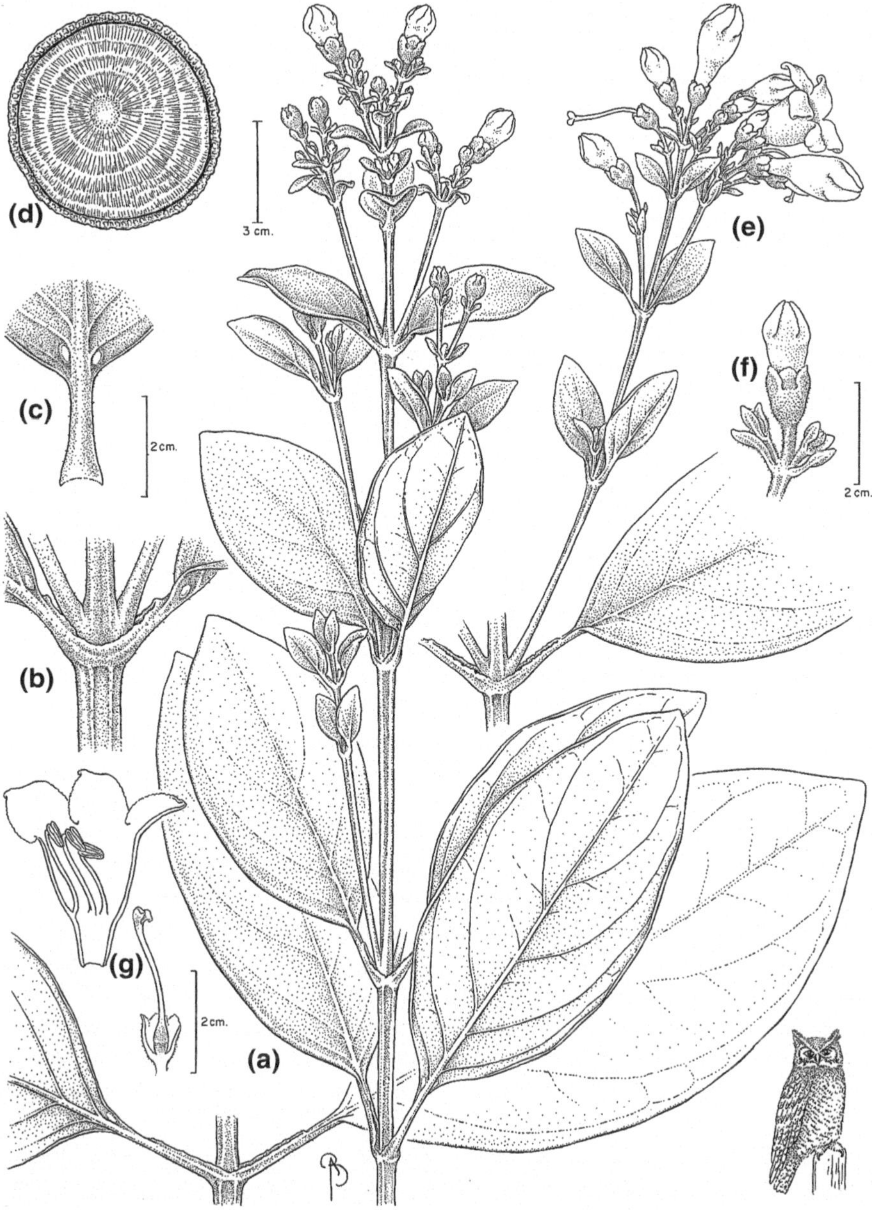

Fig. 3.22 *Macrocarpaea bubops* J.R. Grant & Struwe. **a** habit of flowering stem and leaves; **b** interpetiolar ridge; **c** lower surface of leaf showing two *orange dots*; **d** wood anatomy of cross-section of main trunk; **e** flowering stem; **f** floral bud; **g** cross-section of corolla and pistil. **a–d**, **f** from pickles, *photos* and *specimens* of Grant, J.R & L. Struwe 01-4046; **e**, **g** from Madsen 74050 (AAU)

Macrocarpaea bubops belongs to *Macrocarpaea* section *Choriophylla* and occurs from 1,100 to 2,800 m in the Amotape–Huancabamba region of southern Ecuador and northern Peru. On the underside of most leaves, at the base, on either side of the midrib just above the petiole, are two, orange dots that become translucent when preserved in alcohol. Similar colored dots also occur on the related species *M. noctiluca* and *M.* × *matii*. It is part of a distinctive monophyletic group including *M. apparata*, *M. bubops*, *M. elix*, *M. jensii*, *M.* × *mattii*, and *M. noctiluca*.

Etymology. From the Latin "*Bubo*," owl, and the Greek "*ops*," eye, for the orange dots on the underside of most leaves.

Specimens examined: ECUADOR. **Loja:** En una pista que sale de la carretera de Loja a Zamora, 2,600 m, 28 September 1995, *Garmendia & Paredes 311* (LOJA, QCNE); Loja-Zamora rd, km 11, 79° 11′ W, 03° 59′ S, 2,630 m, 23 February 1988, *Jørgensen 65084* (AAU, QCA); Estación San Francisco, 03° 59′ 08″ W, 079° 04′ 23″ W, 2,300 m, 22 January 2002, *Lozano, P.* et al. *E-337* (LOJA); Loja-Zamora rd., approx. km 15, 79° 08′ W, 03° 58′ S, 2,400–2,700 m, 23–24 April 1988, *Madsen 74050* (AAU, LOJA, MO, QCA); Zamora Huaycu, 04° 02′ 06″ S, 079° 09′ 42″ W, 2,366 m, 7 September 2007, *Rivas, V.M. & E. Guanín A. 21* (LOJA); Zamora Huaycu, 2,600 m, 1983 *Vivar C., F.* et al. *2017* (LOJA). **Morona-Santiago:** Elfin forest and semi-open areas along sandstone escarpment above and west of Tikimints, valley of Río Coangas (on topo maps and known locally as "Ijiach Naimt" in Shuar, meaning "Ridge of the Grubs," 03° 15′ 49″ S, 078° 10′ 12″ W, 2,000 m, 20 March 2001, *Berry, P. & D. Neill 7630* (MO, QCNE); *Berry, P. & D. Neill 7640* (MO, QCNE); Elfin forest and semi-open areas along sandstone escarpment above and west of Tikimints, valley of Río Coangas (on topo maps and known locally as "Ijiach Naimt" in Shuar, meaning "Ridge of the Grubs," 03° 15″49′ S, 078° 10′ 12″ W, 2,000 m, 20 March 2001, *Berry, P. & D. Neill 7665* (type of *M. berryi*, MO, NY, QCNE); Limón Indanza, Cordillera del Cóndor, Centro Shuar Warints, Recorrido desde la Cumbre del Cerro Chankinias hasta la cumbre de la cordillera, 03° 13′ 40″ S, 078° 17′ 29″ W, 2,200 m, 15 October 2002, *Kajekai, C. & Grupo Shuar de Conservación 41* (MO); San Juan Bosco, Cima de la Cordillera del Cóndor, Centro Shuar Numpatkaim. Bosque enano arbustivo y páramo de Bromeliaceae con arbustivos, 03° 16′ 08″ S, 078° 19′ 06″ W, 2,700–2,820 m, 22 July 2005, *Katán, T.*, et al. *367* (LOJA, MO, NY, QCNE); San Juan Bosco, Cima de la Cordillera del Cóndor, Centro Shuar Numpatkaim. Bosque enano arbustivo y páramo de Bromeliaceae con arbustos. Arbolito de 3 m, flores tubulares, 03° 16′ 08″ S, 078° 19′ 06″ W, 2,700–2,820 m, 22 July 2005, *Katán, T.*, et al. *425* (MO, NY, QCNE); San Juan Bosco, Cordillera del Cóndor, Comunidad Shuar Numpatkaim, 03° 15′ 23″ S, 078° 18′ 59″ W, 2,490–2,700 m, 24 July 2005, *Morales, C. & Embrey 1413* (MO, NY, QCNE); Limon Indanza Cantón, Cordillera de Huaracayo, east of Cordillera del Cóndor and Río Coangos, Cerro Ijiach Naimt, flat-topped sandstone mountain, east of Shuar village of Tinkimints, low dense montane forest and shrubby vegetation on summit, 03° 15′ 49″ S, 78° 10′ 13″ W, 2,000 m, 21 March 2001, *Neill, D.* et al. *13102* (MO, QCNE); Cordillera de Cutucu, centro Shuar, Angel Roubi, Pie de la Cordillera, 02° 21′ 41″ S, 078° 02′ 24″ W, 1,100 m, 26 March 2002, *Nicolalde, F.*

et al. *1326* (MO, QCNE); Cordillera del Cóndor, recorrido desde la cumbe del Cerro Chankinias a 2,200 m hasta la cumbre de la Cordillera del Condor a 2,700 m, 03° 14′ 41″ S, 078° 17′ 29″ W, 2,200 m, 15 October 2002, *Tsuink & Grupos Shuar de Conservación 37* (MO); **Zamora-Chinchipe:** Canton Zamora, sector el Tambo, Estación Cientifica "San Francisco," 2,170 m, 17 April 2000, *Cerna 109* (QCNE); Carretera Loja-Zamora, Cuenca del rio San Francisco, Fundacion Arco Iris, 2,100 m, 03° 59′ S, 79° 05′ W, bosque de neblina montano, transectos de 50 × 4 *M.* × 5 (0.1 Ha.), especies de 2.5 cm. de DAP en adelante (modelo radial), arbol de 7 m × 10 cm. de DAP, 27 February 2000, *Cerón* et al. *40136* (QAP); Yantzaza, parroquia las Peñas, Kinross Aurelian, Fruta del Norte, Cordillera del Cóndor, 03° 47′ 56″ S, 078° 28′ 56″ W, 1,900 m, 2 November 2011, *Cerón, C.* et al. *70342* (QAP); *Cerón, C.* et al. *70511* (QAP); *Cerón, C.* et al. *70615* (QAP); Estación Cientifica San Francisco, research station along the highway on the edges of the city of Loja to Zamora at approx. 30 km, 03° 58′ 18″ S, 079° 04′ 44″ W, 1,800–3,200 m, 31 January 1999, *Dziedzioch, C. 163* (LOJA); Parroquia Sabanilla, a 30 km de Loja, Cuesta occidental de la Cordillera del Consuelo, 03° 58′ S, 079° 04′ W, 2,000 m, 3 February 2003, *Fernández, E. & J. Quesada 989* (QCNE); Parque Nacional Podocarpus, San Francisco entrance area, between Loja and Zamora, 03° 59′ 17.1″ S, 079° 05′ 35.4″ W, 2,169 m, 10 February 2011, *Grant, J.R. & J. Vieu 11-4681* (NY); Parque Nacional Podocarpus, San Francisco entrance area, between Loja and Zamora, 03° 59′ 29.1″ S, 079° 05′ 46.2″ W, 2,103 m, 10 February 2011, *Grant, J.R. & J. Vieu 11-4686* (NY); P.N. Podocarpus (San Francisco entrance), trail leading west from San Francisco, 03° 59′ 24″ S, 079° 05′ 48″ W, 2,100 m, 16 February 2001, *Grant, J.R & L. Struwe 01-4046* (G [2 sheets], LOJA, MO, NEU [3 sheets], NY, QCA, QCNE, S, SBBG, US [3 sheets]); Cordillera del Condor, 35 km E of Los Encuentros in direction of Paquisha Alta Military Camp, 1,617 m, 24 February 2006, *Grant, J.R., M.L. Cheung, F. F. Luisier, & N. Villard 06-4349* (G, LOJA, MO, NY, QCA, QCNE); RBSF [Estacion Cientifica San Francisco], transekt 1, above NUMEX plots, 2,340 m, 11 November 2009, *Hillmann, B.M. 50* (QCA); Area of the Estación Cientifica San Francisco, road Loja-Zamora, ca 35 km from Loja, 03° 58′ S, 079° 04′ W, 2,340 m, 13 March 2000, *Homeier, J. 328* (LOJA); Reserva Quebrada de Monos, 04° 01′ S, 079° 01′ W, 2,300 m, 24 January 1998, *Lozano, P.* et al. *920* (LOJA); Estación San Francisco, 03° 59′ 08″ S, 079° 04′ 23″ W, 2,400 m, 22 January 2002, *Lozano, P.* et al. *E-359* (LOJA); Estación San Francisco, 03° 59′ 41″ S, 079° 04′ 07″ W, 2,600 m, 22 January 2002, *Lozano, P.* et al. *E-414* (LOJA); Estación San Francisco, 03° 59′ 41″ S, 079° 04′ 07″ W, 2,200 m, 22 January 2002, *Lozano, P.* et al. *E-445* (LOJA, QCA); Cordillera del Cóndor region, Río Machinaza matershed, easte of Los Encuentros, summit area of Hollín sandstone plateau, northwest of "big bend" of Río Machinaza, 03° 45′ 06″ S, 078° 31′ 23″ W, 1,720 m, 26 June 2009, *Neill, D. & C. Kajekai 16974* (MO, QCNE); Cordillera del Cóndor, El Pangui, summit of the Contraduerte de Wawaime, a sandstone butte isolated by about 2 km from the main sandstone plateau of the Cordillera del Cóndor, 03° 35′ 12″ S, 078° 26′ 21″ W, 1,660 m, 7 December 2005, *Neill, D. & W. Quizhpe 15014* (LOJA, MO, NY, QCNE); El Pangui,

Cordillera del Cóndor, summit of the sandstone sandstone plateau of Cordillera, southeast headwaters of Río Wawaime, 03° 35′ 40″ S, 078° 25′ 11″ S, 1,930 m, 8 December 2005, *Neill, D. & W. Quizhpe 15036* (LOJA, MO, NY, QCNE); El Pangui, Cordillera del Cóndor, summit of sandstone plateau of Cordillera, southeast headwaters of Río Wawaime, 03° 35′ 40″ S, 078° 25′ 11″ W, 1,930 m, 19 September 2006, *Neill, D. & W. Quizhpe 15263* (MO, NY); Centinela del Cóndor, Cordillera del Cóndor, Machinaza plateau area, above Paquisha Alto military post, upper watershed of Río Machinaza, 03° 53′ 54″ S, 078° 28′ 49″ W, 2,350 m, 16 March 2008, *Neill, D. & W. Quizhpe 16177* (LOJA, MO, NY, QCNE); Yantzaza Cantón, Cordillera del Cóndor region, southeast of Los Encuentros, Valley of Río Suárez, tributary of Río Machinaza, along road to Las Peñas camp of Aurelian mine company, 03° 48′ 12″ S, 078° 31′ 21″ W, 1,450 m, 22 March 2008, *Neill, D. & W. Quizhpe 16336* (LOJA, MO, QCNE); Cordillera del Cóndor, carretera desde Los Encuentros hacia el Cerro Machinaza, Parroquia Los Encuentros, Barrio San Antonio, 03° 49′ 39″ S, 078° 31′ 05″ W, 1,470 m, 17 July 2005, *Quizhpe, W. 1563* (F, LOJA, MO, NY, QCNE); Cordillera del Cóndor, carretera desde Los Encuentros hacia el Cerro Machinaza, sendero a Las Peñas, 1,400 m, 20 July 2005, *Quizhpe, W. 1630* (LOJA, MO, NY, QCNE); Region de la Cordillera del Cóndor, Parroquia San Carlos de las Minas, Namibija, sendero al Cerro Colorado, 04° 04′ 44″ S, 078° 47′ 58″ W, 2,100 m, 26 January 2005, *Quizhpe, W.* et al. *751* (LOJA, MO, NY, QCNE); Palanda, Región de la Cordillera del Cóndor, sector sur, Parroquia San Francisco de Vergel, cuenca alta del Rio Vergel, sendero a Sol Naciente, 04° 43′ 33″ S, 078° 57′ 18″ W, 2,010 m, 11 March 2005, *Quizhpe, W.* et al. *962* (LOJA, MO, NY, QCNE); Palanda, Región de la Cordillera del Cóndor, sector sur, Parroquia San Francisco de Vergel, cuenca alta del Rio Vergel, 04° 54′ 55″ S, 078° 58′ 07″ W, 2,200 m, 12 March 2005, *Quizhpe, W.* et al. *974* (LOJA, MO, NY, QCNE); Estación Cientifica San Francisco, bosque nublado del transecto 1, 03° 58′ 59″ S, 079° 04′ 44″ W, 2,200 m, 1 October 1997, *Rbu & SL 2629* (QCA); Cordillera del Condor, destacamento militar Cóndor Mirador (Cóndor Mirador 2), 1 km al sureste del destacamento, 1,962 m, 7 September 2003, *Rodríguez, E.R.* et al. *2641* (NY); Along new road Loja-Zamora, 2,500 m, 25 April 1987, *van der Werff, H. & Palacios, W. 8984* (MO).

PERU. Amazonas: Condorcanqui, Cordillera del Condor, la pared del "tepui" Cerro Machinaza, cabeceras del Rio Comainas, tributario al oeste del Rio Cenepa, arriba de Puesto de Vigilancia Alfonso Ugarte, subiendo atras y norte del campamento, 03° 52.8 S, 078° 26.0 W, 1,950–2,050 m, 21 July 1994, *Beltrán & R. Foster 1197* (F, USM).

5. ***Macrocarpaea claireae*** **J.R. Grant**

Harvard Pap. Bot. 13(2): 253. 2008. TYPE: ECUADOR. Zamora-Chinchipe: 2 km N of Valladolid, 04° 32′ 354″ S, 079° 07′ 908″ W, 1,767 m, 4 January 2008, *Grant, J.R., C. Agier, C. Arnold & M.L. Cheung 08-4528* (Holotype NY; Isotypes G, QCNE) (Figs. 3.23, 3.24).

Fig. 3.23 *Macrocarpaea claireae* J.R. Grant. *Photo* J.R. Grant

Unbranched shrub to small tree, 1.5–2.0 m, glabrous throughout. Trunk to 2 cm diameter, wood hollow in trunk and in branches, rings scarcely visible; bark papery thin to scarcely measurable, 0.05 mm thick, outer surface smooth to rugose, tan to greenish. Stems terete to quadrangular, hollow, 6–8 mm diameter just below the inflorescence. Leaves oval, elliptic to ovate, petiolate, (17) 26–35 cm long. Petioles (20)35–50 mm, slender with slight vagination; interpetiolar ridge 1–3 mm high. Blades (15) 22.5–30 × (8.5) 11.5–17.5 cm, entire, not revolute, dark green though yellowing along the margins of all major veins, with slightly impressed veins above, and slightly raised veins below, glabrous above and below, papery thin; base aequilateral to oblique, rounded to cuneate and slightly decurrent on the petiole; apex acuminate to acute. Inflorescence a few branched open thyrse, 75+ cm long; branches 10–24 cm long; 5–9 flowered per branch. Bracts ovate, sessile to short-petiolate, 10–60–(120) × 5–(65) mm; base aequilateral to oblique, cuneate to rounded; apex acute to acuminate; bract petiole 0–6 mm. Flowers pedicellate, erect to slightly spreading; pedicels 10–22 mm long; bracteoles linear to lanceolate, 3–11 × 0.5–3.0 mm. Calyx campanulate to urceolate, 7–8 × 6–8 mm, glabrous to faintly scabrous, green, ecarinate, reniform to ovate; calyx lobes 3–4 × 4.0–5.5 mm, apex rounded to obtuse. Corolla funnel-shaped, 30–33 mm long, 20–23 mm wide at the apex of the tube, light green, smooth; corolla lobes ovate, 10–12 × 8–10 mm; apex obtuse to rounded. Stamens 25.5–28.0 mm long; filaments 22–24 mm long, filiform, flattened; anthers elliptic to oblong, 3.5–4.0 × 1.5–2.0 mm, sagittate, versatile; pollen glabra-type. Pistil 30-32 mm long; ovary 6–8 × 2–3 mm; style 21–22 × 0.5–1.0 mm; stigma lobes spathulate, 2–3 × 1–2 mm. Capsules and seeds unknown.

Macrocarpaea claireae belongs to *Macrocarpaea* section *Choriophylla* and occurs from 2,300 to 1,800 m in the Amotape–Huancabamba region of southern Ecuador. It is closely related to *M. xerantifulva* and even forms mixed stands together along the roadside near Valladolid. This species belongs to a closely related group of species restricted to the Amotape–Huancabamba zone including

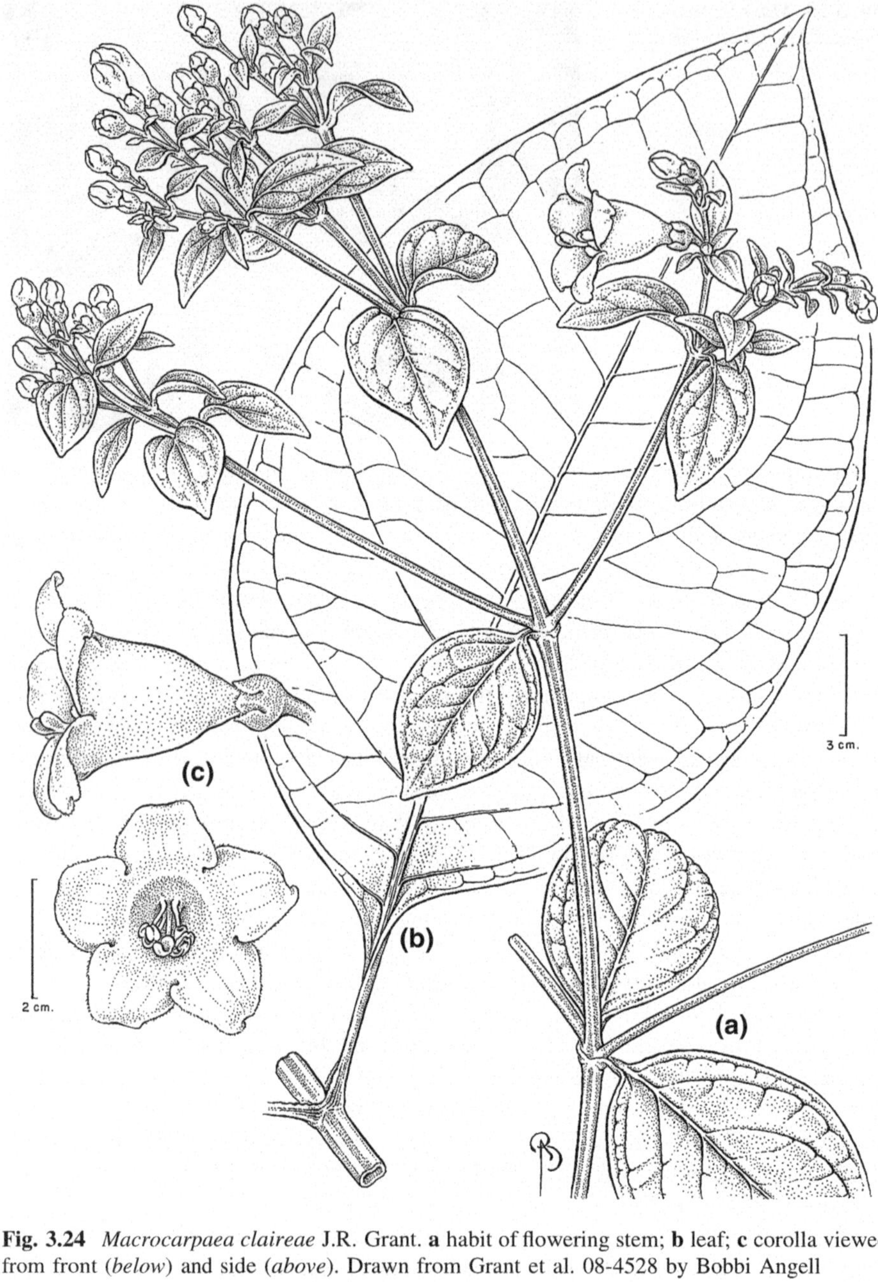

Fig. 3.24 *Macrocarpaea claireae* J.R. Grant. **a** habit of flowering stem; **b** leaf; **c** corolla viewed from front (*below*) and side (*above*). Drawn from Grant et al. 08-4528 by Bobbi Angell

Fig. 3.25 *Macrocarpaea dies-viridis* J.R. Grant. *Photo* J.R. Grant

M. catherineae, M. claireae, M. dies-viridis, M. illuminata, M. lenae, M. quizhpei, and *M. xerantifulva*.

Etymology. Named for Dr. Claire Elvire Arnold (1968–), co-collector of the type, Swiss plant population biologist.

Specimens examined: ECUADOR. **Zamora-Chinchipe:** 1.5 km north of Valladolid, 04° 32′ 26.9″ S, 079° 07′ 48.2″ W, 1,779 m, 6 February 2011, *Grant, J.R. & J. Vieu 11-4666* (NY); Palanda, région del Río Chinchipe, por la carretera entre Palanda y Valladolid, 04° 36′ 47″ S, 079° 07′ 20″ W, 1,300 m, 11 March 2007, *Quizhpe, W. & A. Wisum 2652* (LOJA, MO, QCNE).

6. *Macrocarpaea dies-viridis* J.R. Grant

Harvard Pap. Bot. 11(2): 129 2007. TYPE: ECUADOR. Zamora-Chinchipe: Cordillera del Cóndor, 8.8 km E of Paquisha, 03° 56′ 03″ S, 078° 37′ 32″ W, 1,369 m, 27 February 2006, *Grant, J.R., M.L. Cheung, F. Luisier & N. Villard 06-4352* (Holotype NY; Isotypes G, LOJA, MO, QCA) (Figs. 3.25, 3.26).

Shrub or small tree, 1–4 m, glabrous throughout, though calyces are hyaline scabrous with short simple hairs. Trunk to 3 cm in diameter, wood solid in lower portions to hollow above and in branches, rings scarcely visible; bark papery thin to scarcely measurable, 0.05 mm thick, outer surface smooth to rugose, tan to greenish, blotched with lichens and mosses. Stems terete, hollow, 6–14 mm in diameter just below the inflorescence. Leaves ovate to ovate-elliptic, long-petiolate, (17)–44–51 cm long. Petioles (20)–80–100 mm, slender with slight vagination; interpetiolar ridge 2–3 mm high. Blades (15)–33–41 × (7)–18–32 mm, entire, not revolute, dark green, with slightly impressed veins above, and slightly raised veins below, glabrous above and below, papery thin; base aequilateral to oblique, rounded to cuneate and slightly decurrent on the petiole; apex nearly rounded then acuminate, acute to nearly cuspidate. Inflorescence a much branched open thyrse, 40+ cm long; branches 5–30 cm long; 9–20 flowered per branch. Bracts ovate to lanceolate, sessile to short-petiolate, 10–95 × 3–60 mm; base

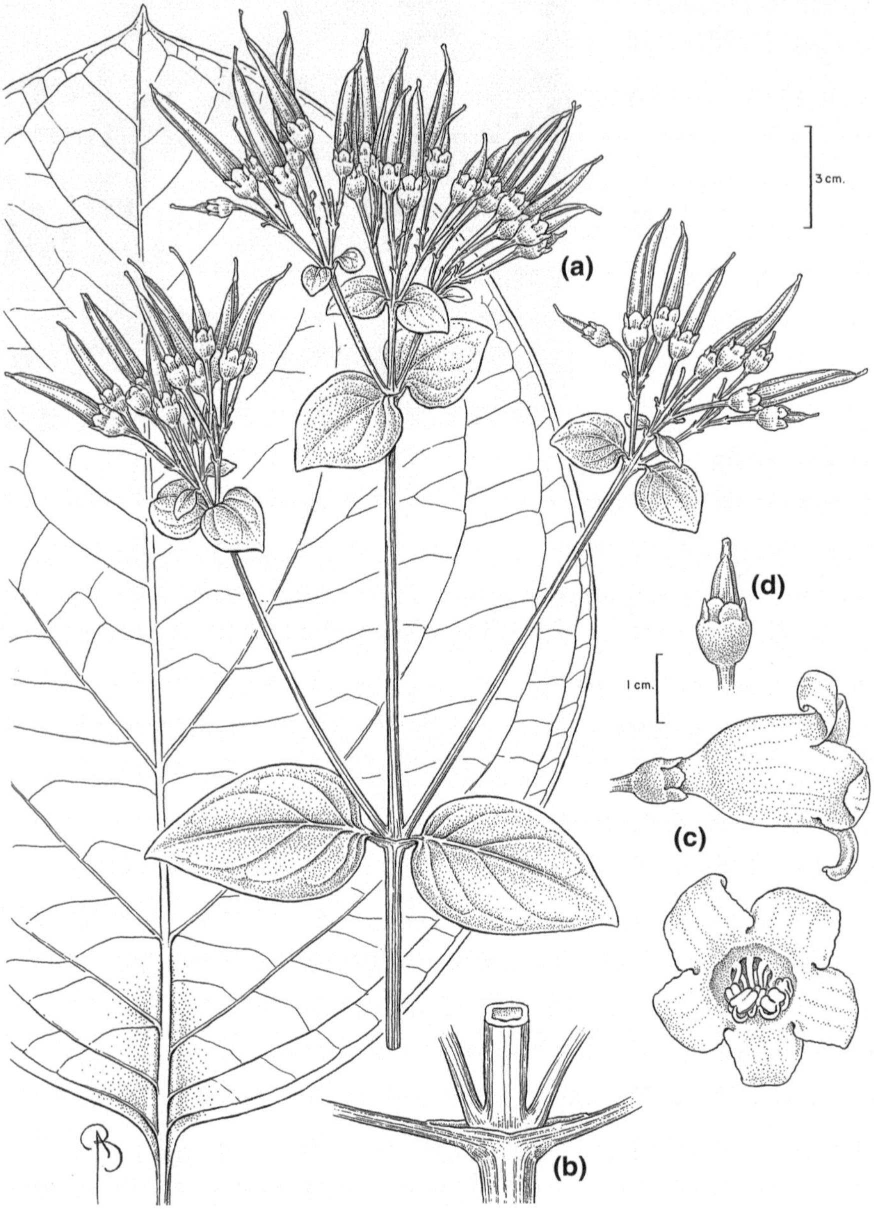

Fig. 3.26 *Macrocarpaea dies-viridis* J.R. Grant. **a–d** *Macrocarpaea dies-viridis*. **a** habit of fruiting stem, and leaf behind; **b** interpetiolar ridge; **c** corolla viewed from front (*below*) and side (*above*); **d** young fruit. **a–d** drawn from Grant et al. 06-4348 and Grant et al. 06-4352

Fig. 3.27 *Macrocarpaea elix* J.R. Grant. *Photo* J. Cortina

aequilateral to oblique, rounded, cuneate, to slightly cordate; apex acute to acuminate; bract petioles 0–5 mm. Flowers pedicellate, erect to spreading; pedicels 14–25 mm long; bracteoles linear to lanceolate, 2–10 × 1–3 mm. Calyx campanulate, 6–8 × 6–7 mm, glabrous to hyaline scabrous with short simple hairs, green, ecarinate; calyx lobes ovate to rotund, 3–4 × 4–5 mm, apex rounded to obtuse. Corolla funnel-shaped, yet measurements unknown. Capsules ellipsoidal to linear-long, 32–45 × 6–8 mm, smooth to faintly ribbed, greenish brown to tan, erect; style remnant 5–12 mm long. Seeds "perimetrically winged type," flattened, roughly 3–4 sided in outline, yet appearing as myriads of different puzzle pieces, 0.9–1.3 × 0.9–1.0 mm, bicolored, testa tanish and reticulate, wings straw-colored and ribbed.

Macrocarpaea dies-viridis belongs to *Macrocarpaea* section *Choriophylla* and occurs from 1,369 to 1,558 m in the Amotape–Huancabamba region of southern Ecuador. This species belongs to a closely related group of species restricted to the Amotape–Huancabamba zone including *M. catherineae, M. claireae, M. dies-viridis, M. illuminata, M. lenae, M. quizhpei*, and *M. xerantifulva.* It occurs in the Cordillera del Condor region.

Etymology. From the Latin "dies," day, and "viridis" green, for the American punk rock music group *Green Day*.

Specimen examined: ECUADOR. **Zamora-Chinchipe:** Cordilllera del Condor, Grant 43.9 km E of Los Encuentros in direction of Paquisha Alta Military Camp, 03° 51′ 02″ S, 078° 33″06′ W, 1,558 m, 21 January 2009, *Grant, J.R., B. Angell, W. Grant & V. Trunz 09-4561* (NY); Cordillera del Condor, 43.9 km E of Los Encuentros in direction of Paquisha Alta Military Camp, 1,558 m, 24 February 2006, *Grant, J.R., M.L. Cheung, F. F. Luisier, & N. Villard 06-4348* (G, LOJA, MO, NY, QCA) Cordilllera del Condor, 9.3 km E of Paquisha, 03° 55′ 31.1″ S, 078° 37′ 27.6″ W, 1,518 m, 9 February 2011, *Grant, J.R. & J. Vieu 11-4679* (NY); Cordilllera del Condor, 44.3 km E of Los Encuentros in direction of Paquisha Alta Military Camp, 03° 54′ 02.8″ S, 078° 30′ 41.9″ W, 1,605 m, 13 February 2011, *Grant, J.R. & J. Vieu 11-4693* (NY).

7. ***Macrocarpaea elix* J.R. Grant**

Harvard Pap. Bot. 9(1): 14. 2004. TYPE: ECUADOR. Azuay: highway Cuenca-Cola de San Pablo, km 88, 2,540 m, 14 February 1977, *J.D. Boeke & H. Loyola 988* (Holotype: NY; Isotypes: AAU, CONN, MO, QCA) (Figs. 3.27, 3.50).

Shrub or small tree to 1–10 m, hyaline hispid to spiculate with short simple hairs on stems, petioles, leaves, inflorescences, bracts, and calyces. Stems terete to slightly quadrangular above, hollow to solid, 5–9 mm in diameter just below inflorescence. Leaves elliptic to ovate, petiolate, (11–) 20–33 cm long. Petioles 10–45 long, robust with strong open vagination or groove nearly equaling the length of the petiole; interpetiolar ridge 2–5 mm high. Blades (10–) 19–28.5 × (5–) 11–18 cm, entire, not revolute, dark green above, lighter below, with slightly impressed veins above, and strongly raised veins below, glabrous above, hyaline hispid on all veins below, papery thin, to thin coriaceous; base aequilateral to oblique, cuneate to rounded; apex acuminate to acute. Inflorescence a much branched open thyrse, often appearing glaucous on dried herbarium specimens due to minute white crystals covering upper stems, petioles and calyces, 37+ cm long (the total inflorescence unknown as has been cut into various segments for separate herbarium sheets); branches 11–22 cm long; 9–15 flowered per branch. Bracts obovate, elliptic, obovate, to oblanceolate, sessile to short-petiolate, 15–120 × 7–68 mm; base aequilateral to oblique, cuneate, rounded to short-attenuate; apex acuminate to acute; bract petioles 2–16 mm long. Flowers pedicellate, erect to slightly spreading; pedicels 9–25 mm long; bracteoles inconspicuous and scabrous, linear, or triangular to ovate, lanceolate to oblanceolate, 1.5–20 × 1–8 mm. Calyx campanulate, 8–10 × 7–8 mm, spiculate to glabrous, smooth, green, ecarinate; calyx lobes ovate, 2–4 × 2.5–4.0, apex rounded to obtuse. Corolla funnel-shaped, (22–35–45 mm long, 10–20 mm wide at the apex of the tube, pale yellow, greenish-yellow, pale greenish, smooth; corolla lobes ovate, 7–12 × 4–9 mm, apex obtuse to rounded. Stamens 24–30 mm long; filaments 20–25 mm long, filiform, terete; anthers linear to linear-elliptic, 4–5 × 1.5–2.0 mm, sagittate, versatile; pollen "glabra-type." Pistil 35–40 mm long; ovary 5–7 × 2–3 mm; style 28–30 × 0.75–1.0 mm; stigma lobes spathulate to elliptic, 2–3 × 1.0–1.5 mm. Capsules ovoid to ellipsoidal, 21–35 (–41) × 7–11 mm, smooth to faintly rugose, light to dark brown, erect to nodding; style remnant 15–23 mm; seeds "perimetrically Cwinged," flattened, roughly 3–4 sided in outline, yet appearing as myriads of different puzzle pieces, 0.8–1.9 × 0.5–1.0 mm, bicolored, testa tan, wings orangish straw-colored, testa reticulate, wings ribbed.

Macrocarpaea elix belongs to *Macrocarpaea* section *Choriophylla* and occurs from 2,100 to 3,100 m in the Amotape–Huancabamba region of southern Ecuador. It is mostly closely related to *M. apparata* with which it shares hairy leaves and calyces. This species is part of a distinctive monophyletic group including *M. apparata*, *M. bubops*, *M. elix*, *M. jensii*, *M.* × *mattii*, and *M. noctiluca*.

Etymology. From the Latin *elix*, a channel, for the large grooves on the leaf petioles.

Specimens examined: ECUADOR. **Azuay:** The eastern Cordillera, 1–8 km north of the village of Sevilla de Oro, 2,440–2,745 m, 27 July 1945, *Camp 4325* (NO, S, U); Carretera Cuenca-Paute-Sevilla de Oro, colecciónes entre Paute-Sevilla de Oro, 2,490 m, 8 August 1986, *Jaramillo 8297* (AAU, QCA); Km 29 Paute-Amaluza road, 02° 41′ S, 79° 36′ W, 2,550 m, 7 August 1986, *P.M. Jørgensen 61429* (AAU, QCNE). **Cañar:** Above Rivera on road to Pindilig, forest remnants, 2,800 m, 9 March 1985, *Harling, G. & Andersson 23031* (GB, MO, QCA); Pass on road Azogues-Taday, dry "ceja" scrub, 3,100 m, treelet 4–5 m tall, 30 March 1985, *Harling, G. & Andersson 23442* (GB, MO, QCA). **Morona-Santiago:** Along road between Gualaceo and Limón, 42.5 km E of Gualaceo, 17.7 km E of summit, 16.5 km E of Azuay border, 11.4 km W of jct to Indanza, 2,072 m, 22 May 2003, *Croat, T.B. & Menke 89230* (MO, NY, QCNE); SW of military checkpoint in Limón (=General L. Plaza Gutiérrez (02° 58′ S, 78° 25′ W), 65 km Sw of Mendez (02° 42′ S, 78° 18′ W), 2,340 m, shrub 1–2 m tall, 27 October 1988, *Dorr, L.& Barnett 5910* (QCA, QCNE, US); Ca 15 km up road from Plan de Milago/Limon to Gualaceo, 03° 00′ 07.5″ S, 078° 32′ 52.7″ W, 2,222 m, 12 February 13, *Grant, J.R & J. Cortina 13-5122* (NY); Road Gualaceo-Limón, Loma del Puerco, mountain forest, 2,500 m, 11 November 1988, *Harling, G. 25814a* (GB); Road Gualaceo-Limón, Loma del Puerco, 2,500 m, ca. 3 m tall, 11 November 1988, *Harling, G. 25814b* (GB); 20 km northwest of Indanza, between Cerro Cruzado and Zapote area, 2,134 m, 2–16 October 1944, *H. Jorgensen 49* (NY); Along Río Tintas below Arenillas, dense forest between Campanas and Arenillas, along Río Tintas, 10 leagues SE of El Pan, 2,195 m, 13 July 1943, *Steyermark, J. 53548* (F [3 sheets], US).

8. *Macrocarpaea ericii* J.R. Grant

Harvard Pap. Bot. 9(1): 23. 2004. TYPE: PERU. Cajamarca: San Ignacio, Huarango, Localidad Romerillo, Cordillera entre Romerillo y Nuevo Mundo, vegetación primaria con predominancia de árboles enanos, 05° 14′ S, 78° 46′ W, 2,300 m, 27 July 1997, *J. Campos, E. Rodríguez, & S. Núñez 4296* (Holotype: MO; Isotype: HUT) (Fig. 3.28).

Unbranched shrub, 1–1.6 m, glabrous throughout. Stems terete to slightly quadrangular, solid, 3–8 mm in diameter just below inflorescence. Leaves obovate, varying to slightly elliptic, short-petiolate, (4–)7–15 cm long. Petioles 6–2 mm long, robust with strong open vagination one half the length of the petiole; interpetiolar ridge 2–5 mm high. Blades (3.4–6–13 × (2.5–)4.5–8.5 cm, entire, slightly revolute, dark above and below, lateral veins scarcely visible either above or below, glabrous, thick, leathery-coriaceous; base aequilateral to oblique, cuneate to rounded; apex rounded to obtuse. Inflorescence a few branched open thyrse, 8–44 cm long; branches 3–28 cm long, 1–3 flowered per cyme. Bracts obovate to oblanceolate, petiolate, 25–55 × 13–23 mm; base aequilateral, cuneate; apex rounded to obtuse; bract petioles 5–8 mm long. Flowers pedicellate, erect to spreading; pedicels 3–22 mm long; bracteoles linear to oblanceolate, 5–20 × 2–5 mm. Calyx campanulate, 8–11 × 6–8 mm, glabrous, faintly

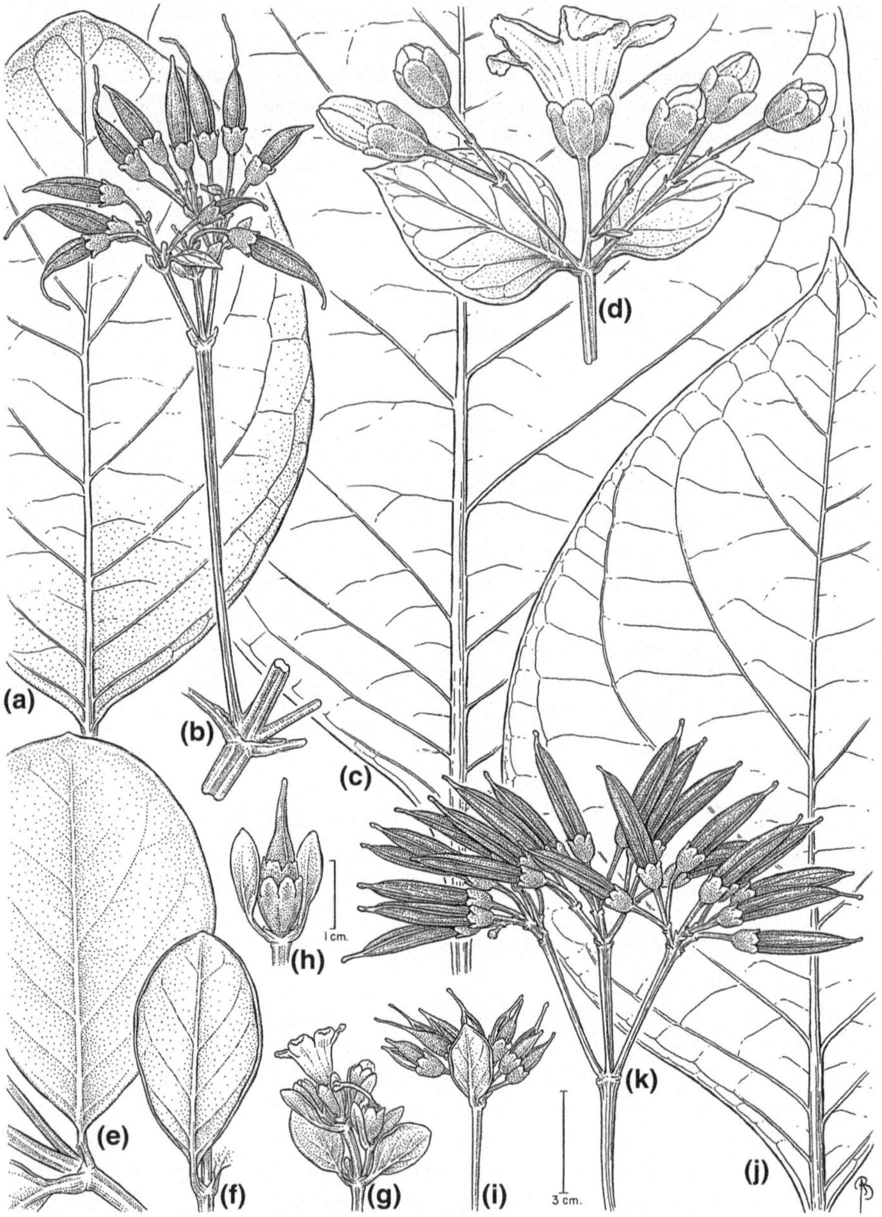

Fig. 3.28 **a–b** Macrocarpaea bubops; **a** leaf; **b** branch of flowering stem and interpetiolar ridge; **c–d** *M. ostentans* from Peru; **c** leaf; **d** habit of flowering stem; **e–i** *M. ericii*. **e** mature leaf; **f** smaller mature leaf; **g** habit of flowering stem; **h** detail of fruiting flower, notice two pronounced bracteoles; **i** habit of fruiting stem. **j–k** *M. dillonii* from Peru. **j** leaf; **k** habit of fruiting stem; **a–b** drawn from Beltran & Foster 1197 (USM); **c–d** drawn from Wallnöfer 12-9688 (U); **e** drawn from Campos et al. 4296 (MO); **f–h** drawn from Weigend 98/500 (M); I drawn from Campos et al. 5564 (U); **j–k** drawn from Sánchez Vega et al. 9572 (HAO)

verrucose, ecarinate, ovate to elliptic; calyx lobes 4–5 × 3.5 mm, rounded to obtuse. Corolla funnel-shaped, 20–30 mm long, 11–14 mm wide at the apex of the tube, yellow to greenish-yellow, smooth; corolla lobes ovate to elliptic, 6–10 × 5–8 mm, apex obtuse to rounded. Gynoecium, androecium and pollen unknown. Capsules ovoid to ellipsoidal, 16–20 × 8–10 mm, smooth, faintly verrucose to pusticulate, straw-colored, erect to slightly nodding; style remnant 5–10 mm. Seeds "perimetrically winged–reversal to rimmed", angular to polyhedral to flattened (pyramids, tetrahedrons, rectangular prisms, to multisided prisms), 2–5-sided, not winged, 0.5–1.0 × 0.3–0.5 mm, straw-colored, rugose-reticulate.

Macrocarpaea ericii belongs to *Macrocarpaea* section *Choriophylla* and occurs from 1,600 to 2,400 m in the Amotape–Huancabamba region of southern Ecuador and northern Peru. It is easily recognized by its thick leathery obovate leaves, and flowers that are subtended by a pair of spathulate bracteoles. This species is part of a group of species with smooth leathery leaves where the secondary leaf veins absent or scarcely visible including *M. arborescens, M. ericii, M. harlingii, M. lilliputiana, M. subsessilis*, and *M. zumbae*.

Etymology. Named for Eric Frank Rodríguez Rodríguez (1963–), Herbarium Truxillense, Universidad Nacional de Trujillo, Peru.

Specimens examined: ECUADOR. **Zamora-Chinchipe:** Nangaritza, summit of tapeui "Cima de la Reserva San Miguel de Las Orquideas", property owned by the Association de Trabajadores Autonomos San Miguel de Las Orquideas, near Cabañas Yankuam, 04° 14′ 54″ S, 078° 38′ 34″ W, 13 May 2009, Clark, J.L. et al. *10779* (NY); Cordillera del Cóndor, Comunidad San Miguel de Las Orquideas, Tepuy 2, vertiente occidental del río Nangaritza, 1,600 m, 1 April 2009, Jadán, O., et al. *292* (LOJA).

PERU. San Ignacio: San José de Lourdes, Localidad Estrella del Oriente "Cerro del Oso", vegetación enana, 04° 46′ S, 78° 59′ W, 1,600–1,700 m, 6 September 1997, *Campos & Díaz 4410* (HUT [not seen], MO, NEU, U [2 sheets]); San Ignacio, near San José de Lourdes, llanos, 1,900–2,100 m, 10 June 1998, *Weigend* et al. *98/500* (M, USM [not seen]). Piura: Cordillera de San José, páramo 1,700 m, flores verdosas, 23 September 1961, *Friedberg 692* (USM).

9. *Macrocarpaea gattaca* J.R. Grant

Harvard Pap. Bot. 9(1): 16. 2004. TYPE: ECUADOR. Pichincha: Vía San Juan-Chiriboga-Empalme, 2,550 m, 10 September 1977, *J. Jaramillo 22* (Holotype: NY; Isotypes: AAU, MO, QCA) (Fig. 3.29).

Unbranched shrub to tree to 2–5 m, hyaline spiculate to scabrous with short simple hairs on stems, petioles, leaves, inflorescences, bracts, and bracteoles. Stems terete to quadrangular, solid to hollow, 4–10 mm in diameter just below inflorescence. Leaves ovate to nearly deltoid, sessile to short-petiolate, 10.5–18.5 cm long. Petioles 0–5 mm long, usually absent, but rarely a short robust petiole with strong open vagination present; interpetiolar ridge 3–5 mm high. Blades 10–18 × 8–11 cm, entire, not revolute, dark green, with slightly impressed

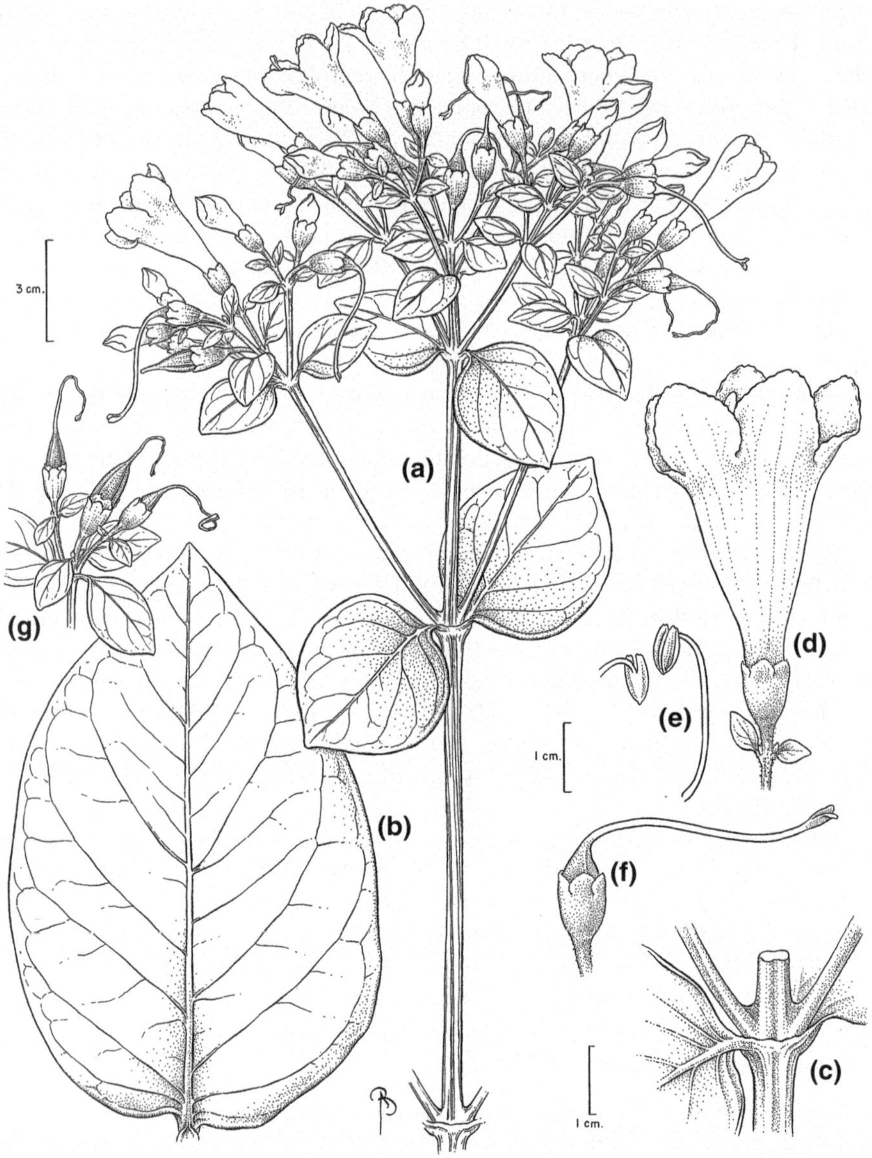

Fig. 3.29 *Macrocarpaea gattaca* J.R. Grant. **a** habit of flowering stem; **b** leaf; **c** detail of interpetiolar ridge; **d** flower; **e** anthers; **f** immature fruit; **g** maturing fruits. **a, c–f** drawn from Asplund 18200 (S); **b** drawn from Jaramillo 22 (QCA); **g** drawn from Freire 456 (AAU)

veins above, and slightly raised veins below, hyaline hispid to spiculate on veins on lower surface, papery thin, to thin coriaceous; base aequilateral to oblique, rounded to nearly truncate; apex acute. Inflorescence a much branched open thyrse, 33+ cm long, the total inflorescence unknown as has been cut into various segments for separate herbarium sheets; branches 9–38 cm long; 3–9 flowered per branch. Bracts ovate to nearly deltoid, inaequilateral, suborbicular to elliptic, sessile to petiolate, 12–77 × 6–54 mm; base aequilateral to oblique, rounded to nearly truncate; apex acuminate to acute; bract petioles 0–5 mm long. Flowers pedicellate, erect to spreading; pedicels 7–20 mm long; bracteoles ovate to obovate, 5–18 × 3.5–14.0 mm. Calyx narrowly campanulate, 8–11 × 5–6 mm, glabrous, smooth, green, ecarinate; calyx lobes ovate to reniform, 1.3–2.5 × 3–5, obtuse, acute to rounded. Corolla funnel-shaped, 45–53 mm long, 18–23 mm wide at the apex of the tube, yellow, yellowish-green, to pale green, smooth; corolla lobes ovate, 9–13 × 8–14 mm, obtuse to rounded, often crisped. Stamens 27–37 mm long; filaments 22–30 mm long, filiform; anthers linear to linear-elliptic, 5–7 × 2–3 mm, sagittate, versatile; pollen "glabra-type". Pistil 47–52 mm long; ovary 10–12 × 2–4; style 34–36 × 1.0; stigma lobes spathulate, 3–4 × 1.0–1.5 mm. Capsules obovoid to ellipsoidal, 25–36 × 7–9 mm, smooth, rugose, to faintly ribbed, greenish to tan, erect to nodding; style remnant 5–17 mm. Seeds "rimmed" type, angular to polyhedral (tetrahedrons, rectangular prisms, to multisided prisms), 2–7-sided, 0.6–1.0 × 0.3–0.7 mm, faint orangish-tan, rugose-reticulate.

Macrocarpaea gattaca belongs to section *Macrocarpaea*, and occurs in the northern Andes, on Pacific slopes on the Cordillera Occidental, the Chocó of Ecuador. It appears to be related to *M. densiflora* (Benth.) Ewan, *M. gaudialis* J.R. Grant, *M. nicotianifolia* Weaver & J.R. Grant, and *M. pachyphylla* Gilg, from southern Colombia. *Macrocarpaea gattaca* is nearly sympatric with *M. sodiroana* yet occur in two distinct elevation bands; 1,000–1,600 m for *M. sodiroana*, and 2,500–2,700 m, for *M. gattaca*.

Etymology: From the DNA nucleotides adenine, cytosine, guanine, and thymidine that were used in the identification of this species.

Specimens examined: ECUADOR. **Pichincha**: Parroquia Pacto-Gualea, sector del Río Pishashi, 1,030–1,070 m, 23 July 1994, *Aimancaña & Guzmán 156* (QAP); Salvador, between San Juan and Chiriboga, shrubby slope, 2,600 m, 28 October 1955, *Asplund 18200* (S); Carretera Quito-Las Palmas-Palmeras, 2,800 m, 29 November 1986, *Freire, A. 456* (AAU, QCA); Cerro Guarumos-La Y sendero al Cerro El Castillo, derecho de vía del Oleoducto de Crudos Pesados, 2,600–2,700 m, 7 September 2001, *Freire A.* et al. *3170* (MO, NY); 67 km up the old road Santo Domingo-Quito from where it merges with the new Santo Domingo-Quito road, 02° 17′ S, 078° 36′ W, 26 February 2002, *Grant, J.R. & J.-M. Torres 02-4209* (NEU, NY, US); Via San Juan-Chiriboga, km 33 desde Quito, 2,700 m, 2 April 1976, *Oldeman 3449* (NY, QCA); Canton Quito, Parroquia Nanegal, Montañas de Maquipucuna, Cerro Sosa and ridge adjacent to Cerro Montecristi, 00° 04–05′ N, 78° 37′ W, 2,100–2,200 m, 9 July 1991, *Webster* et al. *28902* (DAV, MICH, MO, QCNE, TEX).

10. *Macrocarpaea gondoloides* J.R. Grant

Harvard Pap. Bot. 9(2): 320. 2005. TYPE: ECUADOR: Carchi: Tulcan, Reserva Indígena Awá, Comunidad San Marcos, 25 km al NW de El Chical, Parroquia Maldonado, bosque pluvial premontano, 01° 06′ N, 78° 14′ W, 1,500 m, 16–30 November 1990, *D. Rubio, C. Quelal, & J. Pai 1077* (Holotype MO; Isotype NY, QCNE) (Fig. 3.30).

Shrub to liana-like shrub, 0.5–3.0 m, glabrous throughout, cm in diameter. Stems terete, solid, 2–5 mm in diameter just below the inflorescence. Leaves linear-lanceolate, lanceolate, elliptic to ovate, petiolate, (6)10.5–22 cm long. Petioles 2–20 mm long, slender with slight vagination to one quarter the length of the petiole; interpetiolar ridge 1–3 mm high. Blades (5.8)10.3–20 × 3.5–8 cm, entire, not revolute, dark above and below, with slightly impressed veins above, and slightly raised veins below, glabrous, papery thin; base aequilateral to oblique, cuneate to attenuate and decurrent on the petiole; apex acuminate. Inflorescence a few-branched open thyrse, 9–40 cm long; branches 9–22 cm long; 3–9 flowered per branch. Bracts ovate to lanceolate, sessile to short-petiolate, 10–115 × 4–58 mm; base aequilateral to oblique, cuneate to rounded; apex acuminate to acute; bract petioles 0–2 mm long. Flowers pedicellate, erect to slightly spreading, but never nodding; pedicels 8–20 mm long; bracteoles ovate to lanceolate, 5–16 × 1.5–6.0 mm. Calyx campanulate, 11–13 × 5–7 mm, glabrous, vertically striated, straw-colored to tan when dried as herbarium specimen, ecarinate; calyx lobes ligulate to oblong, 6–9 × 3.5–5 mm, apex rounded to obtuse. Corolla greenish-white, greenish-yellow. Stamens and pistil unknown. Capsules ovoid, 9–14 × 5–10 mm, smooth, ribbed along sutures, straw-colored, erect to slightly spreading but not nodding; style remnant 6–8, remaining as a strong lignified tube even when the capsule below deteriorates to release the seeds mm long. Seeds "Rimmed –reversal to Perimetrically winged type," 0.4–0.9 × 0.3–0.5 mm, bicolored, testa golden-tan, wings translucent golden, testa reticulate, wings ribbed.

Macrocarpaea gondoloides belongs to section *Macrocarpaea*, and occurs in the northern Andes, on the Cordillera Occidental, the Chocó of Colombia and Ecuador. It is closely related to several facultatively epiphytic species: *M. gulosa* J.R. Grant and *M. luteynii* of the Cordillera Occidental of Colombia, and *M. browallioides* (Ewan) A. Robyns & Nilsson and *M. subcaudata* Ewan of Costa Rica and Panama.

Etymology. For the oblique shape of the leaves, similar to the outline of an Italian *gondola*, a type of boat.

Specimens examined: ECUADOR. **Carchi:** Tulcan, Reserva Étnica Awá, Parroquia El Chical, Centro Gualpi Medio, 01° 02′ N, 78° 16′ W, 900 m, 18 February 1993, *Aulestia & Grijalva 1122* (MO, QCNE); Tulcan, Reserva Étnica Awá, Parroquia El Chical, Centro San Marcos, bosque muy húmedo premontano, 01° 06′ N, 78° 14′ W, 750 m, 20–30 April 1990, *Méndez* et al. *231* (AAU, MO, QCNE); Tulcan Canton, Chical, Reserva Étnica Awá-Camumbí, 78° 16′ W, 00° 53′ N, 1,700–1,900 m, 20–29 July 1991, *Quelal* et al. *276* (MO, QCNE); Tulcan Cantón, Parroquia Chical, Sector Gualpi medio, Reserva Indígena Awá, Sendero a

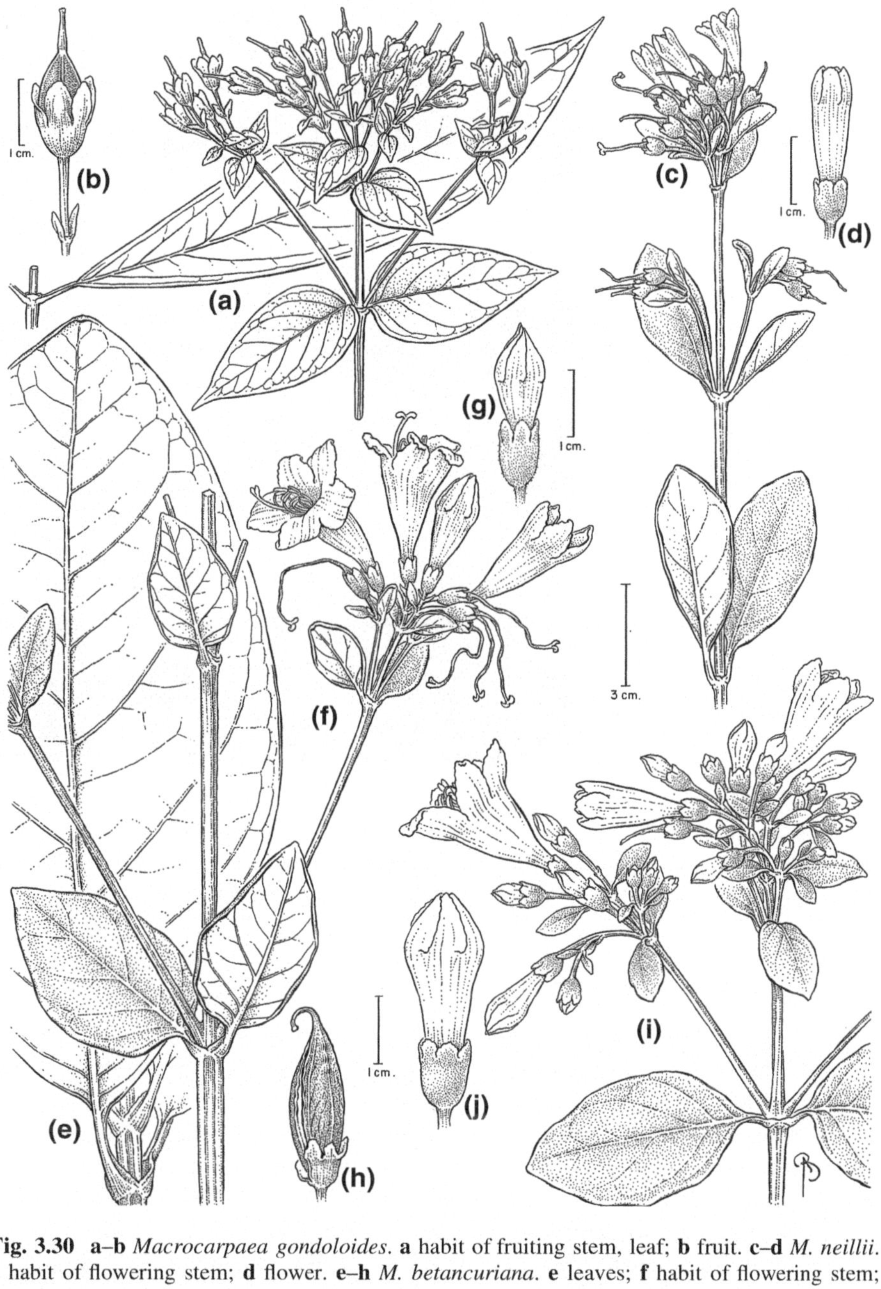

Fig. 3.30 **a–b** *Macrocarpaea gondoloides*. **a** habit of fruiting stem, leaf; **b** fruit. **c–d** *M. neillii*. **c** habit of flowering stem; **d** flower. **e–h** *M. betancuriana*. **e** leaves; **f** habit of flowering stem; **g** bud; **h** fruit. **i–j** *M. voluptuosa*. **i** habit of flowering stem; **j** bud. **a–b** drawn from Méndez et al. 231 (QCNE), **c–d** drawn from Neill, D. et al. 13758 (LOJA, QCNE), **e–h** drawn from Weaver 2644 (GH, NY)

Fig. 3.31 *Macrocarpaea harlingii* J.S. Pringle. *Photo* J.R. Grant

San Marcos al norte de la casa communal, 01° 02′ N, 78° 16′ W, 3,200 m, 23–17 May 1992, *Tipaz* et al. *1051* (MO [2 sheets]). COLOMBIA. Nariño: Mpio. de Ricaurte, Resguardo Indígena Nulpe Medio, Andalucía, camino a Piguantís, 01° 4′ N, 78° 15′ W, 780 m, 13 January 1996, *González,* m.*S.* & *Ramírez P. 1688* (QCA).

11. *Macrocarpaea harlingii* J.S. Pringle

Fl. Ecuador 53(159A): 101. 1995. TYPE: ECUADOR. Loja: N slope of Nudo de Sabanilla, ca. 8 km above Yangana on road to Valladolid, 2,300–2,500 m (Holotype GB; Isotype QCA) (Figs. 3.31, 3.50).

Small tree or large shrub 3–5 m tall, glabroud. Stems quadrangular (older woody portions obscurely so). Internodes mostly much shorter than leaves. Leaves subcoriaceous; venation inconspicuous except for thick midrib, secondary veins branching from midrib at diverse angles, stronger ones mostly at 40–50 (–80) degrees, distally somewhat upcurved, obscure near margins, not brochidodromous. Blades elliptic to broadly oblanceolate, 10–20 cm long, 3–6 cm wide, base gradually tapering to a petiole 1.5–3.0 cm, apex abruptly acute to short-acuminate; margin narrowly cartilaginous, revolute. Inflorescence a large, diffuse dichasium or thyrse, several times divided, with elongate, arcuately spreading lower branches; lower divisions subtended by reduced leaves, upper by minute bractlets; flowers ca. 30–60. Ultimate divisions and pedicels arcuate, ture pedicels 5–15 mm, appearing longer because uppermost bractlets often subtend a single flower. Calyx 9–15 mm, papillose except near margins of lobes; lobes suborbicular, about half to nearly as long as tube, margins broadly hyaline, erose-fimbriate; decurrent ridges extending from sinuses to pedicels (as seen on flowers of the type collection). Corolla 30–52 mm, greenish white, campanulate; tube gradually to rather abruptly flaring above summit of calyx, exterior surface smooth near base, conspicuously papillose above; obles 7–15 mm, ovate-triangular, obtuse. Casule ellipsoid, 18–25 mm, its papillose surface contrasting with the glossy style base. This description is largely from Pringle (1995).

Macrocarpaea harlingii belongs to *Macrocarpaea* section *Choriophylla* and occurs from 900 to 2,500 m in the Amotape–Huancabamba region of southern Ecuador and northern Peru. This species is part of a group of species with smooth leathery leaves where the secondary leaf veins absent or scarcely visible including *M. arborescens, M. ericii, M. harlingii, M. lilliputiana, M. subsessilis*, and *M. zumbae*.

Etymology. Named for Swedish botanist, Gunnar Harling (1920–2010), University of Stockholm, Sweden.

Specimens examined: ECUADOR. **Loja**: W slope of Nudo de Sabanilla, 2,300–2,500 m, 2 April 1985, *Harling, G. & Andersson, 2353* (GB, QCA); San Francisco, 03° 59′ 09″ S, 079° 05′ 38″ W, 2,100 m, 15 January 2002, *Lozano, P.* et al. *E-188* (LOJA). **Morona-Santiago**: Gualaquiza Cantón, région de la Corillera del Cóndor, Valle del Río Quimi, 03° 2954″ S, 078° 25′ 13″ S, 900 m, 10 May 2007, *Quizhpe, W. & A. Wisum 2668* (LOJA, MO, QCNE). **Zamora-Chinchipe**: Cordillera del Cóndor, W of Tikimints, valley of Río Coangas, Ijiach Naimt, 2,000 m, 20 March 2001, *Berry, P. & D. Neill 7651* (MO); Carr. Loja-Zamora, Cuenca del rio San Francisco, Fundacion Arco Iris, 2,100 m, 27 February 2000, *Cerón* et al. *40105* (QAP); Estación Cientifica San Francisco, km 30 de la via Loja-Zamora, 2,100 m, 15 May 2001, *Chimbo & Chambo 87* (LOJA, MO); El Pangui, Vicinity of Ecua-Corrientes copper mine development, valley of Río Waiwaime, along road to mine site, 03° 34′ 44″ S, 078° 26′ 08″ W, 1,312 m, 4 April 2006, *Croat, T.B.* et al. *96578* (LOJA, MO, NY, QCNE); N bank of the Río Zamora, 25 km E of Loja, 2,092–2,125 m, 11 July 1989, *Dorr, L.& Valdespino 6616* (NY, QCA, QCNE); Parroquia Sabanilla, a 30 km de Loja, Cuesta occidental de la Cordillera del Consuelo, 03° 58′ 49″ S, 079° 04′ 18″ W, 2,000 m, 24 September 2003, *Fernández, D.* et al. *1449* (QCNE); P.N. Podocarpus (San Francisco entrance); trail leading west from San Francisco, 2,100 m, 16 February 2001, *Grant, J.R. & L. Struwe 01-4048* (G, LOJA, NY, QCNE, SBBG); P.N. Podocarpus (San Francisco entrance); trail leading west from San Francisco, 2,100 m, 16 February 2001, *Grant, J.R. & L. Struwe 01-4049* (LOJA, NEU, NY, QCA, QCNE, US); Parque Nacional Podocarpus, San Francisco entrance area, between Loja and Zamora, 03° 59′ 22.8″ S, 079° 05′ 43.6″ W, 2,140 m, 10 February 2011, *Grant, J.R. & J. Vieu 11-4682* (NY); Cordillera del Cóndor, Comunidad San Miguel de Las Orquideas, Tepuy 2, vertiente occidental del río

Nangaritza, 1,780 m, 1 April 2009, *Jadán, O.*, et al. *250* (LOJA); Loja-San Francisco, 2,160 m, 17 November 2002, *Merino & Delgado E-1349* (LOJA); Estacion Cientifico San Francisco, Antennae Trail, 03° 58.4892 S, 079° 3.9518 W, 1,910 m, 5 April 2010, *Muchhala, N. 433* (QCA, QCNE); El Pangui, western slopes of Cordillera del Cóndor, Río Wawaime watershed, tributary of Río Quimi, 03° 34′ 43″ S, 078° 26′ 07″ W, 1,320 m, 2 April 2007, *Neill, D. & NSF Dendrology course 15518* (LOJA, MO, NY, QCNE); Cordillera del Cóndor, El Pangui, sandstone plateau of Contrafuerte Tres Patines, west of main Cóndor ridge, above "Jardin Botánico" of EcuaCorriente copper company, south of km 15 of Cóndor Mirador military road, 1,685 m, 9 December 2005, *Neill, D. & W. Quizhpe 15059* (LOJA, MO, QCNE); El Pangui, Cordillera del Cóndor, western slopes, Upper Río Wawaime watershed, tributary of río Quimi, 1,320 m, 4 April 2006, *Neill, D.* et al. *15123* (F, LOJA, MO, NY, QCNE); P.N. Podocarpus, La Esmeralda (Cooperativo San Francisco de Numbala Alto), bosque primario "achaparrado" con dosel de 6 m de altura en cresta de colina, 04° 22′ S, 079° 03′ W, 2,300 m, 1 January 1995, *Palacios, W. & Tirado 13098* (MO, QCNE, U); Trail from the Loja-Zamora road to Río Santiago in P.N. Podocarpus, Sitio San Francisco, 2,000 m, 4 May 1995, *Pedersen* et al. *104170* (LOJA, QCA); Cordillera del Cóndor, vertiente occidental, cuenca del Rio Quimi, 1,180–1,400 m, 5 April 2006, *W. Quizhpe 2115* (LOJA, MO, QCNE); El Pangui, Cordillera del Cóndor, vertiente occidental, cuenca del Río Tundayme, Vale del Río Wawaime, afluyente del Rio Quimi-Tundayme, 1,250 m, 10 April 2006, *Quizhpe, W. & T. Croat 2225* (LOJA, MO, NY, QCNE); El Pangui, Cordillera del Cóndor, vertiente occidental, cuenca del Río Quimi, zona de la futura mina de cobre, 1,400–1,670 m, 19 March 2006, *Quizhpe, W. & F. Luisier 1968* (LOJA, MO, NY, QCNE); El Pangui, Cordillera del Cóndor, vertiente occidental, cuenca del Río Quimi, zona de la futura mina de cobre, 1,400–1,670 m, 19 March 2006, *Quizhpe, W. & F. Luisier 1977* (F, LOJA, MO, NY, QCNE); El Pangui, Cordillera del Cóndor, vertiente occidental, cuenca del Río Quimi, zona de la futura mina de cobre, de la compañia EcuaCorriente, 1,630 m, 20 March 2006, *Quizhpe, W. & F. Luisier 1995* (LOJA, MO, QCNE); El Pangui, Cordillera del Cóndor, vertiente occidental, cuenca del Río Tundayme, carretera hacia el destacamento militar Cóndor Mirador 1,690–2,000 m, 21 March 2006, *Quizhpe, W. & F. Luisier 2016* (LOJA, MO, NY, QCNE); Nangaritza Cantón, Región de la Cordillera del Cóndor, Parrioquia San Francisco del Vergel, Playones, 1,800–2,100 m, 13 March 2005, *Quizhpe, W.* et al. *1040* (LOJA, MO, NY, QCNE); Cordillera del Cóndor, zona del alto Río Nangaritza, reserva "Los Tepuyes," San Miguel de Las Orquídeas, 04° 15′ 32″ S, 078° 41′ 04″ W, 1,620 m, 24 April 2008, *Quizhpe, W.* et al. *3053* (MO); Cordillera del Cóndor, zona del alto Río Nangaritza, reserva "Los Tepuyes," San Miguel de Las Orquídeas, 04° 15′ 32″ S, 078° 41′ 04″ W, 1,620 m, 24 April 2008, *Quizhpe, W.* et al. *3069* (MO); Cordillera del Condor, Destacamento militar Cóndor Mirador (Cóndor Mirador 2), 1 km al norte del destacamento, 03° 37′ 41″ S, 078° 23′ 42″ W, 1,975 m, 6 September 2003, *Rodríguez, E.* et al. *2616* (HUT, LOJA, MO, NY, QCNE, USM); Cordillera del Condor, campamento Militar Pachicutza, 1,849 m, 22 February 1994, *Tirado & Gray 903* (MO, NEU, QCNE, U); P.N. Podocarpus, Via Loja-

Zamora, San Francisco, Apr-May 1992, *Vivar C.* et al. *3968* (LOJA); P.N. Podocarpus, Via Loja-Zamora, San Francisco, Apr-May 1992, *Vivar C.* et al. *3980* (LOJA); Area of ECSF (Estación Cientifica San Francisco), 2,200 m, 7 November 2001, *Wolff, D. & F. Matt, F. 118* (LOJA).

PERU. **Amazonas**: Condorcanqui, Cordillera del Condor, la pared del "tepui" Cerro Machinaza, 1,950–2,050 m, 21 July 1994, *Beltrán & R. Foster 1204* (F, USM); Bongará, 4 km N of Pomacochas on road to Rioja, 2,150–2,200 m, 2 June 1986, *Knapp* et al. *7467* (MO, USM); Bongará, Bosque de Protección Alto Mayo, 2,550 m, 19 March 2002, *Vásquez 27655* (NEU, MO). **Cajamarca**: San Ignacio, Huarango, Localidad Romerillo, Cordillera entre Romerillo y Nuevo Mundo, 2,300 m, 27 July 1997, *Campos* et al. *4284* (MO, U); San Ignacio, Distrito Huarango, Cordillera Huarango, El Romerillo, 05° 16′ 26″ S, 078° 41′ 55″ W, 2,035 m, 19 July 2005, *Rodríguez, E.* et al. *2929* (HUT, LOJA, MO, QCNE, USM); San Ignacio, Distrito Huarango, Cordillera Huarango (El Romerillo), 05° 16′ 26.4 S, 078° 41′ 31.8″ W, 1,885–2,168 m, 2 May 2006, *Rodríguez, E.* et al. *3096* (HUT, LOJA, MO, QCNE, USM). **San Martín**: Along road Rioja-Pedro Ruis, ceja de la montaña, El Mirador, 1,850 m, 25 March 1998, *van der Werff* et al. *15702* (MO).

12. *Macrocarpaea innarrabilis* J.R. Grant

Harvard Pap. Bot. 9(1): 27. 2004. TYPE: PERU. Amazonas: Bagua District, Imaza, Tayu Majaji, Comunidad Nativa de Wawas, 05° 15′ S, 078° 22′ W, 1,100 m, 7 February 1999, *R. Vásquez, C. Vargas, J. Yactayo, & E. Palomino 26057* (Holotype NEU; Isotypes: F, MO, U) (Fig. 3.32).

Shrub or small tree to 1.5, glabrous throughout. Stems terete to slightly quadrangular, solid to hollow, 5–8 mm in diameter just below inflorescence. Leaves obovate to elliptic, petiolate, 6.5–15.5 cm long. Petioles 13–30 mm long, robust with slight vagination one-eighth the length of the petiole; interpetiolar ridge 2–3 mm high. Blades 5.2–12.5 × 3.5–7.5 cm, entire, not revolute, dark above and conspicuously lighter below, with slightly impressed veins above, and slightly raised veins below, leathery-coriaceous to papery thin when dried; base aequilateral to oblique, cuneate to rounded; apex rounded, obtuse, to slightly mucronulate. Inflorescence a much branched open thyrse, 16–32 cm long; branches 9–29 cm long; 1–7 flowered per branch. Bracts obovate to elliptic, petiolate, 25–80 × 14–32 mm; base aequilateral to oblique, cuneate; apex rounded, obtuse, to slightly mucronulate; bract petioles 3–15 mm long. Flowers pedicellate, erect to spreading; pedicels 12–45 mm long; bracteoles varying from inconspicuous and linear to spathulate, 0.5–9.0 × 0.5–3.0. Calyx campanulate, 6–11 × 4–7 mm, glabrous, smooth, ecarinate; calyx lobes ovate, 2–3 × 2.5–4.5 mm, rounded to obtuse. Corolla funnelshaped, 32–43 mm long, 13–17 mm wide at the apex of the tube, whitish, yellow, smooth; corolla lobes ovate to elliptic, 7–12 × 5–7 mm; corolla lobe apex obtuse to rounded. Stamens 26–31 mm long; filaments 22–26 mm long, filiform, flattened; anthers elliptic to oblong, 4–5 × 1.5–2.0 mm, sagittate, versatile; pollen "glabratype". Pistil 32–35 mm long; ovary 5–6 × 2–3 mm; style 26–27 × 0.75–1.0; stigma lobes spathulate, 3 × 1.5–2.5 mm. Capsules ellipsoidal to ovoid,

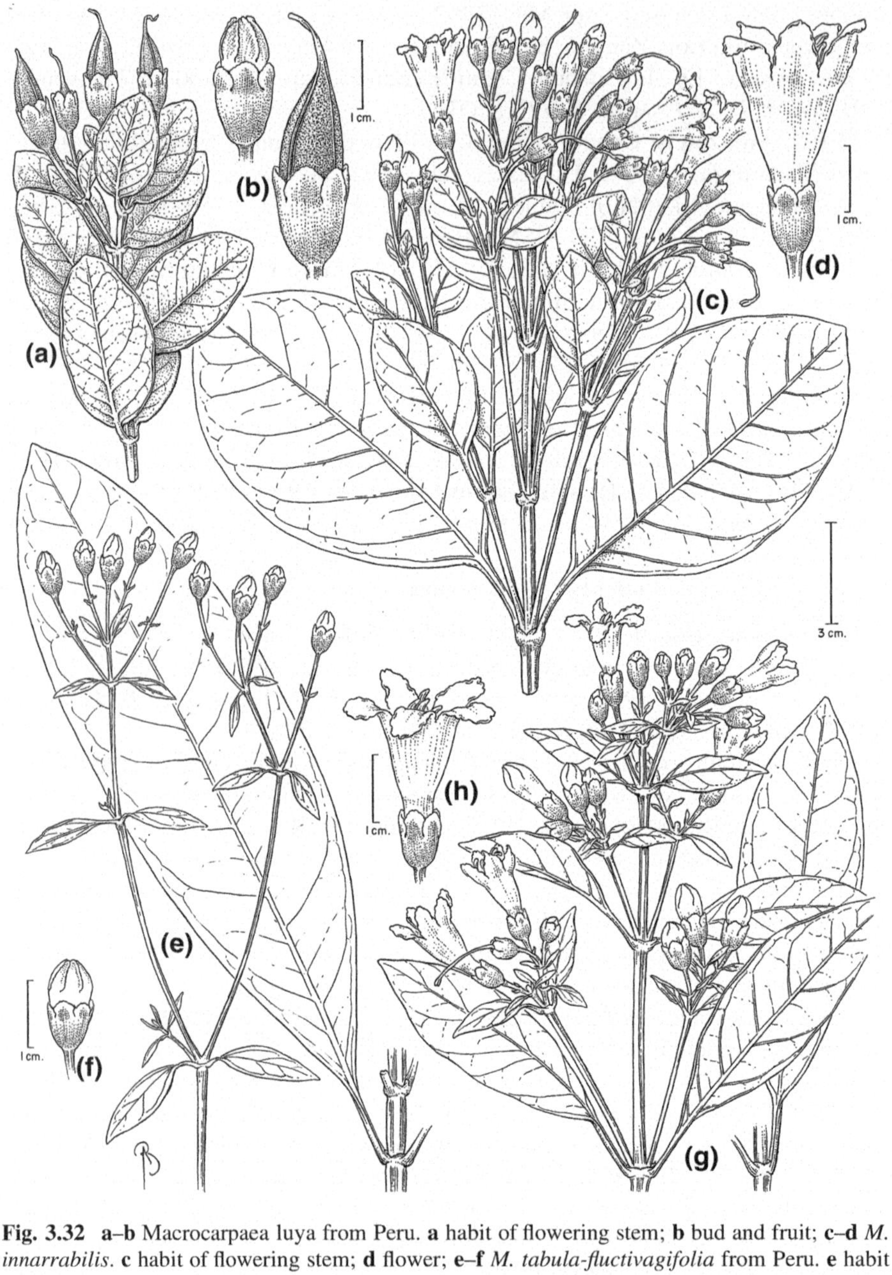

Fig. 3.32 **a–b** Macrocarpaea luya from Peru. **a** habit of flowering stem; **b** bud and fruit; **c–d** *M. innarrabilis*. **c** habit of flowering stem; **d** flower; **e–f** *M. tabula-fluctivagifolia* from Peru. **e** habit of flowering stem, leaf, and inter-petiolar ridge; **f** bud; **g–h** *M. revoluta*. **g** habit of flowering stem, leaf, and interpetiolar ridge; **h** flower. **a–b** drawn from Díaz & Campos 3786 (USM); **c–d** drawn from Matthews 1314 (K); **e–f** drawn from Wolfe 12333 (F); **g–h** drawn from Smith 2748 (NY)

20–21 × 7–8 mm, smooth to pusticulate, shiny, chestnut to brown, spreading to nodding; style remnant 8–9 mm. Seeds "perimetrically winged," flattened with wings all around, 0.4–1.0 × 0.6–1.0, faint orangish-tan, rugose-reticulate.

Macrocarpaea innarrabilis belongs to *Macrocarpaea* section *Choriophylla* and occurs from 1,000 to 2,500 m in the Amotape–Huancabamba region of southern Ecuador and northern Peru. It is a 1.5 m shrub with obovate to elliptic leaves, and ecarinate calyx lobes. It belongs to a group of closely related species including *M. innarrabilis*, *M. jalca*, and *M. kuelap*. They all occur in northern Peru in Amazonas and Cajamarca provinces, and are related to *M. noctiluca* of southern Ecuador.

Etymology. From the Latin *innarrabilis*, indescribable.

Specimens examined: ECUADOR. **Zamora-Chinchipe:** Nangaritza Cantón, Cordillera del Cóndor, comunidad San Miguel de Las Orquideas, Tepuy 1 vertiente oriental del Río Nangaritza, 1,299 m, 1 April 2009, *Jadán, O.*, et al. *214* (LOJA); *Luisier, F. & W. Quizhpe 2* (LOJA, QCNE); Nangaritza Cantón, Región de la Cordillera del Cóndor, cuenca alta del Río Nangaritza, La Wantza, 1,050 m, 13 February 2006, *Quizhpe, W. & F. Luisier 1958* (LOJA, MO, NY, QCNE); Nangaritza Cantón, Región de la Cordillera del Cóndor, cuenca alta del Río Nangaritza, Sector Las Orquídeas, 1,500 m, 23 February 2003, *Quizhpe, W.* et al. *575* (LOJA, MO, NY, QCNE); Nangaritza Cantón, Región de la Cordillera del Cóndor, Parroquia Surmi, Comunidad Yawi, 1,600 m, 14 June 2005, *Quizhpe, W.* et al. *1416* (LOJA, MO, NY, QCNE).

PERU. Amazonas: Yambasbamba, August, *Matthews 1314* (K); Bagua, Imaza, Cerro Tayu-Muyaji, Comunidad Aguaruna Wawas, trocha desde Puente Almendro, cumber, 1,000 m, 16 February 2004, *Rodríguez, E.R.*, et al. *2680* (HUT, NY); upper slopes and summit of Cerro Teyu, 05° 15′ S, 078° 22′ W, 1,030 m, shrub 1.5 m, flower yellow, 22 March 2001, *van der Werff* et al. *16315* (F, MO, NEU); Bagua District, Imaza, Tayu Majaji, 05° 15′ S, 078° 22′ W, 900–1,030 m, 17 February 2002, *Vásquez 27601* (MO). Without locality, "Voyage à l'Équateur et au Pérou," 1876–1877, *Vidal-Sénège s.n.* (P); Rodrigues de Mendoza, low forests of the eastern slopes of the Andes, 1,400–2,500 m, July–August 1963, *Woytkowski 8018* (MO, U). Cajamarca: San Ignacio, Dist. Huarango, alrededores del Caserio Palma, 1,810 m, 26 April 2006, *Ortiz, E.*, et al. *1398* (MO, NY); San Ignacio, Distrito Huarango, Caserio el Porvenir, 05° 06′ 38″ S, 078° 47′ 07″ W, 1,200 m, *Perea, J. & V. Flores 2370* (MO, NY, QCNE); San Ignacio, Distrito Huarango, Poblado Huarandoza, Caserio el Eden, 05° 09′ 38″ S, 078° 44′ 19″ W, 1,598 m, *Perea J. & V. Flores 3534* (MO, NY).

13. ***Macrocarpaea jactans* J.R. Grant**

Harvard Pap. Bot. 9(2): 334. 2005. TYPE: ECUADOR. Napo: P.N. Napo-Galeras, Cordillera de Galeras, Sendero hacia el Río Pucuno, 00° 39′ S, 077° 32′ W, 1,550–1,650 m, 20 March 1997, *A. Alvarez, P. Cerda, & B. Shiguango 1726* (Holotype MO; Isotypes NY, QCNE) (Figs. 3.33, 3.34, 3.35).

Fig. 3.33 *Macrocarpaea jactans* J.R. Grant. *Photo* J.R. Grant

Fig. 3.34 *Macrocarpaea jactans* J.R. Grant. *Photo* J.R. Grant

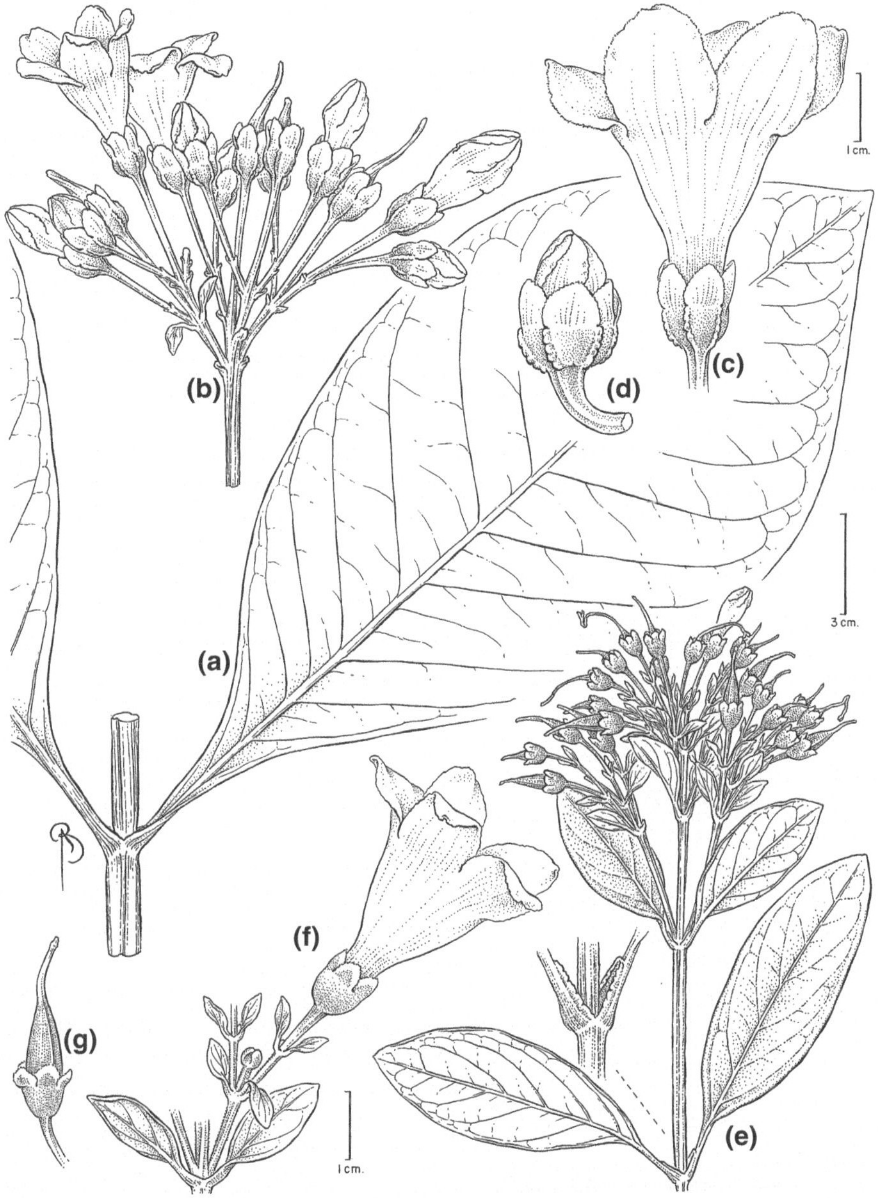

Fig. 3.35 *Macrocarpaea jactans* J.R. Grant. **a–d** *M. jactans*. **a** leaves; **b** inflorescence; **c** flower; **d** bud. **e–g** *Macrocarpaea berryi* (=*M. bubops*) **e** habit of flowering stem; **f** flower; **g** fruit. **a–c** drawn from Alvarez 1726 (MO), **d** drawn from Harling, G. & Andersson 21370 (GB), **e–g** drawn from Berry, P. & D. Neill 7665 (NY)

Shrub or small tree, 1.5–5.0 m, glabrous throughout, cm in diameter. Stems terete to slightly quadrangular, solid to hollow, 10–15 mm in diameter just below the inflorescence. Leaves oval, elliptic to ovate, sessile to short-petiolate, 20.5–35.5 cm long. Petioles 0–25 mm long, robust with strong open vagination one quarter to one-third the length of the petiole; interpetiolar ridge 5–8 mm high. Blades 20.5–33.5 × 12–16.5 cm, entire, not revolute, dark above and conspicuously lighter below, with slightly impressed veins above, and slightly raised veins below, papery thin, to thin coriaceous; base aequilateral to attenuate and decurrent on the petiole; apex rounded, obtuse to acute. Inflorescence a much branched open thyrse, 24–38+ cm long; branches 5–38 cm long; 2–5 flowered per branch. Bracts ovate to elliptic, sessile to short-petiolate, 45–96 × 32–50 mm; base aequilateral to oblique, cuneate, rounded to short-attenuate; apex acute to obtuse; bract petioles 0–3 mm long. Flowers pedicellate, erect to spreading; pedicels 10–34 mm long; bracteoles inconspicuous and scabrous, linear to triangular, 1–13 × 0.5–6.0 mm. Calyx campanulate, 12–16 × 10–16 mm, glabrous, smooth, faintly rugose to pusticulate, ecarinate, yet with pronounced ridges between calyx lobes from their bases and extending to the base of the calyx; calyx lobes rotund, ovate to elliptic, 5–9 × 7–10 mm, apex rounded, obtuse to acute. Corolla funnel-shaped, 45–55 mm long, 20–30 mm wide at the apex of the tube, crema-verdosas, yellow, smooth; corolla lobes ovate, 14–22 × 14–19 mm, obtuse to rounded. Stamens 18–30 mm long; filaments 13–25 mm long, filiform, flattened; anthers elliptic to oblong, 5–9 × 2–4 mm, sagittate, versatile; pollen *Corymbosa*-type. Pistil 40–43 mm long; ovary 7–10 × 2–4 mm; style 33–37 × 1.0 mm; stigma lobes spathulate, 3–7 × 2–4 mm. Capsules ellipsoidal to oblong, 30–45 × 12–16 mm, shiny smooth to pusticulate, tan to light brown, spreading to slightly nodding; style remnant 7–14 mm long. Seeds "Winged type," 1.2–2.2 × 0.3–0.5 mm, straw-colored, testa reticulate, wings ribbed.

Macrocarpaea jactans belongs to section *Magnolifoliae* and occurs from 1,500 to 2,300 m in the Amotape–Huancabamba region of southern Ecuador and northern Peru. It is unique in Ecuador in having bright true yellow colored flowers that are broadly flared open, and leaves with pinnate venation. It is related to *M. ostentans* J.R. Grant, *M. robin-fosteri* J.R. Grant, and *M. tahuantinsuyuana* J.R. Grant of the Cordillera Central of Peru, and *M. cinchonifolia* (Gilg) Weaver of the Cordillera Oriental of Peru and Bolivia. The unique characters that particularly unite these species are *Corymbosa*-type pollen morphology and Winged-type seeds. Despite belonging to two different species groups both *M. jactans* and *M. opulenta* occur in the same geographic region of the Cordillera del Cóndor, and have similar shaped corollas indicating that they may have a common pollinator.

Etymology. From the Latin *jactans*, boastful, for its large showy inflorescences.

Specimens examined: ECUADOR. **Zamora-Chinchipe:** Chinchipe Cantón, Parroquia Zumba, Finca de Sandy León, forest near Río Tarrangami, 2,100–2,300 m, 29 March 2005, *Clark, J.L. 8927* (NY, QCNE, US); Approx. 50 km S Amaluza on road to Zumba that nearly parallels the Ecuador–Peru border, 2,224 m, 1 January 2008, *Grant, J.R., C. Agier, C. Arnold & M.L. Cheung 08-4520* (LOJA, NY); Along road between Zumba and Amaluza that parallels the Ecuador–Peru

Fig. 3.36 *Macrocarpaea jensii* J.R. Grant & Struwe. *Photo* J.R. Grant

border, 04° 47′ 58.6″ S, 079° 18′ 18.4″ W, 2,030 m, 7 February 2011, *Grant, J.R. & J. Vieu 11-4668* (NY); Cordilllera del Condor, 32 km from Los Encuentros in direction of Paquisha Alta Military Camp, north for 1–3 km north on road toward a mine, 03° 50′ 54.3″ S, 078° 31′ 45.6″ W, 1,500 m, 9 February 2011, *Grant, J.R. & J. Vieu 11-4677* (NY); Cordilllera del Condor, 15 km E of Los Encuentros in direction of Paquisha Alta Military Camp, 03° 49′ 18.2″ S, 078° 36′ 47.2″ W, 1,458 m, 9 February 2011, *Grant, J.R. & J. Vieu 11-4678* (NY); Cordilllera del Condor, 46.8 km E of Los Encuentros, entrance to Paquisha Alta Military Camp, 03° 54′ 49.1″ S, 078° 29′ 36.9″ W, 1,834 m, 13 February 2011, *Grant, J.R. & J. Vieu 11-4694* (NY); Above Valladolid on road to Yangana, 2,300 m, 1 February 1985, *Harling, G. & Andersson 21370* (GB); Yantzaza Cantón, Cordillera del Cóndor region, broad ridge on Hollín sandstone substrate, one km north of Las Peñas, above bend of Río Machinaza, "Fruto del Norte" site of Aurelian mine company, 03° 46′ 17″ S, 078° 29′ 28″ W, 1,610 m, 22 March 2008, *Neill, D. & W. Quizhpe 16323* (LOJA, QCNE, MO); Región de la Cordillera del Cóndor, sector sur, Parroquia San Francisco de Vergel, Playones, Cuenca alta del Río Vergel, 04° 43′ 01″ S, 078° 57′ 47″ W, 1,900 m, 10 March 2005, *Quizhpe, W.* et al. *903* (LOJA, QCNE, MO; Cordillera del Condor, campamento Militar Pachicutza, 1,849 m, 22 February 1994, *Tirado, M. & B. Gray 887* (QCNE, MO, U).

14. *Macrocarpaea jensii* J.R. Grant & Struwe

Harvard Pap. Bot. 8(1): 68. 2003. TYPE. ECUADOR. Zamora-Chinchipe: P.N. Podocarpus (San Francisco entrance), trail leading west from San Francisco, 03° 59′ 24″ S, 079° 05′ 48″ W, 2,100 m, 16 February 2001, *Grant, J.R & L. Struwe 01-4047* (Holotype: US [3 sheets]; Isotypes: G [2 sheets], LOJA, NEU, NY, QCNE, SBBG) (Figs. 3.36, 3.37).

Unbranched tree, overall inflorescence triangular in outline, 1–5 m tall, glabrous throughout; trunk to 1.5 cm diameter wood solid to hollow (pith to 3 mm), growth rings scarcely visible; bark papery thin to scarcely measurable, 0.05 mm thick, outer

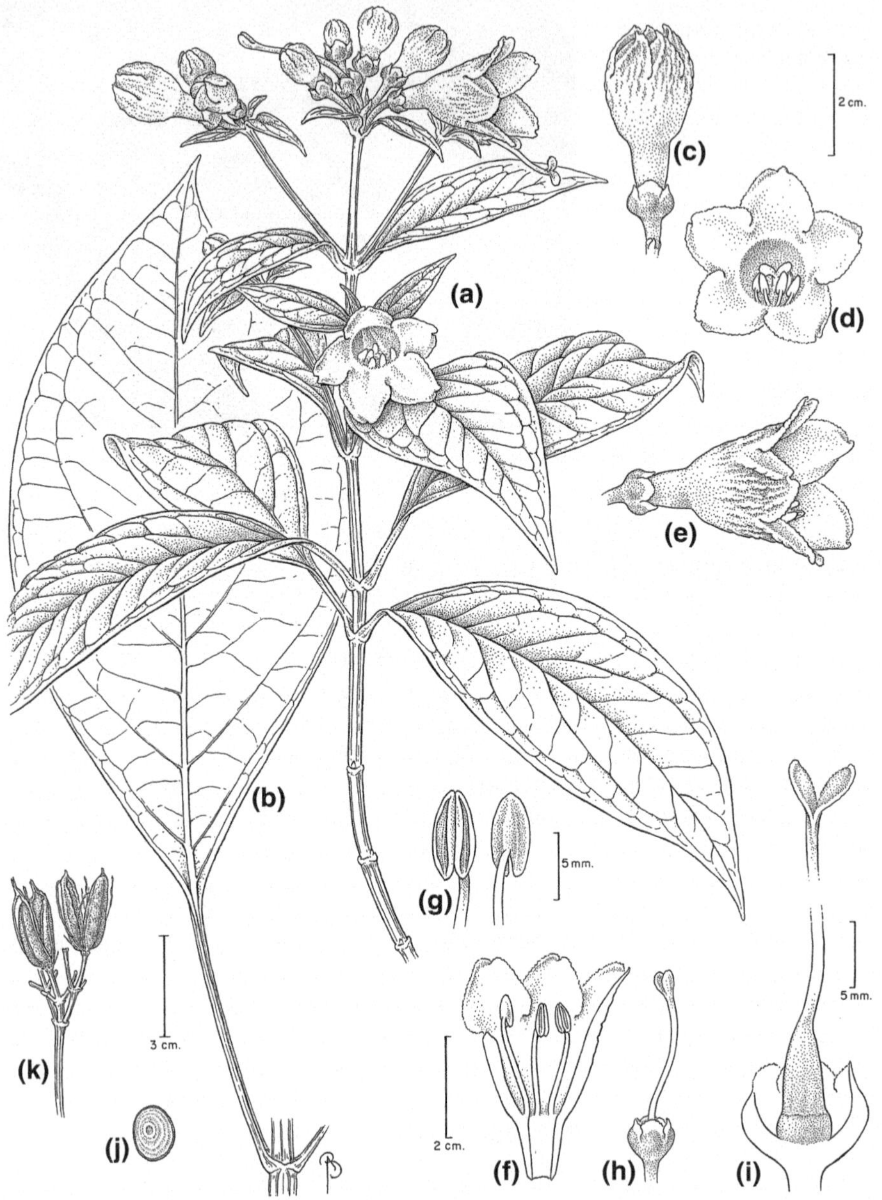

Fig. 3.37 *Macrocarpaea jensii* J.R. Grant & Struwe. **a** habit of flowering stem; **b** leaf; **c** floral bud; **d** corolla viewed from front; **e** side-view of flower at anthesis; **f** cross-section of corolla; **g** anthers; **h** pistil; **i** pistil with sepals removed to show the ovary; **j** wood anatomy of cross-section of main trunk; **k** mature dehisced capsules. **a–k** from pickles, photos and specimens of Grant, J.R & L. Struwe 01-4047

surface smooth to rugose, tan. Stems terete, solid to hollow, 4–6 mm in diameter just below inflorescence. Leaves obovate, elliptic-rhomboid to oval, slightly asymmetric, petiolate, 12–32 cm; blades 14–23 × 4.5–11.5 cm, long attenuate and decurrent on the petiole, entire, not revolute, base aequilateral, apex acute to acuminate, membranaceous, thin, flexible, dark green, with slightly impressed veins above, and slightly raised veins below, glabrous above and typically below, yet often with short tuberculate hairs on lower veins; interpetiolar ridge 1–5 mm high; petioles slender with very slight vagination, 30–90 mm. Branches of the inflorescence 6–23 cm, 3–12 flowered per branch. Bracts elliptic-rhomboid to oval, slightly asymmetric, 18–80(–140) × 9–35(–50) mm, base aequilateral to attenuate, apex acute to acuminate, sessile to petiolate; bract petioles 0–10 mm. Flowers pedicellate, erect to horizontal to oriented in all directions in the inflorescence; pedicels 6–17 mm; bracteoles lanceolate to ovate, 1–12 × 1–9 mm. Calyx urceolate to campanulate, 7–9 × 7–9 mm, glabrous, smooth, green, ecarinate, no ridges extend down from calyx lobes; calyx lobes dividing calyx one-third to one half, apex rounded, obtuse to acute, 2–4 × 3–4 mm. Corolla funnel-shaped, 29–41 mm, 17–25 mm wide at corolla lobe sinuses, yellowish-green, smooth to spongy, fleshy, warty and uneven; corolla lobes ovate, apex obtuse to rounded, 7–14 × 6–13 mm. Stamens 19–21 mm; filaments 16–18 mm, filiform, terete; anthers elliptic to oblong, 3–4 × 2–3 mm, sagittate, versatile. Pistil 26–32 mm; ovary 7–10 × 2.5–3.0 mm; style 17–18 × 0.5–1.0 mm, 2-lobed; stigma 2-lobed, lobes 2–4 × 1–2 mm, spathulate. Capsules dry, bilocular, medially dehiscent, ellipsoidal, 20–26 × 6–11 mm, smooth to very faintly ribbed, faint-orangish tan, erect to slightly nodding, style remnant 2–6 mm. Seeds flattened, angular, roughly triangular in outline, winged, 0.3–1.0 × 0.2–0.5 mm, faint orangish-tan, rugose-reticulate.

Macrocarpaea jensii belongs to section *Choriophylla* and occurs from 1,850 to 2,600 m in the Amotape–Huancabamba region of southern Ecuador. It is unique in having a thick spongy corolla and obovate, elliptic-rhomboid leaves. This species is part of a distinctive monophyletic group including *M. apparata*, *M. bubops*, *M. elix*, *M. jensii*, *M.* × *mattii*, and *M. noctiluca*.

Etymology. Named for Danish botanist Jens Madsen, 1959–, of the University of Aarhus, prolific collector of southern Ecuadorian plants.

Specimens examined: ECUADOR. **Zamora-Chinchipe:** Parque Nacional Podocarpus, San Francisco entrance area, between Loja and Zamora, 03° 59′ 22.8″ S, 079° 05′ 47.1″ W, 2,126 m, 10 February 2011, *Grant, J.R. & J. Vieu 11-4683* (NY); Loja-Zamora rd., E of the pass, disturbed forest, 2,600 m, slender treelet, 4–5 m, 15 February 1985, *Harling, G. & Andersson 22118* (GB, MO, QCA); At 25 km from Loja (road to Zamora), in the forest of the 'Estación Cientifica San Francisco', south of the river San Francisco, beside the path connecting the hydroelectric station and the entrance of the water channel, 1,850 m, January 1998, *Matt, F. 16* (ER); At 25 km from Loja (road to Zamora), in the forest of the 'Estación Cientifica San Francisco', south of the river San Francisco, beside the path connecting the hydroelectric station and the entrance of the water channel, 1,850 m, January 1998, *Matt, F. 17* (ER); Km 17 de la Via Loja-Zamora, Quebrada Navidad, 03° 58′ 42.5″ S, 079° 7′ 3″ W, 1,950 m, 30 January 1996, *Merino,*

Fig. 3.38 *Macrocarpaea lenae* J.R. Grant. *Photo* J.R. Grant

B. et al. *4764* (LOJA); P.N. Podocarpus, La Esmeralda (Cooperativa San Francisco de Numbala Alto), bosque primario alto, 04° 22′ S, 79° 03′ W, 2,250 m, January 1995, *Palacios, W.* & *Tirado 13033* (MO, QCNE, U); Area of ECSF (Estación Cientifica San Francisco), 03° 58′ 18″ S, 079° 04′ 44″ W, 2,300 m, 21 January 2001, *Wolff, D. 49* (LOJA).

15. *Macrocarpaea lenae* J.R. Grant

Harvard Pap. Bot. 8(1): 70. 2003. TYPE: ECUADOR. Zamora-Chinchipe: 5 km S of Zamora toward P.N. Podocarpus (Bombuscaro entrance), 04° 06′ 31″ S, 078° 57′ 49″ W, 1,030 m, 13 February 2001, *Grant, J.R & L. Struwe 01-4013* (Holotype: US [3 sheets]; Isotypes: G, LOJA, MO, NEU [2 sheets], NY, QCA, QCNE, S, SBBG) (Figs. 3.38, 3.39).

Unbranched tree, overall inflorescence triangular in outline, 2–3 m tall, glabrous throughout; trunk to 2.7 cm diameter wood always hollow (pith 5–9 mm), growth rings scarcely visible, bark papery thin to scarcely measurable (to 0.05 mm), outer surface smooth to rugose, tan. Stems terete to slightly quadrangular, hollow, 4–6 mm in diameter just below inflorescence. Leaves elliptic, oblong to oval, petiolate, 12–45 cm; blades 10.5–37 × (4.5–) 11.5–19 cm, cuneate to rounded, entire, not revolute, base aequilateral to oblique, apex acute to acuminate, papery thin, dark green, with slightly impressed veins above, and slightly raised veins below, glabrous above and below; interpetiolar ridge 2–8 mm high; petioles slender with very slight vagination, 15–70 mm. Branches of the inflorescence 11–26 cm, 6–20 flowered per branch. Bracts ovate to elliptic, 9–70 × 5–32 mm, base aequilateral, apex acute to acuminate, short-petiolate; bract petioles 1–5 mm. Flowers pedicellate, erect to horizontal to oriented in all directions in the inflorescence; pedicels 7–25 mm; bracteoles linear to ovate, 1–15 × 1–9 mm. Calyx urceolate to campanulate, 8–10 × 6–8 mm, glabrous, smooth, glaucous green, ecarinate; no ridges extend down from calyx lobes; calyx lobes dividing calyx one-third to one half, rounded to obtuse, 3–4 × 3.0–4.5 mm.

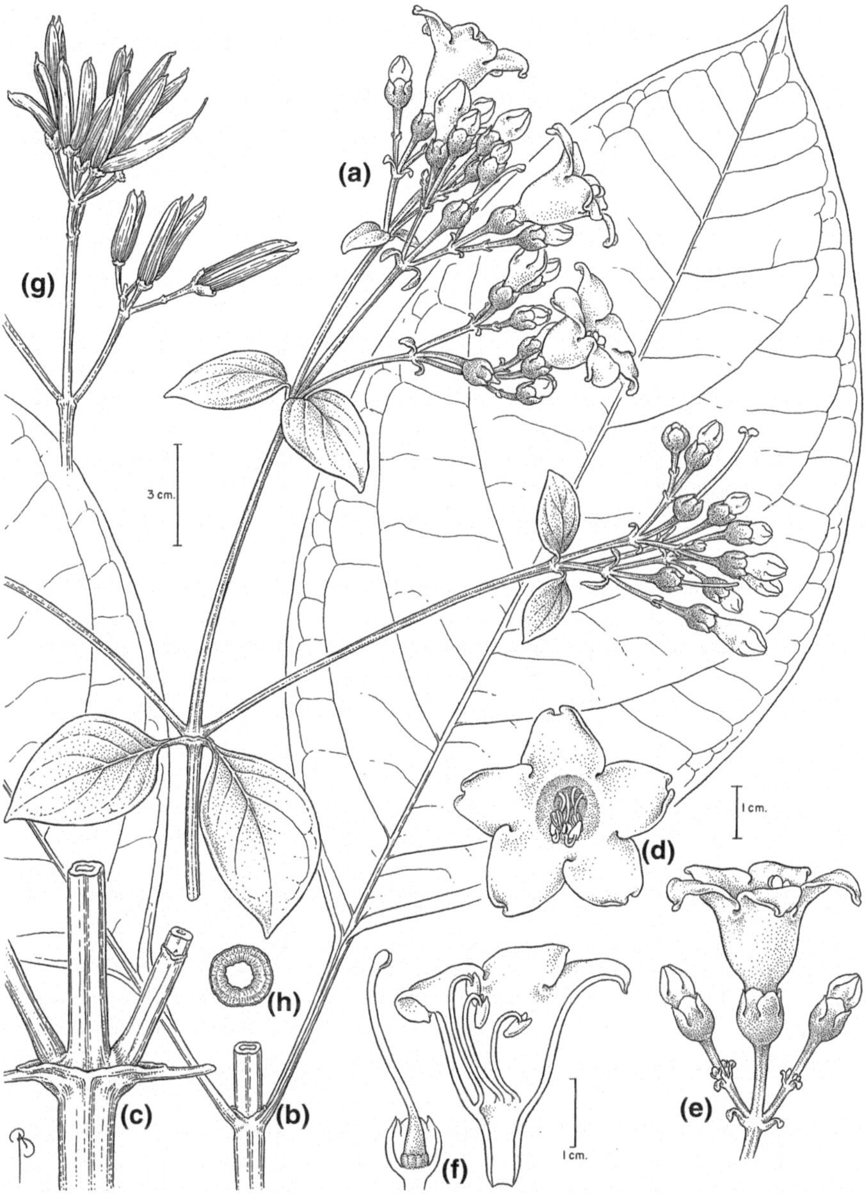

Fig. 3.39 *Macrocarpaea lenae* J.R. Grant. **a** habit of flowering stem; **b** leaves; **c** interpetiolar ridge; **d** corolla viewed from front; **e** flower and floral buds; **f** cross-section of corolla; **g** mature dehisced capsules; **h** wood anatomy of cross-section of main trunk. **a–h** from pickles, photos and specimens of Grant, J.R & L. Struwe 01-4013

Corolla funnel-shaped, 30–37 mm, 12–21 mm wide at corolla lobe sinuses, light green, smooth; corolla lobes ovate, apex obtuse to retuse, 9–12 × 4–9 mm. Stamens 24–28 mm; filaments 20–24 mm, filiform, terete; anthers elliptic to oblong, 3–4 × 2–3 mm, sagittate, versatile. Pistil 25–31 mm; ovary 5–7 × 2.5–3.5 mm; style 17–20 × 0.5–1.5 mm; stigma 2-lobed, lobes 3–4 × 2.0–2.5 mm, rounded to spathulate. Capsules dry, bilocular, medially dehiscent, ellipsoidal, 24–29 × 8–11 mm, smooth, faint orangish-tan, erect to slightly nodding, style remnant 1–2 mm. Seeds flattened, angular, roughly triangular in outline, winged, 0.3–1.1 × 0.4–0.6, faint orangish-tan, rugose-reticulate.

Macrocarpaea lenae belongs to section *Choriophylla* and occurs from 800 to 1,700 m in the Amotape–Huancabamba region of southern Ecuador. It is best known as a lownlad species around Zamora, ranging northwards to Chuchumblezta. At the Bombuscaro entrance to the P.N. Podocarpus and it is sympatric with *M. micrantha*. This species belongs to a closely related group of species restricted to the Amotape–Huancabamba zone including *M. catherineae, M. claireae, M. dies-viridis, M. illuminata, M. lenae, M. quizhpei*, and *M. xerantifulva*.

Etymology. Named for Swedish botanist Karin Lena Elisabet Struwe, 1967-, Gentianaceae specialist, Rutgers University–New Brunswick, NJ, USA.

Specimens examined: ECUADOR. **Loja:** Chaguarpamba Cantón, Alrededores de Buenavista, 1,220 m, Oct-Nov 2004, *Sanchez, O. & Gonzaga 193* (LOJA); 3 km S of Zamora toward the Bombuscaro entrance of P.N. Podocarpus, 1,019 m, 12 January 2008, *Grant, J.R., C. Agier, C. Arnold & M.L. Cheung 08-4546* (NY, LOJA, MO); **Zamora-Chinchipe:** 5 km S of Zamora toward P.N. Podocarpus (Bombuscaro entrance), 04° 06′ 31″ S, 078° 57′ 49″ W, 1,030 m, 1 November 2002, *Grant, J.R., A. Roguenant & A. Raynal-Roques 02-4250* (MO, NEU, NY, SEL, US); P.N. Podocarpus (Bombuscara entrance), 5.7 km S of Zamora, 1,030 m, 22-Feb-2006, *Grant, J.R., M. Cheung, F. F. Luisier, & N. Villard 4346* (CHRB, LOJA, NY); Property of Cabañas Ecológicas Copalinga, between Zamora and the Bombuscaro entrance to P.N. Podocarpus, 04° 05′ 30.1″ S, 078° 57′ 39.4″ W, 994 m, 8 February 2011, *Grant, J.R. & J. Vieu 11-4675* (NY); 1 km south of Chuchumblezta on road toward the River Zamora, 03° 32′ 12.4″ S, 078° 31′ 39.9″ W m, 786 m, 15 February 2011, *Grant, J.R. & J. Vieu 11-4700* (NY); 1 km from Zamora along main road to Loja, 03° 58′ 17.7″ S, 079° 01′ 15.9″ W, 1,578 m, 16 February 2011, *Grant, J.R. & J. Vieu 11-4703* (NY); P.N. Podocarpus, Romerillo, trail at limit of P.N. Podocarpus, mountain rainforest, 78° 56′ W, 04° 13′ S, 1,650–1,700 m, 14 February 1990, *Madsen & Knudsen 86854* (AAU, LOJA); hills and pasture and disturbed forest immediately S and SE of Zamora, 78° 57′ W, 04° 04′ S, 1,000–1,250 m, 14 June 1988, *Øllgaard* et al. *74846* (AAU, LOJA, QCA); P.N. Podocarpus, Guardería Río Bombuscaro, Sendero al Mirador, bosque primario alto, 1,100 m, January 1995, *Palacios, W. & Tirado 13301* (MO, QCNE, U); Zamora, Región de la Cordillera del Cóndor, Parroquia San Carlos de Las Minas, Namibija, Sendero al Cerro Colorado, 2,100 m, 27 January 2005, *Quizhpe, W.* et al. *786* (LOJA, MO, NY, QCNE); Región de la Cordillera del Cóndor, Parroquia San Carlos de Las Minas, Namibija, Sendero al Cerro Colorado, 1,520–1,700 m, 29 January 2005, *Quizhpe, W.* et al. *834* (LOJA, MO, NY, QCNE).

Fig. 3.40 *Macrocarpaea luna-gentiana* J.R. Grant & Stuwe. *Photo* J.R. Grant

16. *Macrocarpaea luna-gentiana* J.R. Grant & Struwe

Harvard Pap. Bot. 8(1): 72. 2003. TYPE: ECUADOR. Loja: Km 21 on road from Yangana to Cerro Toledo, then trail from Cerro Toledo (ca. 300 m below antennas) toward Numbala, (to 1 km down the trail), cool, very rainy, windy páramo, 04° 24′ 01″ S, 079° 06′ 42″ W, 3,350 m, 14 February 2001, *Grant, J.R & L. Struwe 01-4028* (Holotype: US [2 sheets]; Isotypes: G, LOJA, NEU, NY, QCA, QCNE) (Figs. 3.40, 3.41).

Dichotomously branched to unbranched tree, overall branching pattern obtriangular and flat-topped in outline, 2–3 m tall, glabrous to hyaline puberulent with short simple hairs on stems, petioles, leaves, inflorescences, calyces, and corolla lobes; trunk to 7.7 cm diameter wood solid in trunk, hollow in younger branches (to 0.5), growth rings prominent; bark thin to thick, (to 0.5–4.0 mm), brown background, mottled with distinctive tan rippling and furrowing patterns. Stems terete to slightly quadrangular, solid to hollow, 5–10 mm in diameter just below inflorescence. Leaves broadly ovate, sessile to short-petiolate, 3–12 cm; blades 10–11.5 × 2.5–8.0 cm, entire, slightly revolute, base aequilateral to slightly oblique to cuneate, apex acute, thick, leathery-coriaceous, light to dark glossy green, bullate, with few to strongly impressed veins above, and strongly raised veins below, glabrous above and typically below yet often with long pubescent hairs on the veins of young leaves; interpetiolar ridge 1–5 mm high; petioles flattened to concave, 3–5 mm. Branches of the inflorescence 5–23 cm, 1–10 flowered per branch. Bracts ovate to lanceolate, 33–82 × 10–42 mm, base aequilateral to slightly oblique or rounded, apex acute, sessile to short-petiolate; bract petioles 0–3 mm. Flowers pedicellate, erect to horizontal to oriented in all directions in the inflorescence; pedicels 3–30 mm; bracteoles ovate, 5–18 × 3–7 mm. Calyx broadly campanulate, 9–18 × 13–14 mm, hyaline puberulent to glabrous, rugose, green, ecarinate, no ridges extend down from calyx lobes; calyx lobes dividing calyx one-third to one half, rounded, 7–9 × 6–12 mm. Corolla funnel-shaped, 341–62 mm, 15–40 mm

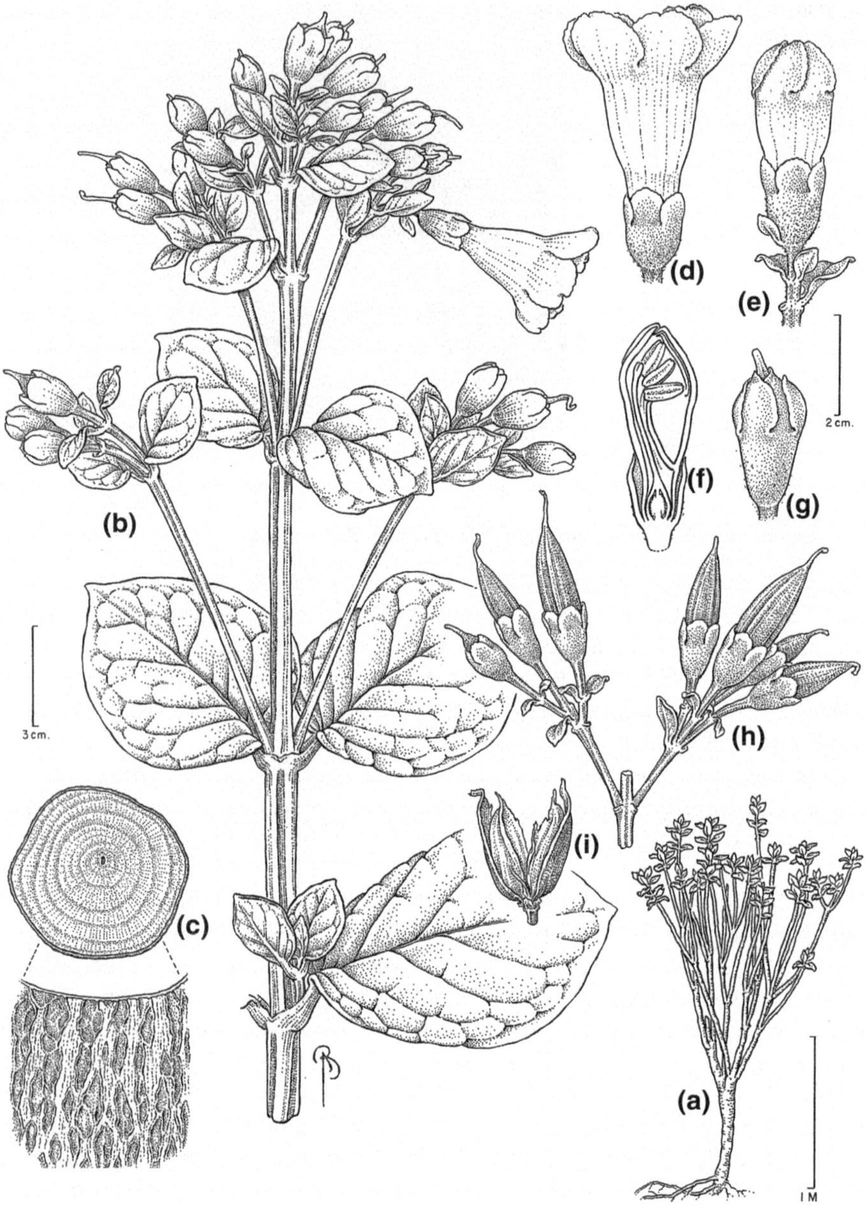

Fig. 3.41 *Macrocarpaea luna-gentiana* J.R. Grant & Stuwe. **a** habit of tree in the field; **b** habit of flowering stem; **c** wood anatomy of cross-section of main trunk, and bark; **d** flower; **e** floral bud; **f** cross-section of floral bud; **g** flower just after anthesis; **h** immature capsules; **i** mature dehisced capsule. **a** from a *photo* of Grant, J.R & L. Struwe 01-4027; **b** from Matt, F. 1 (ER); **c** from Grant, J.R & L. Struwe 01-4027; **d–i** from Grant, J.R & L. Struwe 01-4028

wide at corolla lobe sinuses, light green, smooth; corolla lobes half moon-shaped to broadly ovate, apex rounded, 5–21 × 9–24 mm. Stamens 23–35 mm; filaments 18–29 mm, filiform, flattened; anthers elliptic to oblong, 5–6 × 2.5–3.5 mm, sagittate, versatile. Pistil 41–42 mm; ovary 7–8 × 3–4 mm; style 29–30 × 1.0–1.5 mm; stigma 2-lobed, lobes 4–5 × 2.0–2.5 mm, spathulate. Capsules dry, bilocular, medially dehiscent, ellipsoidal to ovoid, 30–49 × 13–19 mm, wrinkled to ribbed, dark brown, erect to slightly nodding; style remnant 4–15 mm. Seeds unknown.

Macrocarpaea luna-gentiana belongs to section *Magnolifoliae* and occurs from 2,500 to 3,800 m in the Amotape–Huancabamba region of southern Ecuador. Its probable sister species *M. pachyphylla* is restricted to Nariño and Putumayo provinces of Colombia at similar elevations of 3,000–3,400 m. *Macrocarpaea luna-gentiana* It has a generally robust, stout appearance, a few-flowered inflorescence, broad campanulate calyx, rounded calyx lobes, and a broad funnel-shaped corolla with broad rounded to half-moon-shaped corolla lobes.

Etymology: From the Latin "*luna*," moon, and "*gentiana*," gentian, for the half-moon shaped petal lobes.

Specimens examined: ECUADOR. **Loja:** Loja-Zamora, 3,800 m, 1 December 1876, *André 4513* (F, K, NY); Loja-Zamora, 3,500 m, 1 December 1876, *André 4536* (K, NY); El Tiro, en el límite entre las provincias de Loja y Zamora, a 12 km de Loja, Parque Nacional Podocarpus, 04° 00′ 17″ S, 079° 08′ 51″ W, 2,980 m, 9 June 1999, *Cabrera, O.* et al. *379* (LOJA); P.N. Podocarpus, Sendero a Cajanuma-Laguna del Compadre-Mirador-Cajanuma, 2,750–3,200 m, 04° 04′ S, 79° 09′ W, Bo25 February 2000, *Cerón & Curso de Dendrologia 40077* (QAP); Km 21 on road from Yangana to Cerro Toledo, then trail from Cerro Toledo (ca. 300 m below antennas) toward Numbala, (to 1 km down the trail), cool, very rainy, windy páramo, 04° 24′ 01″ S, 079° 06′ 42″ W, 3,350 m, 14 February 2001, *Grant, J.R & L. Struwe 01-4027* (LOJA, NEU, QCNE, US); Km 21 on road from Yangana to Cerro Toledo, 2 November 2002, *Grant, J.R., A. Roguenant & A. Raynal-Roques 02-4274* (NEU, NY, US); Km 21 on road from Yangana-Cerro Toledo, 3,350 m, 19 February 2006, *Grant, J.R., M.L. Cheung, F. F. Luisier, & N. Villard 06-4341* (G, LOJA, NY); Around km 21 on road from Yangana up Cerro Toledo, near the summit of Cerro Toledo, 04° 22′ 32.9″ S, 079° 06′ 43.3″ W, 3,122 m, 3 February 2011, *Grant, J.R. & J. Vieu 11-4658* (NY); P.N. Podocarpus, Cerro Toledo, 79° 07′ W, 04° 23′ W, 2,500–3,400 m, 30 October 1989, *Madsen 86290* (AAU [2 sheets], LOJA); Parque Nacional Podocarpus, vicinity of Lagunas de Compadre, 04° 10′ S, 079° 07′ W, 3,000–3,400 m, 21 November 1989, *Madsen, J. 86430* (LOJA); P.N. Podocarpus, above the refuge of Cajanuma, beside the trail from 'Mirador' on the ridge eastward to the cordillera, 3,200 m, November 1997, *Matt, F. 1* (ER); Cerro Toledo, km 15, Cerretera Yangana-Antenas, 04° 23′ S, 079° 7′ W, 2,920 m, 12 December 1995, *Merino, B.* et al. *4695* (LOJA); P.N. Podocarpus, S of Loja, above "Centro de información," E of Nudo de Cajanuma, 79° 10′ W, 04° 05′ S, 3,050–3,420 m, 3 m tall shrub, flowers green, 24 February 1985, *Øllgaard* et al. *58114* (AAU, MO, NY); P.N. Podocarpus, Yangana-Cerro Toledo, at entrance to crest, subpáramo scrub and bogs in the pass, 70° 06′ W, 04° 23′ S, 3,100 m, 26 February 1985, *Øllgaard* et al.

Fig. 3.42 The hybrid *Macrocatpaea* × *mattii* and its parents. *Macrocarpaea subsessilis* (*left*), *M.* × *mattii* (*middle*), and *M. noctiluca* (*right*). *Photo* J.R. Grant

58258 (AAU, LOJA, MO, QCA). **Zamora-Chinchipe:** Palanda, Valladolid, Reserva Biologica Tapichalaca (Fundación de Conservación Jocotoco), Senderl El Paramo (toward plane wreckage), 04° 29′ 45″ S, 079° 7′ 55″ W, 2,469 m, 1 June 2011, *Clark, J.L.* et al. *12423* (NY).

17. *Macrocarpaea* × *mattii* J.R. Grant

(*Macrocarpaea noctiluca* J.R. Grant & Struwe × *M. subsessilis* Weaver & J.R. Grant), Harvard Pap. Bot. 11(2): 138. 2007. TYPE: ECUADOR. Loja: Beside the road from Yangana to 'Cerro Toledo', 04° 22′ 6.94″ S, 079° 08′ 0.35″ W, 2,500 m, 1 January 1998, *F.G. Matt 14* (Holotype ER) (Fig. 3.42).

Shrub, likely between 1 and 2 m, glabrous throughout. Stems terete, solid, 5–6 mm in diameter just below the inflorescence. Leaves broadly elliptic, petiolate, 9.5–10.5 cm long. Petioles 10–12 mm long, slender with slight vagination; interpetiolar ridge 2–3 mm high. Blades 8.5–9.5 × 5.5–6.5 cm, entire, not revolute, dark above and conspicuously lighter below, with slightly impressed veins above, and slightly raised veins below, glabrous above and below, thin coriaceous; base aequilateral to oblique, cuneate; apex rounded to obtuse. Inflorescence a much branched open thyrse to 15 cm long; branches 5–14 cm long; 3–12 flowered per branch. Bracts obovate to oblanceolate, petiolate, 18–60 × 10–32 mm; base aequilateral to oblique, cuneate; apex obtuse, acute to rounded; bract petioles 4–7 mm long. Flowers pedicellate, erect to slightly spreading; pedicels 2–15 mm long; bracteoles spathulate to oblanceolate, 5–18 × 1–7 mm. Calyx narrowly campanulate, 9–11 × 4.5–6.0 mm, glabrous, smooth, green, ecarinate; calyx lobes ovate to reniform, 3–4 × 4–5 mm, rounded. Corolla funnel-shaped, 28–32 mm long, 12–14 wide at the apex of the tube, yellow, smooth; corolla lobes ovate to elliptic, 7–12 × 4–7 mm, apex obtuse to rounded. Stamens 22–25 mm long; filaments 18–20 mm long, filiform, flattened; anthers linear to linear-elliptic, 4–5 × 1.5–2.0 mm, sagittate, versatile. Pollen glabra-type. Pistil 27–28 mm long;

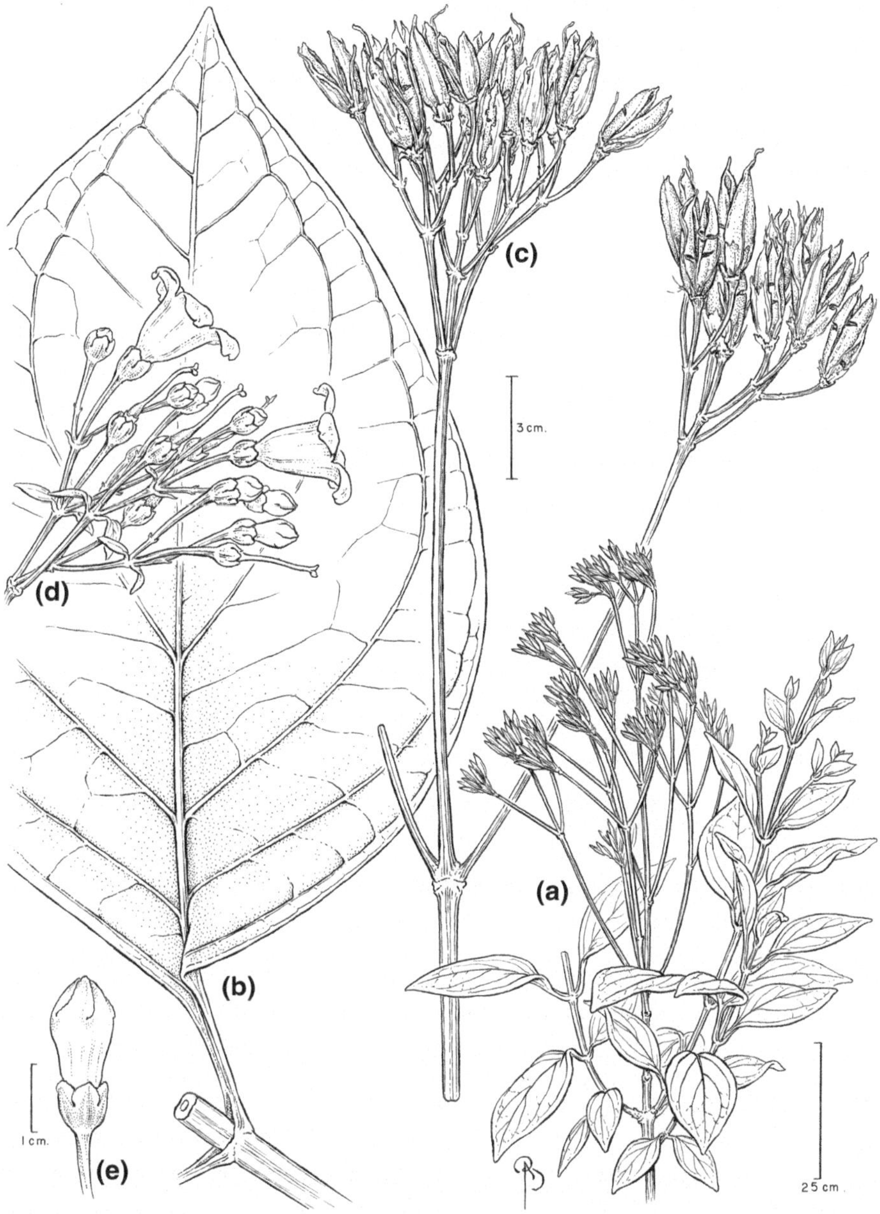

Fig. 3.43 **A**-**C** Macrocarpaea micrantha. **A** leaves; **B** habit of flowering stem; **C** details of flowers;**A** drawn from Grant 11-4691 (NY). **B**-**C** drawn from Quizhpe & Wisum 2673 (MO). (*Please note the Erratum at the end of the book*)

ovary 4–5 × 2–3 mm; style 20–22 × 0.75–1.0 mm; stigma lobes spathulate, 2–3 × 1–2 mm. Capsules and seeds unknown.

Macrocarpaea × *mattii* belongs to section *Choriophylla* and occurs from 2,450 to 2,500 m in the Amotape–Huancabamba region of southern Ecuador. It is a natural hybrid between sympatric *M. noctiluca* and *M. subsessilis*, and is intermediate in its morphological characters between its parents. While *M. subsessilis* has small rounded thick leathery-coriaceous leaves, *M. noctiluca* has much larger broad and papery thin leaves. The leaves of *Macrocarpaea* × *mattii* are intermediate, being elliptic to slightly obovate in shape. It also has the "dots" on the lower backside of leaves as in *M. noctiluca* and *M. bubops*. This species is part of a distinctive monophyletic group including *M. apparata*, *M. bubops*, *M. elix*, *M. jensii*, *M.* × *mattii*, and *M. noctiluca*.

Etymology. Named for German biologist Felix G. Matt, researcher of bats in Ecuador and their pollination of *Macrocarpaea*.

Specimens examined: ECUADOR. **Loja:** Around km 9 on road from Yangana up Cerro Toledo, 04° 22′ 50.5″ S, 079° 08′ 58.1″ W, 2,450 m, 3 February 2011, *Grant, J.R. & J. Vieu 11-4662* (NY); Beside the road from Yangana to 'Cerro Toledo', 2,530 m, January 1998, *Matt, F. 15* (ER).

18. *Macrocarpaea micrantha* Gilg

Bot. Jahrb. Syst. 22: 338. 1896. TYPE: PERU. San Martín: Propre Tarapoto, 1855–1856, *Spruce 4618* (lectotype NY; isolectotypes BM, BR, C, E, G [3 sheets, G, G-BOISS, G-DC], GH, GOET, K, LD, OXF, P, W, designated by Grant 2004) (Fig. 3.43).

Shrub 1–4 m tall, glabrous. Stem terete or nearly so; internodes various, generally shorter than leaves, usually 1.5–5.0 cm. Leaves membranous; venation slender but conspicuous below, larger (lower) secondary veins mostly branching from midrib at 35–50 degrees, strongly upcurved, brochidodromous. Blades of lower and middle cauline leaves elliptic, 12–25 cm long, 4–9 cm wide, base tapering rather abruptly to a distinctively slender petiole 2–4 cm, apex strongly acuminate. Uppermost cauline leaves rather abruptly smaller, often more ovate, all petiolate. Inflorescence a compact dichasium, often small and only 2–4 times divided, or occarionally larger with elongated primary branches; first 1 or 2 divisions subtended by leaves 2.0–6.5 cm long, further divisions my minute or rudimentary bractlets; flowers usually 7–25, occasionally more numerous. Pedicels erect, 15–25 mm. Calyx 7–10 mm; lobes semicircular, 0.4–0.8 times as long as tube, hyaline-margined, erose. Corolla 20–30 mm, pale greenish yellow, narrowly campanulate; lobes 6–8 mm, ovate-triangular, obtuse. Capsule narrowly conic-ovoid, 16–35 mm. This description is largely from Pringle (1995).

Macrocarpaea micrantha belongs to section *Choriophylla* and occurs from 550 to 1,700 m in the Amotape–Huancabamba region of southern Ecuador and northern Peru. It is perhaps the most widespread species in lowlands of the Huancabamba region of southern Ecuador and northern Peru. It is often found growing with other species, such as *M. harlingii*, *M. lenae*, *M. noctiluca* and *M.*

quizhpei. Often its presence may indicate the occurrence of other more rare species of *Macrocarpaea*.

Specimens examined: ECUADOR. **Morona-Santiago:** Cantón Tiwintza, Cordillera del Cóndor, Centro Shuar Kaputna, laderas del Cerro kampa Naint, 550 m, 3 July 2003, *Kajekai, C. & Grupo Shuar de Conservación 68* (LOJA, MO); Gualaquiza Cantón, région de la Corillera del Cóndor, vertiente occidental, Valle del Río Quimi, 03° 2954″ S, 078° 25′ 13″ S, 900 m, 10 May 2007, *Quizhpe, W. & A. Wisum 2673* (LOJA, MO QCNE); Cordillera del Cóndor, Cerro Maka Naint, 6 km al suroeste del Centro Shuar Warints, 1,700 m, 7 October 2002, *Wisum & Grupo Shuar de Conservación-1 18* (MO, NY, QCNE). **Zamora-Chinchipe:** Estación Padmi, 4 km de Los Encentros en la carretera Zamora-Gualaquiza, 880 m, 12 June 1997, *Aguirre* et al. *71* (LOJA); El Panqui, Tundayme, Cordillera del Cóndor region, Río Wawaime watershed, tributary of Río Quimi, below site of proposed EcuaCorriente coppermine, on ridge near 1-ha forest inventory "Wawaime" plot, 03° 34′ 18″ S, 078° 26′ 46″ W, 15 May 09, *Clark, J.L.* et al. *10819* (NY); Along road from Zamora to Romerillos, 13.3 km N of bridge over Rio Bombuscaro at Zamora, 0.4 km N or Pituca along river, 975 m, 30 May 2003, *Croat, T. & Menke 89733* (MO, NEU, NY, QCNE); P.N. Podocarpus, Guardería Bombuscara, ca. 6 km S de Zamora, camino que va desde el final del Carretero Zamora-P.N. Podocarpus hasta la Guardería Bombuscara, 1,050 m, 11 July 1992, *Freire-Fierro* et al. *2284* (QCA, QCNE, U); Miazi, Río Nangaritza, hillside behind camp, 930 m, 27 July 1993, Gentry 80502 (MO, QCNE); Miazi, Río Nangaritza, hillside behind camp, 930 m, 27 July 1993, *Gentry, A. 80502* (MO, QCNE); Región de la Cordillera del Cóndor, cuneca del alto Río Nangaritza, Las Orquídeas, 877 m, 8 January 2008, *Grant, J.R. & C. Arnold 4538* (LOJA, MO, NY); 3 km S of Zamora toward the Bombuscaro entrance of P.N. Podocarpus, 1,019 m, 12 January 2008, *Grant, J.R., C. Agier, C. Arnold & M.L. Cheung 08-4547* (G, LOJA, MO, NY); Nangaritza Valley, south of Guayzimi, Cordillera del Condor region, Las Orquídeas, 04° 15′ 314″ S, 078° 40′ 087″ W, 1,180 m, 24 January 2009, *Grant, J.R., B. Angell, W. Grant & V. Trunz 09-4556* (NY); Property of Cabañas Ecológicas Copalinga, between Zamora and the Bombuscaro entrance to P.N. Podocarpus, 04° 05′ 30.1″ S, 078° 57′ 39.4″ W, 994 m, 8 February 2011, *Grant, J.R. & J. Vieu 11-4674* (NY); Forest trail west of Cabañas Yankuam, south of Las Orquideas, 04° 14′ 58.8″ S, 078° 39′ 46.6″ W, 966 m, 11 February 2011, *Grant, J.R. & J. Vieu 11-4691* (NY); Road from Chuchumbleza, Quimi and Tundayme toward the Condor Mirador, 14 km from end of road at Military camp 'Condor Mirador', 03° 37′ 23.5″ S, 078° 27′ 32.7″ W, 1,210 m, 15 February 2011, *Grant, J.R. & J. Vieu 11-4696* (NY); Road from Chuchumbleza, Quimi and Tundayme toward the Condor Mirador, 14 km from end of road at Military camp 'Condor Mirador', 03° 37′ 23.5′ S, 078° 27′ 32.8″ W, 1,212 M, 15 February 2011, *Grant, J.R. & J. Vieu 11-4697* (NY); Parque Nacional Podocarpus, bosque subtropical-nublado-andino, 03° 09′ S, 079° 01′ W, 1,100–2,560 m, 21 July 2002, *Jaramillo, J. 23479* (QCA); Nangaritza, Parroquia Pachicutza, colecciónes en el sector noreste del Campamento military, 900 m, 6 December 1990, *Jaramillo & Grijalva 13411* (NY, QCA); Yacuambi, Parroquia La Paz, Centro Shuar Kurints, en ladera arriba del Río Yacuambi, 03° 46′ S, 078° 54′ W, 920 m, 7 September 2006,

Kajekai, C. & A. Wisum 746 (MO, NY); El Padmi, 6 June 1995, *Merino* et al. *4526* (LOJA); Nangaritza, lower slopes of Cordillera del Condor, above Río Nangaritza at Pachicutza, tropical premontane wet forest, 04° 07′ S, 78° 37′ W, 950 m, ere4 December 1990, *Neill, D. & W. Palacios 9496* (MO, QCNE); El Pangui, Cordillera del Cóndor region, Río Wawaime watershed, tributary of Río Quimi, 03° 34′ 18″ S, 078° 26′ 46″ W, 1,130 m, 22 September 2007, *Neill, D.* et al. *15919* (MO, NY, QCNE); P.N. Podocarpus, Guardería Bombuscara, 1 January 2000, *Piguet s.n.* (NEU); Cordillera del Cóndor, vertiente occidental, Cuenca del Río Quimi, 03° 34′ 45″ S, 078° 25′ 20″ W, 1,180–1,400 m, 5 April 2006, *Quizhpe, W. 2117* (LOJA, QCNE); Nangaritza Cantón, Región de la Cordillera del Cóndor, cuenca alta del Río Nangaritza, Sector Las Orquídeas, 920 m, 11 May 2006, *Quizhpe, W. 2266* (LOJA, MO, NY, QCNE); Cordillera del Cóndor, vertiente occidental, Cuenca alta del Río Wawaime, afluyente del Río Quimi, 03° 35′ 06″ S, 078° 26′ 18″ W, 1,500 m, 28 September 2006, *Quizhpe, W. & A. Wisum 2320* (LOJA, MO, NY, QCNE); El Pangui, Cordillera del Cóndor, vertiente occidental, cuenca del Río Machinaza, Comunidad Machinaza, 1,050 m, 22 March 2006, *Quizhpe, W. & F. Luisier 2056* (LOJA, MO, NY, QCNE); Yantzaza Cantón, Cordillera del Cóndor, vertiente occidental, carretera de Los Encuentros a La Zarza, 3 km al noroeste de Jardéin del Cóndor, 03° 48′ 40″ S, 078° 36′ 28″ W, 1,400 m, 27 March 2008, *Quizhpe, W. & O. Jaramillo 2968* (LOJA, MO, QCNE); El Pangui, Cordillera del Cóndor region, vertiente occidental, Cantón El Pangui, Parroquia Tundayme, Valle del Río Quimi, 03° 33′ 59″ S, 078° 26′ 29″ W, 900 m, 17 May 2008, *Quizhpe, W. & R. Steeves 3078* (MO); Regíon de la Cordillera del Cóndor, cuenca alta del Río Nangaritza, destacamento militar de Miazi, rio abajo de la confluencia del Rio Cumbiriatza y el Rio Nangaritza, 900 m, 1 January 2002, *Quizhpe, W.* et al. *702* (MO, NY, QCNE); P.N. Podocarpus, near park refugio, southwest of Zamora, 1,100 m, 10 July 1990, *Rome, A.* et al. *1003* (MO, QCNE); Parroquia La Paz, Centro Shuar Kiim, Terraza aluvial del Río Kiim, 03° 47′ 02″ S, 078° 54′ 15″ W, 900 m, 18-Jun-2006, *Wisum, A.* et al. *598* (MO, NY, QCNE).

PERU. Amazonas: Bagua, Imaza, Imazita, Wawiko, primary forest trail from Wawiko to Achu community, 04° 56′ 7″ S, 078° 12′ 11″ W, 400 m, 21 November 1996, *Albán, J.* et al. *9313* (MO); Bagua, Sukutim, annex of Wawas; Aguaruna name: uchi tsuwim, 21 July 1998, *Apanú & E. Chávez 19262* (MO): Condorcanqui, Cordillera del Condor, Puesto de Vigilancia Alfonso Ugarte (PV3), 1,300–1,500 m, 16 July 1994, *Beltrán, H. & R. Foster 845* (F, USM); Bagua, Imaza, Aguaruna Putuim (anexo Yamayakat), 760 m, 24 September 1994, *Díaz, C.* et al. *7227* (MO, U); Bagua, Imaza, Comunidad Aguaruna de Wanas (Km 92 Carretera Bagua-Imacita) 700–800 m, 29 August 1996, *Díaz, C.* et al. *8040* (MO, U, USM); Bagua, Imaza, Reg. Nororiental del Marañon, Kampaenza, 320 m, 23 September 1994, *Jaramillo* et al. *468* (MO, NY); Bagua, Imaza, Aguaruna de Putuim, 780 m, 13 June 1996, *Rodríguez, E.* et al. *1000* (AMAZ, HUT, MO, U, USM); Bagua, Imaza, Tayu Mujaji, Comunidad de Wawas, 800 m, 23 October 1997, *Rojas, R.*, et al. *443* (MO, U); Quebrado El Almendro, 430 m, 8 March 1998, *van der Werff* et al. *14518* (MO, U); Bagua, Imaza, Tayu Mujaji, Comunidad de Wawas, 600 m, 25 October 1997, *Vásquez* et al. *24690* (MO, U). Loreto: Loreto, 1 September 1904, *Ule 6392* (G,

Fig. 3.44 *Macrocarpaea noctiluca* J.R. Grant & Struwe. *Photo* J.R. Grant

MG). San Martín: Pedro Ruíz-Moyobamba road, km 399, 1,410 m, 29 January 2001, *Grant, J.R. & Rodríguez 01-3966* (G, HUT, NEU, NY, US); Pumayacu, between Balsapuerto and Moyobamba, 600–1,200 m, 1 August 1933, *Klug, G. 3138* (F, G, GH, K, MO, NY, S, US, WIS); San Martín, trail to television antenna, km 17.5 of Tarapoto-Yurimaguas road (2.5 km N of Cataratas de Ahuashiyacu), 850–1,200 m, 7 August 1986, *Knapp, S. 7911* (MO, NEU, U, USM); Km 399 of Carretera Marginal, trail to Quebrada Venceremos and Río Serranoyacu, 67 km E of Pomacochas, 8 km W of bridge over Río Serranoyacu, 1,400–1,500 m, 13 June 1986, *Knapp, S. & P. Alcorn 7743* (MO, NY, USM); San Martín, Dtto Tarapoto, Carretera de Tarapoto a Yurimaguas, from km 12 to km 14, 3 December 1979, *Rimachi Y., M. 4751* (MO); Dtto Banda de Silcayo, Carretera de Tarapoto-Yurimangas km 16, altura de las cataratas del Ahuashiyacu, 800 m, 10 September 1992, *Rimachi Y., M. 10316* (US); Tarapoto, 1 November 1982, *Rios T., J. 1010* (MO); Rioja, salida a Mashoyacu-Schacucaqui, Bosque de Protección Alto Mayo, Toma de Agua, Quebrada Cuchachi, 1,000 m, 15 July 1995, *Sánchez Vega, I. & m.O. Dillon 8048* (F); Rioja, Pardo Miguel, El Afluente y la Marginal, 1,450 m, 25 June 1997, *Sánchez Vega, I. & M.O. Dillon 9038* (F, U); Rioja, Pardo Miguel, Centro Poblado Aguas Verdes, fila entre La Perla y Nuevo Edén, 1,350 m, 25 June 1998, *Sánchez Vega, I.* et al. *9520* (F, U); Between Balsafuerte & Mayobamba, 1 August 1938, *Sandeman s.n.* (BM).

19. ***Macrocarpaea neillii*** **J.R. Grant**

Harvard Pap. Bot. 9(2): 336. 2005. TYPE: ECUADOR. Zamora-Chinchipe: Nangaritza Cantón, Cordillera de Nanguipa, Cerro Colorado, about 8 km SSE of Nambija, 20 km ESE of Zamora, montane forest and dense shrubby vegetation on exposed ridges, 04° 07′ 29″ S, 78° 46′ 25″ W, 2,700 m, 18 February 2002, *D. Neill, W. Quizhpe, J. Manzanares, A. Hirtz, T. DeLinks, & C. Cole 13758* (Holotype QCNE; Isotypes LOJA, MO, NY) (Fig 3.30).

Unbranched shrub, 2 m, glabrous throughout, cm in diameter. Stems terete, solid, 3–4 mm in diameter just below the inflorescence. Leaves obovate varying to elliptic, petiolate to nearly sessile in upper leaves, 4–8 cm long. Petioles 0–10 mm long, slender with very slight vagination; interpetiolar ridge 1–4 mm high. Blades 4–7 × 2.0–3.5 cm, entire, slightly revolute, dark above and conspicuously lighter below, with slightly impressed veins above, and slightly raised veins below, leathery-coriaceous; base aequilateral to oblique, cuneate to attenuate and decurrent on the petiole; apex acute, obtuse to rounded. Inflorescence a few-branched open thyrse, 4–10 cm long; branches 3–4 cm long; 3–9 flowered per branch. Bracts obovate nearing on spatulate, sessile, 11–21 × 4–17 mm; base aequilateral to oblique, cuneate; apex acute to obtuse; bract petioles 0 mm long. Flowers pedicellate, erect to slightly spreading, but never nodding; pedicels 9–15 mm long; bracteoles inconspicuous and scabrous, to ovate to obovate, 1–3 × 0.5–1.0 mm. Calyx campanulate, 7–8 × 5–6 mm, glabrous, faintly rugose to pusticulate, ecarinate; calyx lobes ovate to rotund, 2–3 × 2.5–3.0 mm, apex rounded to obtuse. Corolla funnel-shaped, 21–27 mm long, 7–10 mm wide at the apex of the tube, greenish-yellow (*Neill, D.* et al. *13758*), smooth; corolla lobes ovate, 5–6 × 4–5 mm, obtuse to rounded. Stamens 13.0–15.5 mm long; filaments 10–12 mm long, filiform, flattened; anthers elliptic to oblong, 3.0–3.5 × 1.5–2.0 mm, sagittate, versatile; pollen *Glabra*-type. Pistil 22–29 mm long; ovary 5–6 × 2.0–2.5 mm; style 15.0–20.5 × 0.5 mm; stigma lobes spathulate, 2.0–2.5 × 1.0–1.5 mm. Capsules and seeds unknown.

Macrocarpaea neillii belongs to section *Choriophylla* and occurs at 2,700 m in the Amotape–Huancabamba region of southern Ecuador. It is related to *M. wurdackii* Weaver & J.R. Grant and *M. stenophylla* Gilg, both of the Huancabamba region of Peru.

Etymology. Named for David A. Neill (1953–), collector of the type, specialist of the flora of Ecuador.

20. *Macrocarpaea noctiluca* J.R. Grant & Struwe

Harvard Pap. Bot. 8(1): 76. 2003. TYPE: ECUADOR. Loja: Nudo de Sabanilla, 15 km S of Yangana, 04° 25′ 22″ S, 079° 09′ 04″ W, 2,486 m, 10 February 2001, *Grant, J.R & L. Struwe 01-3977* (Holotype: US [2 sheets]; Isotypes: LOJA, QCA, QCNE, NY) (Figs. 3.44, 3.45).

Dichotomously branched to unbranched tree, overall branching pattern triangular in outline, 1–6 m tall, glabrous throughout; trunk to 4.0 cm diameter wood solid to hollow (pith to 5 mm), growth rings prominent, bark thin 0.005–0.1, outer surface rugose, brown. Stems terete, solid to hollow, 6–12 mm in diameter just below inflorescence. Leaves broadly ovate, oblong to elliptic, sessile to short-petiolate, 8–43 cm; blades 7–35 × 4.5–16.0 cm, entire, slightly revolute, base aequilateral to slightly oblique or rounded, apex acute, papery thin, shiny to opaque, light green to dark green, with slightly impressed veins above, and slightly raised veins below, glabrous above and below; interpetiolar ridge 3–6 mm high; petioles slender with very slight vagination, 10–90 mm. Branches of the

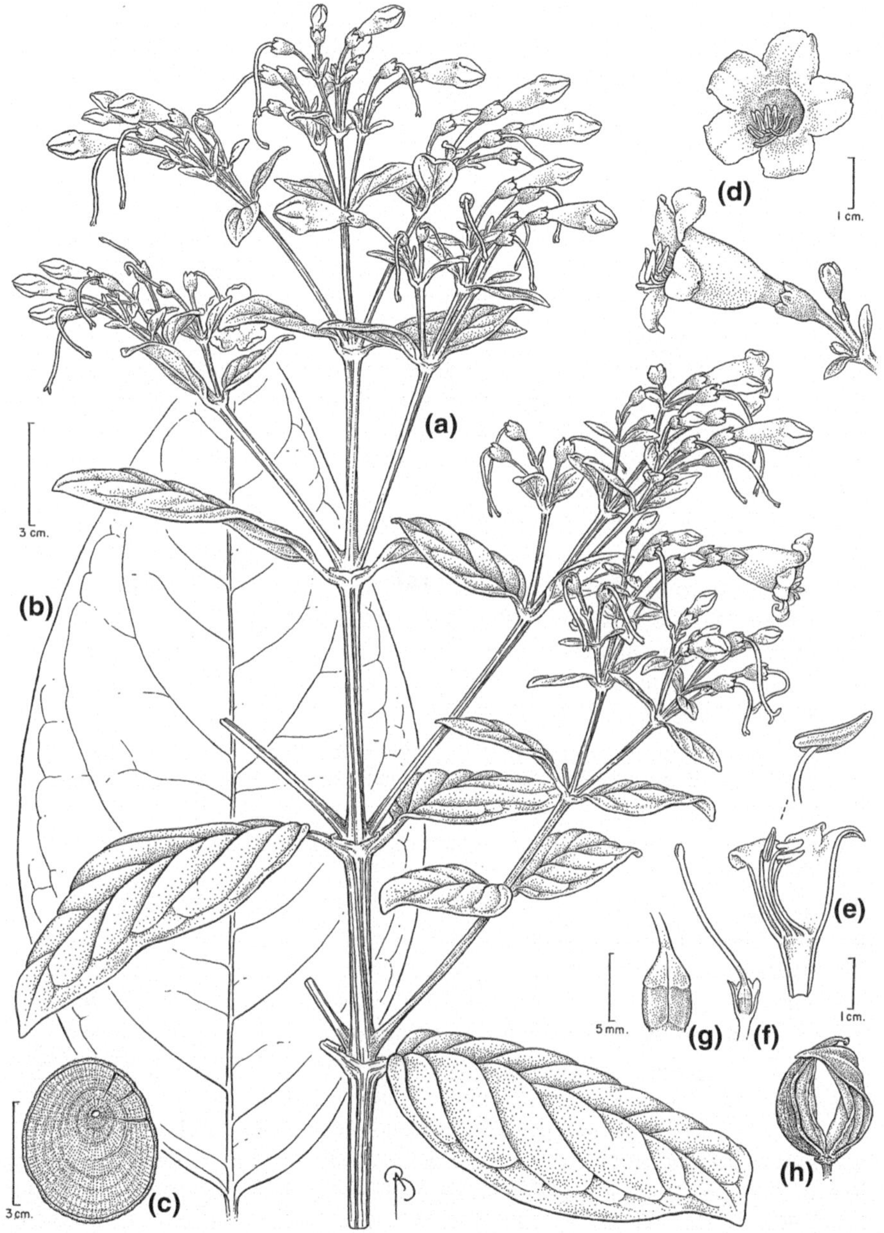

Fig. 3.45 *Macrocarpaea noctiluca* J.R. Grant & Struwe. **a** habit of flowering stem; **b** leaf; **c** wood anatomy of cross-section of main trunk; **d** corolla viewed from front, and side-view of flower; **e** cross-section of corolla, and anther; **f** pistil; **g** ovary; **h** mature dehisced capsule. **a** from field *photo* Grant, J.R & L. Struwe 01-4021; **b, g** from Grant, J.R & L. Struwe 01-4074; **c** from Grant, J.R & L. Struwe 01-3994; **d–f** from pickles of Grant, J.R & L. Struwe 01-3979

inflorescence 17–34 cm, 3–18 flowered per branch. Bracts ovate to lanceolate to oblong to elliptic, 9–135 × 5–75 mm, base aequilateral to attenuate, apex acute to obtuse, sessile to short-petiolate; bract petioles 1–7 mm. Flowers pedicellate, erect to horizontal to oriented in all directions in the inflorescence; pedicels 3–31 mm; bracteoles linear to lanceolate to obovate, 1–13 × 1–8 mm. Calyx narrowly to broadly campanulate, 7–10 × 5–9 mm, glabrous, smooth, green, ecarinate, to sometimes a small keel on the back of calyx lobe, often with pronounced ridges between calyx lobes extending to the base of the calyx; calyx lobes dividing calyx to one-third, obtuse, rounded to acute, 2–4 × 2.5–3.0. Corolla funnel-shaped, 30–50 mm, 11–24 mm wide at corolla lobe sinuses, creamy white to yellow to yellowish-green, smooth; corolla lobes ovate to elliptic, apex obtuse to rounded, 7–15 × 4–13 mm. Stamens 22-34 mm; filaments 18-28 mm, filiform, terete; anthers linear to linear elliptic, 4–6 × 2.0–2.5 mm, sagittate, versatile. Pistil 25–43 mm; ovary 4–7 × 2.5–3.0 mm; style 18–31 × 1.0–1.5 mm; stigma 2-lobed, lobes 3–5 × 2–3 mm, rounded to spathulate. Capsules dry, bilocular, medially dehiscent, ellipsoidal to ovoid, 29–34 × 10–14 mm, smooth to rugose, dark brown, erect to slightly nodding; style remnant 3–6 mm. Seeds unknown.

Macrocarpaea noctiluca belongs to section *Choriophylla* and occurs from 2,100 to 3,200 m in the Amotape–Huancabamba region of southern Ecuador and northern Peru. It is the most common species of the genus in Ecuador, occurring in Azuay, Loja, Morona-Santiago, and Zamora-Chinchipe, and subsequently, the mostly frequently collected. It is mostly commonly observed as a primary colonizer in roadside secondary vegetation. It occurs at different sites with *M. angelliae*, *M. apparata*, *M. arborescens*, *M. bubops*, *M. harlingii*, and *M. jensii*. Often its presence may indicate the occurrence of other more rare species of *Macrocarpaea*.

This species is part of a distinctive monophyletic group including *M. apparata*, *M. bubops*, *M. elix*, *M. jensii*, *M.* × *mattii*, and *M. noctiluca*.

Etymology. From the Latin "*noctis*", night, and "*lucis*," light, or glowing for flowers that look like small glowing lights.

Specimens examined: ECUADOR. **Azuay:** Km 74 de Cuenca, carretera Zigzig-Molón-Gualaquiza, 2,790 m, 6 August 1986, *Jaramillo* et al. *8866B* (AAU, GB, HUA, QCA); Road Sigsig-Gualaquiza, km 25.6, at the pass on military post road, 3,200–3,330 m, 11 January 2000, *Jørgensen* et al. *1832* (CHRB, MO, NY, QCNE). **Loja:** 15 km S of Yangana on the road Loja-Zumba (under construction), 04° 30′ S, 79° 8′ W, 2,500 m, 24 September 1983, *Brandbyge 42299* (AAU, MO, NY, Q); Nudo de Guagrauma, slopes of the Lora de Oro (about 6 km south of Zaraguro), 3,000 m, 2 August 1944, *Camp 275* (NY); P.N. Podocarpus, Sendero a Cajanuma, 2,550–2,700 m, 04° 04′ S, 79° 09′ W, 23 February 2000, *Cerón & Curso de Dendrologia 40065* (QAP); Cerro Villonaco, remanente del bosque Andino en el lado sur oriental de la cumbre 03° 58′ S, 079° 16′ W, 2,900 m, 12 September 1997, *Cerón, C. & L.E. Ocampo 35126* (QAP); Saraguro Canton, along road between Cuenca and Loja, 58.6 km N of Loja (first roundabout after entering the main entrance road), ca. 5 km S of Saraguro), 03° 39′ S, 79° 15′ W, 2,875 m, 4 March 1992, *Croat, T. 72660* (MO, QCNE); Loja-Zamora, 12 km from Loja, on the finca of

Dr. David Espinosa, 03° 55′ S, 79° 09′ W, 2,600 m, 1 October 1988, *Ellemann 66529* (AAU, LOJA, QCA); Loja-Zamora, 12 km from Loja, on the finca of Dr. David Espinosa, 03° 55′ S, 79° 09′ W, 2,400–2,600 m, shrub, flowers yellow, 17–18 November 1988, *Ellemann 75387* (AAU, LOJA); 12 km NW of Saraguro on Loma Paredones, 03° 36′ S, 79° 10′ W, 2,800 m, shrub 2 m high, 9 March 1989, *Ellemann 91665* (AAU, LOJA, QCA); Namandu, S. Loja, 2,400–2,500 m, 18 April 1946, *Espinosa, R. 198* (LOJA, US); Quedrada Honda, en la carretera Yangana-Valladolid, 2,470 m, 9 October 1995, *Garmendia & Paredes 639* (QCNE); De la carretera Yangana-Valladolid, antes del refugio de Quebrada Honda, sale un camino de herradura que va a unos potreros, 04° 28′ 59″ S, 79° 9′ 20″ W, 2,720 m, 9 October 1995, *Garmendia* and *Paredes 688* (QCNE); Quebrada Honda, cuesta carrizal, 04° 28′ 55″ S, 079° 7′ 36″ W, 2,520 m, 26 September 1996, *Garmendia, A. & M. Igual 1464* (QCA); 8 km E of Loja on Loja-Zamora road, 31 October 2002, *J.R. Grant 02-4245* (NEU, NY, US); Km 8.9 on road from Yangana-Cerro Toledo, 2,450 m, 29 December 2007, *Grant, J.R., C. Agier, C. Arnold & M.L. Cheung 07-4508* (LOJA, MO, NY); Km 8.9 on road from Yangana-Cerro Toledo, 2,450 m, 19 February 2006, *Grant, J.R., M.L. Cheung, F. F. Luisier, & N. Villard 06-4342* (CHRB, G, LOJA, MO, NY); Km 10 on road from Yangana–Cerro Toledo, 2 November 2002, *Grant, J.R., A. Roguenant & A. Raynal-Roques 02-4275* (MO, NEU, NY, SEL, US); Km 10 on road from Yangana–Cerro Toledo, 2 November 2002, *Grant, J.R., A. Roguenant & A. Raynal-Roques 02-4276* (MO, NEU, NY, SEL, US); 30.1 km past Yangana toward Valladolid, 5 November 2002, *Grant, J.R., A. Roguenant & A. Raynal-Roques 02-4295* (NEU, NY, US); Nudo de Sabanilla, 20 km S of Yangana, 04° 27′ 12″ S, 079° 09′ 11″ W, 2,485 m, 10 February 2001, *Grant, J.R & L. Struwe 01-3979* (LOJA, QCA, QCNE, NEU, NY, S, US); Nudo de Sabanilla, between 20 and 28.8 km S of Yangana, 04° 27′ 59″ S, 079° 08′ 44″ W, 2,550 m, 11 February 2001, *Grant, J.R & L. Struwe 01-3992* (LOJA, QCA, QCNE, NEU, NY, MO, S, US); Nudo de Sabanilla, between 20 and 28.8 km S of Yangana, 04° 27′ 59″ S, 079° 08′ 44″ W, 2,550 m, 11 February 2001, *Grant, J.R & L. Struwe 01-3993* (LOJA, NEU, QCNE); Nudo de Sabanilla, between 20 and 28.8 km S of Yangana, 04° 27′ 59″ S, 079° 08′ 44″ W, 2,550 m, 11 February 2001, *Grant, J.R & L. Struwe 01-3994* (LOJA, NEU, QCNE, US [2 sheets]); 28.8 km S of Yangana, 04° 27′ 59″ S, 079° 08′ 44″ W, 2,560 m, 11 February 2001, *Grant, J.R & L. Struwe 01-3996* (LOJA, QCNE, US); 10 km from Loja on road to Zamora, 03° 59′ 28″ S, 079° 09′ 57″ W, 2,550 m, 13 February 2001, *Grant, J.R & L. Struwe 01-4003* (LOJA, NEU, NY, QCA, QCNE); Km 8.9 on road from Yangana to Cerro Toledo, 04° 23′ 08″ S, 079° 09′ 03″ W, 2,450 m, 14 February 2001, *Grant, J.R & L. Struwe 01-4021* (LOJA, NEU, NY, QCA, QCNE,); From Loja-Saraguro road, then 200 m on road toward Fierro Urco, 03° 40′ 58″ S, 079° 16′ 22″ W, 2,900 m, 17 February 2001, *Grant, J.R & L. Struwe 01-4074* (G, LOJA, NEU [3 sheets], NY, QCA, QCNE, US [2 sheets]); Around km 9 on road from Yangana up Cerro Toledo, 04° 22′ 59.6″ S, 079° 07′ 41.0″ W, 2,826 m, 3 February 2011, *Grant, J.R. & J. Vieu 11-4660* (NY); Around km 9 on road from Yangana up Cerro Toledo, 04° 22′ 59.6″ S, 079° 07′ 41.0″ W, 2826, 3 February 2011, *Grant, J.R. & J. Vieu 11-4660-6* (NY); Around km 9 on road from Yangana up Cerro Toledo, 04° 22′ 59.6″ S, 079° 07′ 41.0″ W, 2,826 m, 3 February 2011, *Grant,*

J.R. & J. Vieu 11-4660-58 (NY); Km 8.9 on road from Yangana–Cerro Toledo, 2 November 2002, *J.R. Grant* et al. *02-4269* (NEU); Road between Loja and Zamora, 2,600–2,700 m, 20 March 1972, *Harling, G. 11322* (GB, MO); Loma de Loro, 6 km S of Saraguro on road to Loja, 3,200 m, 11 February 1985, *Harling, G. & Andersson 21910* (GB, MO, QCA); W slope of Nudo de Sabanilla, ca. 10 km above Yangana on road to Valladolid, 2,500 m, shrub 2.0–2.5 m tall, 3 April 1985, *Harling, G. & Andersson 23,594* (GB, QCA); Loja-Zamora rd, 2,700–2,900 m, 8 February 1982, *Harling, G.* et al. *20386* (GB); Nudo de Sabanilla, N part, 2,400–2,600 m, 10 February 1982, *Harling, G.* et al. *20558* (GB); Carretera Loja-La Palma, bosque húmedo, suelo cascajosa, 2,850 m, 27 December 1988, *Jaramillo 10514* (QCA); Carretera Saraguro-Loja, Loma del Oro, 03° 31′ 38″ S, 079° 43′ 41″ W, 2,525 M, 24 June 2008, *Jaramillo, J.* et al. *26788* (QCA); Km 10–15 Yangana-Toledo road, subpáramo, 04° 24′ S, 79° 6′ W, 3,000–3,300 m, 1 August 1986 *Jørgensen 61374B* (AAU, MO, NY, QCA, QCNE); Loja-Saraguro, km 58, turnoff toward Fierro Urco, km 1–2, 03° 41′ 49″ S, 79° 16′ 22″ W, 3,000 m, 21 April 1994, *Jørgensen* et al. *499* (LOJA, MO, QCA, QCNE); Cerro Uritusinga, Loja-La Palma, km 18–20, 04° 05′ 03″ S, 79° 13′ 40″ W, 2,910–3,000 m, 30 November 1994, *Jørgensen* et al. *1050* (LOJA, MO, NEU, QCA, QCNE); 4 km S of Loja, 04° 0′ 40″ S, 79° 2′ 15″ W, 2,600 m, 17 April 1966, *Knight 494* (WIS); Slopes of Cerro Villonaco, ca. 10 km west of Loja, 03° 11′ 80″ S, 79° 3′ 10'' W, 2,850 m, small tree 4 m tall, with yellow flowers, 6 March 1966, *Knight 517* (WIS); Road Loja-Zamora, km 10, 03° 58′ S, 79° 08′ W, 2,600 m, 24 October 1996, *Lewis 2709* (K, LOJA, MO, NY, QCNE); Road Loja-Saraguro, km 26, bosque Cofragia, 03° 49′ 437″ S, 79° 17′ 592″ W, 2,750 m, 17 January 1997, *Lewis* et al. *2930* (K, LOJA, MO, QCNE); Road between Loja and Zamora ca. Km 9, 2,500–2,750 m, 31 December 1978, *Luteyn* et al. *6542* (AAU, CAS, GH, MO, NY, QCA); Cordillera de Sabanilla, ca. 15 km S of Yangana, 2,480 m, 30 December 1980, *Madison & Besse 7498* (QCA, SEL); Rd. Loja-Zamora km 18 (El Tiro), 03° 58′ S, 079° 08′ W, 2,650–2,800 m, 21 February 2001, *Madsen, J. 7724* (LOJA, NY, QCNE); Rd. Universidad Nacional de Loja - Las Achira (Uritusinga), km 10, 2,800 m, 25 October 2001, *Madsen, J. and Chimbo 8610* (AAU); P.N. Podocarpus, about 6.5 km from the park entrance to the refuge Cajanuma, 2,500 m, November 1998, *Matt, F. 4* (ER); *Matt, F. 5* (ER); *Matt, F. 6* (ER); Beside the road from Loja to Zamora, still on the eastern slope of the cordillera, 2 km before the former rubbish dump of Loja, 2,550 m, January 1998, *Matt, F. 7* (ER); *Matt, F. 8* (ER); *Matt, F. 9* (ER); *Matt, F. 10* (ER); *Matt, F. 11* (ER); Amaluza, 5–10 km ENE of the village (Pasaje del Romerillo), 2,400–2,700 m, tree 3 m high, 79° 23′ W, 04° 34′ S, 23 September 1976, *Øllgaard* and *Balslev 9740* (AAU, NY); La Palma, bosque alto andino, 2,850 m, 27 December 1988, *Rios, M.* et al. *209* (AAU, QCA); Carretera Valladolid-Loja, km 4–21, 04° 20′ S, 79° 15′ W, 1,900–2,700 m, 18 February 1993, *Romoleroux* et al. *1536* (QCA, QCNE); Cerro Uritusinga (north of Loja), 2,920 m, 14 January 2003, *Smith, S.D. & O. Jadan 229* (MO); western slopes of Cordillera de Condor and northwest slopes of Nudo de Sabanillas, around Tambo Cachiyacu, along Río Cachiyacu, about 2 leagues southeast of Yangana, 2,000–3,000 m, 19 October 1943, *Steyermark, J. 54812* (F); Canton Loja, Hoya de Loja, Cordillera Oriental, Sitio via a Zamora, 2,646 m, 14

Fig. 3.46 *Macrocarpaea opulenta* J.R. Grant. *Photo* J.R. Grant

Fig. 3.47 *Macrocarpaea opulenta* J.R. Grant. *Photo* J.R. Grant

July 2008, *Villa, N. 93* (LOJA); Canton Loja, Hoya de Loja, Cordillera Occidental, sitio Villonaco, 2,818 m, 15 July 2008, *Villa, N. 130* (LOJA); Canton Loja, Hoya de Loja, Cordillera Oriental, Sitio Z. Huayco Alto, 2,363 m, 22 July 2008, *Villa, N. 155* (LOJA); Canton Loja, Hoya de Loja, Cordillera Sur-Oriental, sitio Q. Monica, 2,547 m, 25 August 2008, *Villa, N. 240* (LOJA); Canton Loja, Hoya de Loja, Cordillera Oriental, Sitio Q. Volcan, 2,622 m, 26 August 2008, *Villa, N. 277* (LOJA). **Morona-Santiago:** Path Campamento San Miguel (on road in construction Sigsig-Gualaquiza)–Gualaquiza, 9 April 1968, *Harling, G.* et al. *8115* (GB). **Zamora-Chinchipe:** Palanda Cantón, Parroquia Vallodolid, southern slopes of the Cordillera de Sabanilla (headwaters of Río Chinchipe), Tapichalaca Reserve (Fundación Jocotoco), 2,500–2,700 m, 3 April 2005, *Clark, J.L. 9022* (NY, QCNE, US); Quebrada Honda, cuesta carrizal, 2,520 m, 26 September 1996, *Garmedia & Igual 1464* (QCNE); Approx. 40 km S Amaluza on road to Zumba that nearly parallels the Ecuador–Peru border, 3,016 m, 1 January 2008, *Grant, J.R., C. Agier,*

C. Arnold & M.L. Cheung 08-4515 (LOJA, NY); P.N. Podocarpus (San Francisco entrance), trail leading west from San Francisco, 03° 59′ 24″ S, 079° 05′ 48″ W, 2,100 m, 16 February 2001, *Grant, J.R & L. Struwe 01-4062* (LOJA, QCNE); Along road between Zumba and Amaluza that parallels the Ecuador–Peru border, 04° 47′ 18.1″ S, 079° 22′ 16.6″ W, 2,507 m, 7 February 2011, *Grant, J.R. & J. Vieu 11-4671* (NY); Along road between Zumba and Amaluza that parallels the Ecuador–Peru border, 04° 46′ 47.0″ S, 079° 23′ 50.0″ W, 2,988 m, 7 February 2011, *Grant, J.R. & J. Vieu 11-4672* (NY); Above Valladolid on rd to Yangana, montane rain forest, 2,700 m, 2 February 1985, *Harling, G. & Andersson 21462* (GB); Nudo de Sabanilla-Valladolid, horse trail to Caserío Quebrada Honda, montane rainforest, 2,400–2,600 m, ca. 2.0–2.5 m, 12 February 1993, *Harling, G. & Ståhl 26325* (GB, MO, S); P.N. Podocarpus, road Yangana-Valladolid, km 26, montane forest along ravine, 04° 29′ S, 79° 09′ W, 2,550 m, 2 December 1988, *Madsen* et al. *75753* (AAU, LOJA, QCA, QCNE); Palanda, Tapichalaca Reserve, south of Podocarpus National Park, in upper Río Chinchipe watershed, east of the Yangana-Valladolid road, at the head of the mule trail to Quebrada Honda, 04° 29′ 04″ S, 079° 07′ 59″ W, 2,570 m, 4 December 2006, *Neill, D. 15368* (NY, QCNE); Palanda, Tapichalaca Reserve, south of Podocarpus National Park, east of road between Yangana and Valladolid, upper Río Chinchipe watershed, near beginning of mule trail to Quebrada Honda, 04° 29′ 23″ S, 079° 08′ 11″ W, 2,550 m, 25 September 2007, *Neill, D.* et al. *16001* (LOJA, MO); P.N. Podocarpus, road Yangana-Valladolid, just S of the pass (Nudo de Sabanilla), quebrada with wet montane forest, and low dense scrub on ridgetop, 04° 27′ S, 78° 08′ W, 2,640–2,770 m, 16 February 1989, *Øllgaard* et al. *90616* (AAU, LOJA, QCA); Valladolid, cruz del soldado, 04° 28′ S, 079° 05′ W, 9 December 2000, *Sánchez, O. 17* (LOJA); Palanda, Reserva Tapichalaca, near road between Yangana and Valladolid, south of Podocarpus National Park, trails in Tapichalaca Reserve, Upper Río Chinchipe watershed, 04° 29′ 44″ S, 079° 07′ 55″ W, 2,700 m, 30 October 2006, *van der Werff, H.* et al. *21810* (MO); Area of ECSF (Estación Cientifica San Francisco), 03° 58′ 18″ S, 079° 04′ 44″ W, 2,200–2,600 m, 24 January 2001, *Wolff, D. 98* (LOJA).

21. *Macrocarpaea opulenta* J.R. Grant

Harvard Pap. Bot. 11(2): 135. 2007. TYPE: ECUADOR. Zamora-Chinchipe: Cordillera del Cóndor, 28.1 km E of Los Encuentros (as measured from the bridge crossing the Rio Nangaritza E of the city center) in direction of the Paquisha Alta military camp, 1,441 m, 24 February 2006, *Grant, J.R., M.L. Cheung, F. Luisier & N. Villard 06-4347* (Holotype NY!; Isotypes G, LOJA, MO, QCA, QCNE) (Figs. 3.46, 3.47, 3.48).

Shrub 2–4 m, glabrous throughout. Trunk to 2 cm diameter, wood hollow, rings absent; bark papery thin to scarcely measurable, 0.05 mm thick, outer surface smooth to rugose, tan to greenish, blotched with lichens and mosses. Stems terete, hollow, 8–11 mm in diameter just below the inflorescence. Leaves elliptic, oval to ovate, long-petiolate, (20)–37–47 cm long. Petioles (20)–65–85 mm, slender with slight vagination; interpetiolar ridge 2–3 mm high. Blades (18)–29–40 × (8.5)–

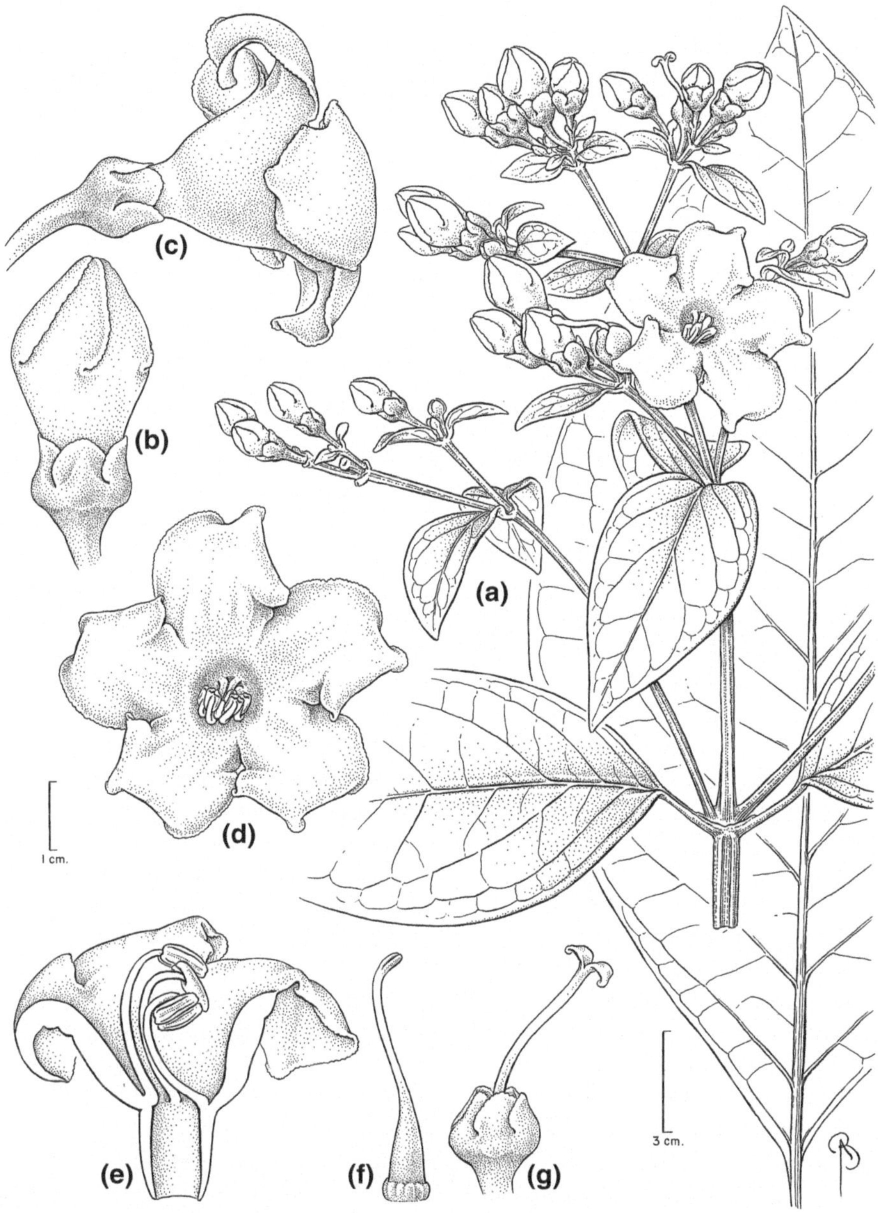

Fig. 3.48 *Macrocarpaea opulenta* J.R. Grant. **a** habit of flowering stem; **b** bud; **c** corolla viewed from side; **d** corolla viewed from front; **e** dissected corolla, noticed thickened petals; **f** young pistil; **g** calyx and pistil once corolla has dropped. **a–g** drawn from Grant et al. 06-4347

14–19 cm, entire, not revolute, dark green, with slightly impressed veins above, and slightly raised veins below, glabrous above and below, membranaceous, thin, flexible; base aequilateral to oblique, cuneate to attenuate; apex acuminate to acute. Inflorescence much branched open thyrse, 29–35+ cm long; branches 5–35 cm long, 3–12 flowered per branch. Bracts ovate, elliptic, to lanceolate, petiolate, 17–130 × 9–65 mm; base aequilateral to oblique, cuneate; apex acuminate to acute; bract petioles 0–5 mm. Flowers pedicellate, erect to slightly spreading; pedicels 7–30 mm long; bracteoles inconspicuous and scabrous, linear, triangular to ovate, 1–5 × 1–2 mm. Calyx campanulate, 13–15 × 11–17 mm, glabrous, smooth, green, ecarinate, ovate, reniform to rotund; calyx lobes 4–9 × 6–10 mm, apex obtuse to rounded. Corolla broadly funnel-shaped with petal lobes slightly spiraled to the right such as to give the impression of a pinwheel, 47–59 mm, 40–45 mm wide at the apex of the tube, greenish-yellow, smooth, spongy, significantly thickened and fleshy; corolla lobes ovate, slightly imbricate (lobes spiraling, curving and overlapping to the right), an additional curling of the right side of the petal lobe gives each petal the shape of a horse's head; corolla lobes 28–32 × 22–30 mm, obtuse to rounded. Stamens 27–33 mm long; filaments 22–27 mm long, filiform, terete; anthers elliptic to oblong, 5–6 × 3.0–3.5 mm, sagittate, versatile; pollen glabra-type. Pistil 38–40 mm long; ovary 11–13 × 3–4 mm; style 23–25 × 1.5–2.0; stigma lobes oblong to spathulate, 3–4 × 2–3 mm. Capsules and seeds unknown.

Macrocarpaea opulenta belongs to section *Choriophylla* and occurs from 1,100 to 1,700 m in the Amotape–Huancabamba region of southern Ecuador. Its flowers are green and broadly flared, having a broad mouth, petals that have a flattened face before curling backwards. The petals that are significantly thickened and fleshy, a slight turn in the angle of the petal lobes so they appear as curving toward the right, and unique curving of the right side of the petal lobes that give each petal the shape of a horse's head. Despite belonging to two different species groups both *M. jactans* and *M. opulenta* occur in the same geographic region of the Cordillera del Cóndor, and have similar shaped corollas indicating that they may have a common pollinator.

Etymology. From the Latin "opulentus," splendid or opulent.

Specimens examined: ECUADOR. **Zamora–Chinchipe:** 13 km E of Los Encuentros (as measured from the bridge crossing the Rio Nangaritza E of the city center) in direction of the Paquisha Alta military camp, 1,444 m, 9 January 2008, *Grant, J.R., C. Agier, C. Arnold & M.L. Cheung 08-4539* (NY, LOJA); 13 km E of Los Encuentros (as measured from the bridge crossing the Rio Nangaritza E of the city center) in direction of the Paquisha Alta military camp, 1,444 m, 9-Jan-2008, *Grant, J.R., C. Agier, C. Arnold & M. Cheung, 08-4540* (NY, LOJA); Cordilllera del Condor, 18.4 km E of Los Encuentros in direction of Paquisha Alta Military Camp, 03° 30′ 28″ S, 078° 35′ 59″ W, 1,433 m, 21 January 2009, *Grant, J.R., B. Angell, W. Grant & V. Trunz 09-4551* (NY); Cordilllera del Condor, 32 km from Los Encuentros in direction of Paquisha Alta Military Camp, north for 1–3 km north on road toward a mine, 03° 50′ 56.2″ S, 078° 31′ 49.8″ W, 1,506 m, 9 February 2011, *Grant, J.R. & J. Vieu 11-4676* (NY); Road from Chuchumbleza, Quimi and Tundayme toward the Condor Mirador, 12 km from end of road at

Military camp 'Condor Mirador', 03° 37′ 07.0″ S, 078° 26′ 41.7″ W, 1,427 m, 15 February 2011, *Grant, J.R. & J. Vieu 11-4698* (NY); Cordillera del Cóndor, vertiente occidental, cuenca del Río Quimi 03° 34′ 45″ W, 078° 25′ 20″ W, 1,180–1,400 m, 5 April 2006, *Quizhpe, W. 2141* (LOJA, MO, NY, QCNE); Yantzaza Cantón, Cordillera del Cóndor, vertiente occidental, carretera de Los Encuentros a La Zarza, 03° 48′ 40″ S, 078° 36′ 28″ W, 1,400 m, 27 March 2008, *Quizhpe, W. & O. Jaramillo 2990* (LOJA, MO, QCNE); Yantzaza Cantón, Cordillera del Cóndor, vertiente occidental, Vía Pindal, al noreste de Los Encuentros, 03° 46 S, 078° 35′ W, 1,100 m, 30 March 2008, *Quizhpe, W. & O. Jaramillo 3011* (MO, QCNE); Yantzaza Cantón, Cordillera del Cóndor, vertiente occidental, Vía Pindal, al noreste de Los Encuentros, 03° 46′ S, 078° 35′ W, 1,100 m, 30 March 2008, *Quizhpe, W. & O. Jaramillo 3016* (MO, QCNE); Región de la Cordillera del Cóndor, Parroquia San Carlos de Las Minas, Namibija, sendero al Cerro Colorado, bosque muy húmedo montano bajo, 04° 04′ 52″ S, 078° 47′ 51″ W, 1,520–1,700 m, 29 January 2005, *Quizhpe, W.* et al. *830* (LOJA, MO, QCNE); Cordillera del Condor, Destacamento militar Cóndor Mirador, ruta hacia el hito limítrofe Perú–Ecuador, 1,770 m, 8 September 2003, *Rodríguez, E.R.* et al. *2669* (NY).

22. *Macrocarpaea pringleana* J.R. Grant

Harvard Pap. Bot. 9(1): 18. 2004. TYPE: ECUADOR. Pastaza: Puyo–Puerto Napo Road, 14–18 km N of Puyo, heavily cut over forest, now pasture with remnant trees, 1,000–1,025 m, 25 April 1978, *J.L. Luteyn & M. Lebron–Luteyn 5830* (Holotype: NY; Isotype: GH [2 sheets]) (Figs. 3.49, 3.50).

Unbranched shrub to tree, 1–5 m, glabrous throughout. Stems terete to slightly quadrangular, hollow, 5–9 mm in diameter just below inflorescence. Leaves ovate to ovate-elliptic, petiolate, 12–29(–38) cm long. Petioles 5–50 mm long, slender with slight vagination to one quarter the length of the petiole; interpetiolar ridge 2–5 mm high. Blades 11.5–26.5(–33) × 6–14.5(–18), entire, not revolute, bright green above, lighter below, with slightly impressed veins above, and slightly raised veins below, glabrous above and below, papery thin; base aequilateral to slightly oblique, cuneate, rounded to attenuate and decurrent on the petiole; apex acute to long acuminate (sometimes nearly apiculate). Inflorescence a much branched open thyrse, 46 + cm long, as the inflorescence of the single plant has been cut into multiple lengths for separate herbarium sheets; branches 6–40 cm long; 3–18 flowers per branch. Bracts ovate, elliptic, to lanceolate, sessile to short-petiolate, 12–113 × 4–63 mm; base aequilateral to oblique, cuneate to rounded; apex acuminate; bract petioles 0–4 mm long. Flowers pedicellate, erect; pedicels 11–17 mm long; bracteoles linear to lanceolate to ovate, 0.5–10 × 0.5–3.0 mm. Calyx campanulate, 6–10 × 6–8 mm, glabrous, smooth, ecarinate; calyx lobes ovate to elliptic, 2–4 × 2–4; apex rounded, obtuse to acute. Corolla funnel-shaped, 30–40 mm long, 15–25 mm wide at the apex of the tube, yellowish-green, greenish yellow to green, smooth; corolla lobes ovate to elliptic, 9–12 × 7–12, apex obtuse to rounded. Stamens 21–28 mm long; filaments 17–22 long, filiform, flattened; anthers elliptic to oblong, 4–6 × 2–3, sagittate, versatile; pollen

Fig. 3.49 *Macrocarpaea pringleana*. *Photo* J. Cortina

"glabra-type". Pistil 29–34 mm long; ovary 8–9 × 2–3 mm; style 19–22 × 0.5–1.5 mm; stigma lobes spathulate, 2–3 × 1.0–1.5 mm. Capsules ellipsoidal to oblong, 23–32 × 6–9, smooth to ribbed, chestnut-tan, erect to slightly spreading; style remnant 2–9 mm. Seeds "perimetrically winged," flattened, roughly 4 sided in outline, yet appearing as myriads of different puzzle pieces, 0.8–1.1 × 0.5–0.9, bicolored, testa brown, wings straw-colored, testa reticulate, wings ribbed.

Macrocarpaea pringleana belongs to section *Choriophylla* and occurs from 500 to 2,200 m in the Amotape–Huancabamba region of southern Ecuador. It occurs exclusively on Amazon-facing slopes of Ecuador, ranging from about Baeza (Napo) to Limón (Morona–Santiago). This species belongs to a closely related group of species with leaves that are long-decurrent on the petiole including *M. cortinae*, *M. pringleana*, *M. sodiroana*, and *M. umbellata*.

Etymology. Named for James S. Pringle (1937–), Royal Botanic Gardens, Hamilton, Ontario, Canada, specialist of North and South American Gentianaceae, and author of the treatment of the family in the *Flora of Ecuador* (1995).

Vernacular name: tabaquillo, *Vargas 3634*.

Specimens examined: **Morona–Santiago:** El Partidero, between the ríos Paute and Negro, eastern slope of the cordillera, valley of the ríos Negro and Chupianza (on the trail from Sevilla de Oro to Mendez), 640–945 m, 14 December 1944, *Camp 1533* (NO, NY, S); Carretera Macas-Riobamba, Parroquia 9 de octobre,

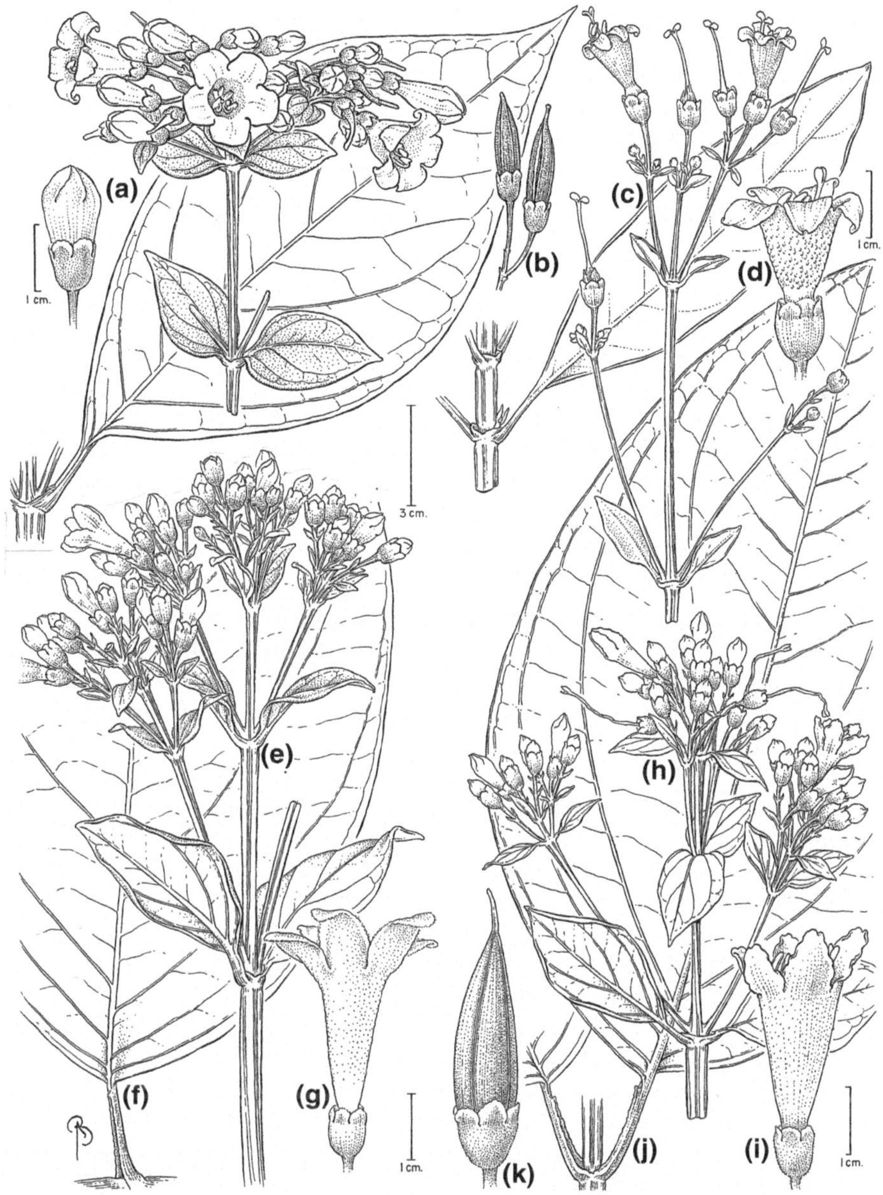

Fig. 3.50 *Macrocarpaea pringleana* J.R. Grant, *M. harlingii* J.R. Grant, and *M. elix* J.R. Grant. **a–b** *Macrocarpaea pringleana*. **a** bud, habit of flowering stem; **b** fruit. **c–d** *M harlingii*. **c** habit of flowering stem; **d** flower. **e–k** *M. elix*. **e** habit of flowering stem; **f** leaf; **g** flower, **h** habit of flowering stem; **i** flower; **j** interpetiolar ridge and leaves; **k** fruit. A drawn from Luteyn 5830 (NY); **b** drawn from Clark et al. 5480 (QCNE); **c–d** drawn from Knapp et al. 7467; **e** drawn from Harling, G. & Andersson 23442; **f–g** drawn from Harling, G. & Andersson 23031; **h–i** drawn from Boeke & Loyala 988; **j–k** drawn from Camp 4325

vegetación aledaña del río Salado, entrada al P.N. Sangay, 1,665 m, 23 August 2002, *Caranqui* et al. *790* (MO, NY, QCNE); Loma San José Grande-Sendero San José-Río Bomboiza, bosque húmedo premontano, vegetación de potreros y bosques secundarios, 02° 38′ S, 78° 27′ W, 1,300–1,600 m, 21 April 1991, *Cerón 14358* (MO, QAP, QCNE); Along road between Limón (Gen. Plaza Guttiérrez) and Gualaceo, 1.2 km N of Limón disturbed roadside banks, 02° 58′ 36″ S, 078° 26′ 24″ W, 1,211 m, 11 August 2002, *Croat, T.B.* et al. *86472* (LOJA, MO, NY); Just north of Limon along highway, 02° 57′ 20.0″ S, 078° 25′ 25.6″ W, 1,059 m, 12 February 2013, *Grant, J.R & J. Cortina 13-5121* (NY); environs of Santa Susana de Chiviaza, rastrojos, 1,200 m, 1 March 1993, *Harling, G. & Ståhl 26808* (GB, S); Environs of Pan de Azúcar, along road Indanza–Don Bosco, 1,000–1,100 m, 2 March 1993, *Harling, G. & Ståhl 26853* (S); Limón Indanza, Región de la Cordillera del Cóndor, Centro Shuar Maikuants, Cerro Wishiwishi Naint, 03° 06′ 49″ S, 078° 16′ 26″ W, 1,140 m, 17 March 2005, *Kajekai, C.* et al. *323* (LOJA, MO, NY, QCNE); Limón Indanza, Región de la Cordillera del Cóndor, Centro Shuar Maikuants, Cerro Wishiwishi Naint, 03° 06′ 49″ S, 078° 16′ 26″ W, 1,140 m, 17 March 2005, *Kajekai, C.* et al. *325* (LOJA, MO, NY, QCNE); 2–4 km N of Arapicos, 800–900 m, 4 April 1981, *Lugo S. 5980* (GB, K, MO, NY); 2–6 km S of Arapicos, 800–900 m, 6 April 1982, *Lugo S. 6003* (GB, K, MO, NY); Macas–Alshi, forest of small trees nearby a small brook, 1,360 m, 6 February 2006, *Luisier, F., D. Neill & W. Quizhpe 1* (CHRB, G, LOJA, MO, NEU, NY, QCA, QCNE); Limón Indanza, Región de la Cordillera del Cóndor, al oeste del Río Zamora, Parroquia Santa Susana de Chiviaza, en la trocha hacia el poblado de Las Orquídeas, 02° 53′ 50″ S, 078° 18′ 42″ W, 1,200 m, 22 February 2006, *Morales, C. & D. Reyes 1628* (MO, QCNE); Limón Indanza, Región de la Cordillera del Cóndor, al oeste del Río Zamora, cerca a la población de Santa Susana de Chiviaza, 1,560 m, 24 February 2006, *Morales, C. & D. Reyes 1672* (MO, QCNE); Taisha, Centro Shuar Wisui, summit area of Cerro Wisui, easte of Río Macuma, the northeasternmost summit of the Cordillera de Cutucú, 02° 05′ 25″ S, 077° 45′ 57″ W, 1,360 m, 24 June 2007, N*eill, D. & U. Politec. Salesiana students 15664* (MO, NY, QCNE); Río Cuyes y vía Bomboiza–Gualaquiza, 03° 25′ S, 78° 35′ W, 800 m, 1 November 1986, *Palacios, W. 1430* (MO, QCNE, U); Limón Indanza, Región de la Cordillera del Cóndor, Santa Susana de Chiviaza, al oeste del Río Zamora, Cerro Chiviaza, 1,150 m, 24 March 2006, *Reyes & C. Morales 1075* (QCNE); Parque Nacional Sangay, Bosque Sardnayacu, 1,500 m, 7 February 1999, *Toasa* et al. *5290* (QCNE). **Napo:** Tena-Pano road, 550 m, 18 July 1982, *Besse* et al. *1668* (SEL); Reserva Ecológica Antisana, comunidad Shamato, entrada por km 21 –Shamata, 1,700 m, 23 April 1998, *Clark, J.L.* et al. *5029* (MO, QCNE); Reserva Ecológica Antisana, Sector Guacamayos, Sendero Jumandy–La Virgen (km 30), path follows potential oil pipeline, ends near Río Urcusiqui, 00° 38′ S, 77° 49′ W, 1,700–1,800 m, 1 May 1998, *Clark, J.L.* et al. *5480* (COL, MO, QCNE); Narupa to Coca Road, 4.5 km E of man Baeza–Archidona, 1,189 m, 19 April 2003, *Croat, T.B.* et al. *87811* (QCNE); P.N. Sumaco, ca. km 239 E 45 Baeza–Tena road, ca. 20 km north of Narupo (turnoff to Hollín-Coca), 00° 38′ 57″ S, 077° 47′ 57.9″ W, 1,866 m, 8 February 2013, *Grant, J.R & J. Cortina 13-5118*

(NY); new road to Loreto (56 km S of Baeza), 28 km E of junction with Baeza–Tena road, i00° 50′ S, 77° 33′ W, 1,100 m, 21 December 1988, *Hammel, B. & Wilder 17267* (MO, QCNE); Carretera Hollín-Loreto, km 40–50, alrededores de la comunidad Huamaní y del Río Pucuno, 1,200 m, 10 October 1988, *Hurtado 674* (MO, QCNE); Carretera Hollín-Loreto, km 25, Centro Challuayacu, 1,230 m, 10 November 1988, *Hurtado & Alvarado 955* (AAU, MO, QCNE); Crescit in silvis, Cuyuyua Baeza, 1 February 1905, *Sodiro s.n.* (QPLS [2 sheets]); Carretera Tena-Baeza, Cordillera de Guacamayos, Reserva Ecológica Antisana, 16 km al norte del desvio a Loreto, bosque pluvial premontano, 00° 40′ S, 77° 50′ W, 1,800 m, 18 June 1995, *Tirado & Murillo 1563* (MO). **Pastaza:** Mera, forest on precipitous bank of Río Pastaza, 1,050 m, 21 December 1955, *Asplund 18939* (S [2 sheets]); Hacienda San Antonio de Baron von Humboldt, 2 km al NE de Mera, 1,300 m, 5 March 1985, *Baker* et al. *5509* (QCNE); Road N from Mera known as the "via Anzu," 01° 20′ S, 78° 10′ W, 1,100 m, National Cancer Institute voucher specimen, 1 January 1988, *Boom* et al. *7884* (MO, NY, QCA, QCNE); Along road from Mera to Río Anzu, 7 km N of Río Alpayacu, 01° 25′ 56″ S, 078° 04′ 53″ W, 1,267 m, 8 May 2003, *Croat, T.B.* et al. *88765* (QCNE); Along road from Mera to Río Anzu, 7 km N of Río Alpayacu, 01° 25′ 56″ S, 078° 04′ 53″ W, 1,267 m, 8 May 2003, *Croat, T.B.* et al. *88795* (QCNE); Just south of Puyo along highway, 01° 31′ 14.5″ S, 077° 55′ 18.8″ W, 992 m, 12 February 2013, *Grant, J.R & J. Cortina 13-5120* (NY); El Topo, 1,250 m, 28 November 1959, *Harling, G. 3413* (S); Mera, along Río Allpayacu, riverside thickets, 1,000 m, 6 March 1980, *Harling, G. & Andersson 16984* (GB [2 sheets], MO); Mera, 1,100 m, 25 May 1968, *Harling, G.* et al. *10208* (GB, MO); Mera, trail to Pindo, 1,000 m, 16 January 1982, *Harling, G.* et al. *19621* (GB); Cabeceras on the Río Bobonaza, ca. 12 km E of Puyo, 16 November 1974, *Lugo S. 4618* (GB, NY, U); Veracruz (Indillama), 22 November 1974, *Lugo S. 4656* (GB, MO); Between Mera and Río Negro, 27 km from Puyo on road to Baños, 1 km N of the road, primary mountain rainforest, 01° 25′ S, 78° 11′ W, 1,300–1,400 m, 14 January 1988, *Molau* et al. *2469* (AAU, GB, HAM, QCA, QCNE); Hacienda San Antonio de Baron von Humboldt, 2 km al NE de Mera, 01° 27′ S, 78° 06′ W, 1,100 m, 27 February –19 March 1985, *Neill, D.* et al. *5809* (MO, QCNE); Munic. Puyo, Canton Puyo, Los Vencedores, Estación Experimental E.S.P.O.C.H., 01° 30′ S, 77° 56′ W, 800–1,000 m, 4 October 1995, *Pinus RBS-0023* (MO); Munic. Puyo, Canton Puyo, Los Vencedores, Estación Experimental E.S.P.O.C.H., 01° 30′ S, 77° 56′ W, 800–1,040 m, 14 December 1995, *Soejarto* et al. *9424* (F [2 sheets], HUA, QCNE); Puyo, Los Vencedores, km 32 Puyo-Macas, Estación Experimental E.S.P.O.C.H., trail from Field Station to river leading to waterfall, 8 June 1996, *Soejarto* et al. *9645* (F); Río Cayabe, between Mera and Río Topo, 1,100 m, 17 July 1967, *Sparre 17619* (S). **Tungurahua:** Parque central de la Parroquia de Río Negro via al Puyo, 18 January 2003, *Alcocer, V. 17* (QCNE); Río Topo, 1,220 m, 12 November 1955, *Asplund 18364* (S); Baños, Parroquia Río Verde, sector Machay, collections made on forested trail starting from Machay (Baños-Puyo road) to Río Verde, 01° 23′ 05″ W, 78° 16′ 50″ W, 1,800–2,200 m, tropical montane wet forest, 24 December 2000, *Clark, J.L.* et al. *5724* (QCA, QCNE, US);

Fig. 3.51 *Macrocarpaea quizhpei* J.R. Grant. *Photo* J.R. Grant

Between Ambato and Baños, 10 January 1981, *D'Arcy 14014* (MO, NY); Baños, Río Negro, 1,500 m, 18 January 2003, *Guamán, N. 12* (QCNE); Cashurco, near Río Negro, northern side of Rio Pastaza, ca. 1.0–1.5 m high, 18 March 1969, *Lugo S. 813* (GB,MO); Trail between Baños and Mera, near Rio Negro, in border low woods, 400–500 m, 23 February 1935, *Mexia 6977* (NA, UC, US); Baños, Zona de amortiguamiento del P.N. Llanganates, Río Verde, canon a las cascadas de Machay, bosque húmedo montaño bajo, 01° 23′ S, 78° 17′ W, 1,630 m, 29 July 1999, *Vargas, H. 3634* (MO, QCA, QCNE).

23. *Macrocarpaea quizhpei* J.R. Grant

Harvard Pap. Bot. 13(2): 256. 2008. TYPE: ECUADOR. Zamora-Chinchipe: Nangaritza, Región de la Cordillera del Cóndor, cuenca del alto Río Nangaritza, Las Orquídeas, bosque muy húmedo, en colina, selo arcilloso, 04° 13′ 15″ S, 078° 40′ 54″ W, 1,100 m, 11 February 2006, *W. Quizhpe & F. Luisier 1913* (Holotype LOJA; Isotypes MO, NY, QCNE) (Figs. 3.51, 3.52).

Unbranched shrub to small tree, 1–3 m, glabrous throughout. Trunk to 2 cm diameter, wood hollow in trunk and in branches, rings scarcely visible; bark papery thin to scarcely measurable, 0.05 mm thick, outer surface smooth to rugose, tan to greenish; stems terete to quadrangular, hollow, 6–13 mm diameter just below the inflorescence. Leaves broadly ovate to elliptic, long-petiolate, (16)-39–46 cm long. Petioles 50–110 mm, slender with slight vagination; interpetiolar

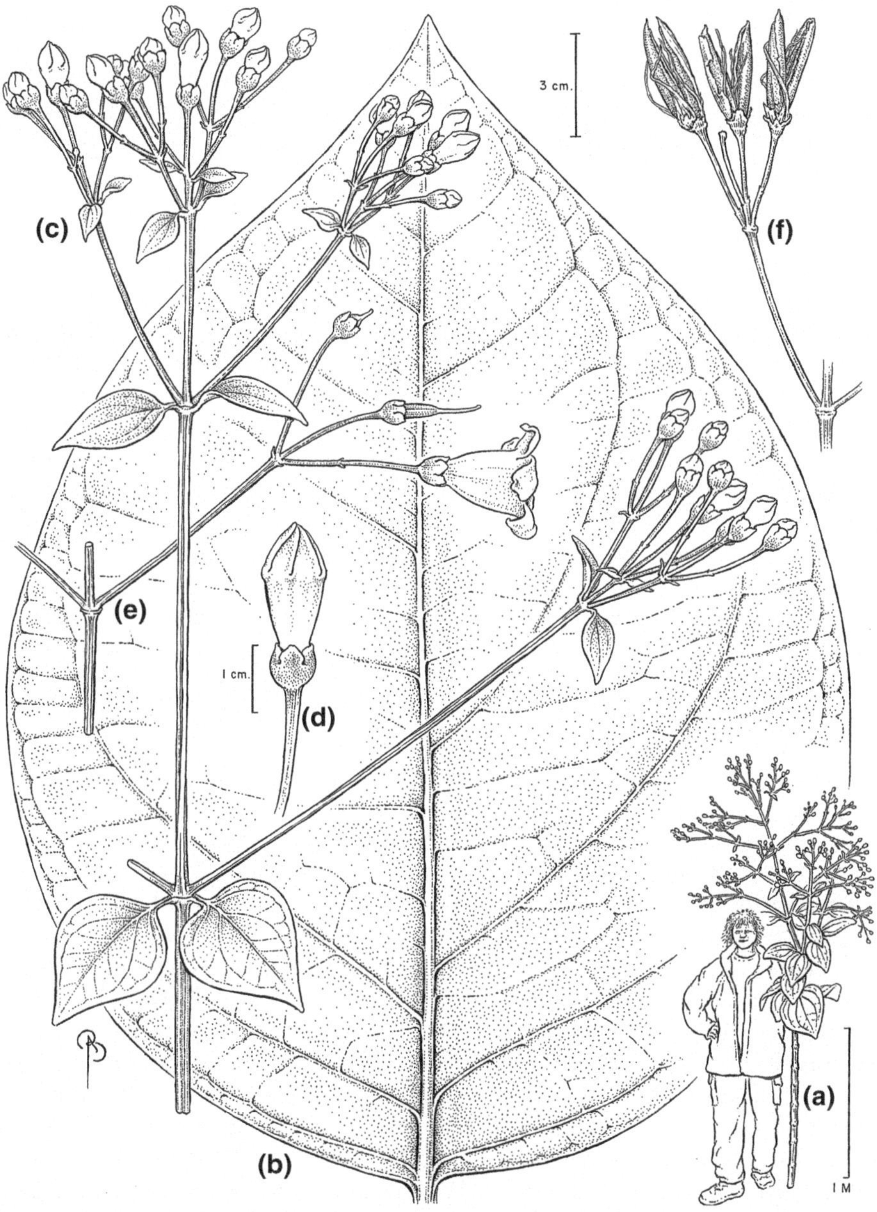

Fig. 3.52 *Macrocarpaea quizhpei* J.R. Grant. **a** habit of tree in the field (Claire Arnold holding plant); **b** leaf; **c** habit of flowering stem; **d** bud; **e** flowering stem; **f** fruits. **a–d**, **f** drawn from Grant et al. 08-4537, **e** drawn from Quizhpe & Luisier 1913 by Bobbi Angell

ridge 2–3 mm high. Blades (14)–30–36 × (9)–23–26.5 cm, entire, not revolute dark green, with slightly impressed veins above, and slightly raised veins below, glabrous above and below, papery thin; base aequilateral to oblique, rounded to nearly truncate; apex acute to acuminate. Inflorescence a much diffusely branched open thyrse to 1.3 m; branches 10–60+ cm long, 3–9 flowered per branch. Bracts broadly ovate, ovate, to lanceolate, short-petiolate, 18–62–(160) × 5–45–(90) mm; base aequilateral to oblique, rounded to nearly truncate; apex acuminate to acute; bract petioles 1–5 mm. Flowers pedicellate, erect to slightly spreading; pedicels 20–52 mm, long; bracteoles linear to lanceolate, 1–10 × 0.5–2.0 mm. Calyx campanulate, 7–9 × 7–8 mm, glabrous, smooth, green, ecarinate, ovate; cayx lobes 2.5–5.0 × 3.5–4.0 mm, apex rounded to obtuse. Corolla funnel-shaped, 38–40 mm long, 18–20 mm wide at the apex of the tube, light green, smooth; corolla lobes ovate, 10–12 × 8–12 mm, obtuse to rounded. Stamens 30.0–32.5 mm long; filaments 28–30 mm long, filiform, flattened; anthers elliptic to oblong, 5–6 × 2.0–2.5 mm, sagittate, versatile; pollen glabra-type. Pistil 31–33 mm long; ovary 8–11 × 3–4 mm; styel 19–21 × 1.0–1.5; stigma lobes spathulate, 2–3 × 1–2 mm. Capsules ellipsoidal to linear-long, 35–38 × 6–8 mm, smooth to faintly ribbed, faint-orangish tan, erect to slightly nodding; style remnant 1–2 mm long. Seeds "perimetrically winged type," flattened, roughly 3–4 sided in outline, yet appearing as a myriads of different puzzle pieces, 0.8–1.2 × 0.8–1.0 mm, bicolored, testa brown, wings straw-colored, testa reticulate, wings ribbed.

Macrocarpaea quizhpei belongs to section *Choriophylla* and occurs from 863 to1,500 m in the Amotape–Huancabamba region of southern Ecuador. It has a large diffuse inflorescence with long pedicels (20–52 mm). This species belongs to a closely related group of species restricted to the Amotape–Huancabamba zone including *M. catherineae, M. claireae, M. dies-viridis, M. illuminata, M. lenae, M. quizhpei,* and *M. xerantifulva.*

Etymology. Named for Wilson Quizhpe (1970–), Ecuadorian botanist of the Herbario Loja, Universidad Nacional de Loja, Ecuador.

Specimen examined: ECUADOR. **Zamora-Chinchipe:** Nangaritza Valley, south of Guayzimi, Cordillera del Condor region, Las Orquídeas, 04° 15′ 314″ S, 078° 40′ 087″ W, 1,180 m, 24 January 2009, *Grant, J.R., B. Angell, W. Grant & V. Trunz 09-4555* (NY); Nangaritza, Región de la Cordillera del Cóndor, cuenca del alto Río Nangaritza, Las Orquídeas, 04° 09′ 222″ S, 078° 38′ 555″ W, 877 m, 8 January 2008, *Grant, J.R. & Claire Arnold 08-4537* (G, LOJA, MO, NY); Between Wants and Las Orquideas, 04° 10′ 51.1″ S, 078° 38′ 38.6″ W, 863 m, 11 February 2011, *Grant, J.R. & J. Vieu 11-4688* (NY); Forest trail west of Cabañas Yankuam, south of Las Orquideas, 04° 15′ 03.1″ S, 078° 39′ 37.7″ W, 925 m, 11 February 2011, *Grant, J.R. & J. Vieu 11-4689* (NY).

24. ***Macrocarpaea sodiroana*** **Gilg**

Bot. Jahrb. Syst. 25: 724. 1898. TYPE: ECUADOR. Pichincha. In regione temperate secus fluv. Pilatón, 800–1,600 m, August flor., *Sodiro 101/1* (Type: B?

Fig. 3.53 *Macrocarpaea sodiroana* Gilg. *Photo* J.R. Grant

[destroyed]; Lectotype: QPLS; isolectotypes: QPLS [2 sheets], designated by Grant (2003a, 2003b) (Fig. 3.53).

Shrub, sometimes largely herbaceous but at least basally woody, usually 1–5 m tall, sometimes up to 10 m and arborescent, glabrous or inflorescence branches and calyx tubes occasionally sparsely pubescent. Stem terete or nearly so; internodes various, often exceeding leaves. Leaves membranous; venation conspicuous below, larger (lower) secondary veins mostly brancing from midrib at 30–50 degree, strongly upcurved, distinctly brochidodromous. Blades of lower and middle cauline leaves elliptic, mostly 20–40 cm long, 9–22 cm wide, base tapterig gradually or rather abruptly to a petiole 2–5 cm, apex acuminate. Upper cauline leaves gradually smaller, more ovate, subsessile or sessile. Inflorescence a dichasium, 3–6 times divided, primary branches often much elongated; first 2–3 divisions subtended by leaves 1.5–12 cm long, ultimate divisions by minute bractlets; flowers 16–100. Pedicels erect or arcuate, 7–30 mm. Calyx 6–12 mm; lobes semicircular, 0.5–0.8 times as long as tube, hyaline-margined, erose. Corolla 24(30–50 mm, pale green to greenish yellow, funnelform to campanulate; lobes 7–12 mm, ovate-triangular, obtuse. Capsule narrowly ovoid-cylindric, 28–50 mm. This description is largely from Pringle (1995).

Macrocarpaea sodiroana belongs to section *Choriophylla*, and occurs in the northern Andes, on Pacific slopes on the Cordillera Occidental, the Chocó of Ecuador. *Macrocarpaea sodiroana* is nearly sympatric with *M. gattaca* yet occur in two distinct elevation bands; 1,000–1,600 m for *M. sodiroana*, and 2,500–2,700 m, for *M. gattaca*. This species belongs to a closely related group of species with leaves that are long-decurrent on the petiole including *M. cortinae*, *M. pringleana*, *M. sodiroana*, and *M. umbellata*.

Specimens examined. ECUADOR. **Imbabura**: Cotacachi, Parroquia García Moreno, Reserva Biológica Los Cedros, sendera Y de la Cascada-Estación, 00° 18′ 13″ N, 078° 47′ 21″ W, 1,550 m, 2 November 2005, *Vargas, H.* et al. *6495* (MO, NY). **Pichincha**: Vía a Mindo, 1,400 m, 17 May 1997, *Avila, M.E. 11* (QCA); On road La Palma-Chiriboga (old road to Quito), 1,200 m, 21 December 1981, *Berg, C.C. 1285* QCA); Carr. Calacalí –Los Bancos, en Mindo, 1,500 m, 17 May 1997, *Carrera 23* (QCA); Quito, trek from Lloa to Mindo Mature Forest, between Hacienda Pacay and main bridge to Mindo, 1,400 m, 8 February 1998, *Clark, J.L. 4521* (MO, QCNE); Ayapi Pachijal, 18 November 2010, *Cornejo, X. & T. Montenegro 8349* (QCA); Along road from La Unión de Toachi to Quito via Chiribogo, 2.2 km up road to Chiriboga from the bridge over the Río Toachia, 10.9 km W of Allurgin, 00° 18′ 57″ S, 078° 55′ 29″ W, 984 m, 24 February 2005, *Croat, T. 95277* (MO, QCNE); Maquipucuna Reserve, vicinity of Nanegal 6.7 km NE of Nanegalito-Nanegal Road (turnoff ca. 3 km S of Nanegal), 19 km N of Nanegalito, 1,331 m, 11 March 2006, *Croat, T.* et al. *95730* (MO, QCNE); Along old road from Santo Domingo de los Colorados to Quito via Chiriboga and San Juan, 10.9 km NE of La Unión and Río Pilaton, 00° 18′ 21″ S, 078° 53′ 03″ W, 1,233 m, 17 March 2006, *Croat, T.B.* et al. *96020* (MO, NY); Along road from Santo Domingo de Los Colorados via Chiriboga and San Juan, 14.3 m NE of La Union and Río Pilaton, 1,432 m, 23 March 2006, *Croat, T.* et al. *96289* (MO, NY, QCNE); Bosque Protector Mindo, 1,325–1,400 m, 15 July 1992, *Delprete* et al. *6100* (TEX); Km 87–84 old road Quito-Santo-Domingo, 1,200–1,300 m, 21 March 1980, *Dodson, C.H. & Gentry 9725* (QCNE, SEL); Dos Rios, km 90 old road Quito-Santo Domingo, 1,100 m, 7 April 1984, *Dodson, C.H. & Thurston 14162* (MO, QCNE); Km 87–84 old road Quito-Santo-Domingo, 1,200–1,300 m, 21 March 1980, *Dodson, C.H., C.H. & Gentry 9725* (MO); 27 km from Chiriboga, *Fagerlind & Wibom 1925a* (S); 27 km from Chiriboga, *Fagerlind & Wibom 1925b* (S); Maquipucuna, 5 km E of Nanegal, 1,630 m, 9 May 1990, *Gentry* et al. *69921* (MO, QCNE); 15 km up road toward Chiriboga from the intersection of the new highway, on the old Quito-Sto. Domingo Road, 00° 18′ 11.8″ W, 078° 52′ 08.2″ W, 1,450 m, 27 January 13, *Grant, J.R. 13-5103* (NY); Mindo, Cordillera San Lorenze, 00° 04′ 35.4″ S, 078° 45′ 49.2″ W, 1,456 m, 5 February 13, *Grant, J.R. & J. Cortina 13-5106* (NY); 15 km up the old road Santo Domingo-Quito from where it merges with the new Santo Domingo-Quito road, 1,461 m, 26 February 2002, *Grant, J.R. & J.-M. Torres 02-4210* (G, NEU, NY, US); Road Nanegal-Nanegalito, 1,200–1,550 m, 24 January 1974, *Harling, G. & Andersson 11575* (GB); Old road Santo Domingo-Quito, 1,500 m, 8 May 1968, *Harling, G.* et al. *9221* (GB); Old road Quito-Santo Domingo, 2–9 km NE of turn off to old road out of Alluriquin, 1,100–1,230 m, 2 February 1982, *Luteyn* et al. *8733* (QCA, QCNE); Km 96 Carr. Quito-Santo Domingo, 1 February 1975, *Ortíz* 20 (MO, QCA); Carretera Cotocollao-Tandayapa-Nanegalito-Palmito-Pamba, 1,500 m, 26 January 1984, *Rodríguez, J. & C. Ulloa 154* (QCA); Mindo, camino a Mindo Gardens, vegetación secundaria cerca de las orillas del río Mindo, 078° 46′ 21″ W, 00° 2′

54″ W, 1,500 m, 17 October 1999, *Romero-Saltos, H. 1507* (QCA); Vía a Mindo, 1,400 m, 17 May 1997, Sáenz, F. 6 QCA); Camino de Quito a Santo Domingo de Los Colorados, entre Chiriboga y Dos Rios, 951–1,863 m, 4 February 1944, *Scolnik 1616* (PH); Nanegal, Res. Biol. Maquipucuna, 1,200–1,700 m, 20 May 1991, *Tipaz & Quelal 154* (MO, QCNE); Reserva Maquipucuna, secondary rainforest along Rio Tulambi W of Hacienda El Carmen, 1,250 m, 18 September 1989, *Webster* et al. *27733* (DAV, MO); Parroquia Nanegal, Bosque Protectora Maquipucuna, disturbed riparian woods on N bank of Río Umachaca, NE of Hacienda El Camen, 1,250 m, 7 July 1992, *Webster* et al. *29065* (DAV); Ridge between Río Mindo and Río Bagasal, 1,325–1,450 m, 14 July 1992, *Webster* et al. *29399* (DAV); Parroquia Nanegal, Reserva Maquipucuna, secondary riparian forest along Río Umachaca across from Hacienda El Carmen, 1,250 m, 15 January 1995, *Webster* et al. *31346* (DAV); Parroquia Nanegal, Bosque Protector Maquipucuna, steep slope in secondary rainforest along Río Umachaca, 1,250 m, 17 June 1996, *Webster* et al. *31550* (DAV, QCNE); 13.5 km by road W of Pacto, 1,400 m, 17 November 1998, *Webster 32889* (QCNE); Carr. Quito-Chiriboga-Empalme, km 92, desvio a Mulaulte, en borde del carretero y en quebrado de Mulaulte, 1,200–1,300 m, 13 December 1987, *Zak & Jaramillo 3169* (MO); Carretera Quito-Aloag-Santo Domingo, 1,100–1,400 m, 15 April 1988, *Zak & Jaramillo 3458* (AAU, G, GB, K, MO, NY, US).

25. ***Macrocarpaea subsessilis*** **Weaver & J.R. Grant**

Harvard Pap. Bot. 8(1): 91. 2003. TYPE: ECUADOR. Loja: Between Tambo Cachiyacu, La Entrada, and Nudo de Sabanilla, 2,500–3,500 m, 7 October 1943, *J. Steyermark, J. 54436* [Holotype: US; Isotypes: F, NY (2 sheets)] (Figs. 3.54, 3.55).

Shrub much branching from the base, the middle or above, 0.5–1.5(–3.0) m, glabrous throughout. Trunk to 1.5 cm in diameter, wood always solid. Stems terete to slightly quadrangular, solid, 2–5 mm in diameter just below inflorescence. Leaves obovate varying to elliptic, petiolate, 2.5–9.0 cm long; blades 2.0–8.5 × (1.5–)2.5–5.0 cm, not revolute, thick, leathery-coriaceous, dark green above, light green below, drying yellowish-green in herbarium material, midrib yellow to orange, lateral veins scarcely visible either above or below, glabrous; base aequilateral to oblique, rounded, apex obtuse to rounded, to rarely slightly mucronulate. Petioles 3–9 mm long, robust with short open vagination one quarter to one-third the length of the petiole; interpetiolar ridge 1–3 mm high. Inflorescence a few-branched nearly sessile thyrse, 3–11 cm long; branches 1–7 cm long, 2–8 flowered per branch. Bracts obovate, elliptic to rotund, petiolate, 12–30 × 6–19 mm; base aequilateral to oblique, rounded, apex obtuse to rounded, to rarely slightly mucronulate; bract petioles 2–4 mm long. Flowers pedicellate, erect to slightly spreading, but never nodding; pedicels 1–6 mm long; bracteoles spathulate, 10–22 × 4–11 mm. Calyx narrowly campanulate, 8–13 × 4–7 mm, glabrous, smooth, green, ecarinate; calyx lobes elliptic, 3–6 × 3–4 mm, apex rounded. Corolla funnel-shaped, 21–35 mm long, 11–25 mm wide at the apex of the tube, pale yellow, yellowish-green, greenish-yellow, cream to white, smooth;

Fig. 3.54 *Macrocarpaea subsessilis* Weaver & J.R. Grant. *Photo* J.R. Grant

corolla lobes ovate to elliptic, 7–12 × 4–10 mm, apex obtuse, rounded to retuse. Stamens 9–24 mm long; filaments 11–21 mm long, filiform, flattened; anthers linear to linear-elliptic, 3–4 × 1.5–2.0 mm. Pistil 13–23 mm long; ovary 4–6 × 1–3 mm; style 14–16 × 0.75–1.0 mm; stigma lobes spathulate to elliptic 2–3 × 1–2 mm. Capsules ovoid to ellipsoidal, 11–15 × 6–10 mm, smooth, pusticulate to rugose, straw-colored, erect to nodding; style remnant 3–9 mm long. Seeds angular to polyhedral to flattened (pyramids, tetrahedrons, rectangular prisms, to multisided prisms), 2–5-sided, not winged, 0.3–1.3 × 0.3–0.6 mm, dark orangish-tan, rugose-reticulate to wrinkled.

Macrocarpaea subsessilis belongs to section *Choriophylla* and occurs from 2,350 to 3,000 m in the Amotape–Huancabamba region of southern Ecuador. Its most frequently collected site, the hillside road from Yangana to Cerro Toledo has been seriously degraded by recent successive years of burning. It can be easily identified by its short, subsessile inflorescence, short-pedicellate flowers, large spathulate bracteoles, and leaves that become light-greenish yellow with yellow to orange midribs when dried as herbarium specimens. This species is part of a group of species with smooth leathery leaves where the secondary leaf veins absent or scarcely visible including *M. arborescens, M. ericii, M. harlingii, M. lilliputiana, M. subsessilis*, and *M. zumbae*.

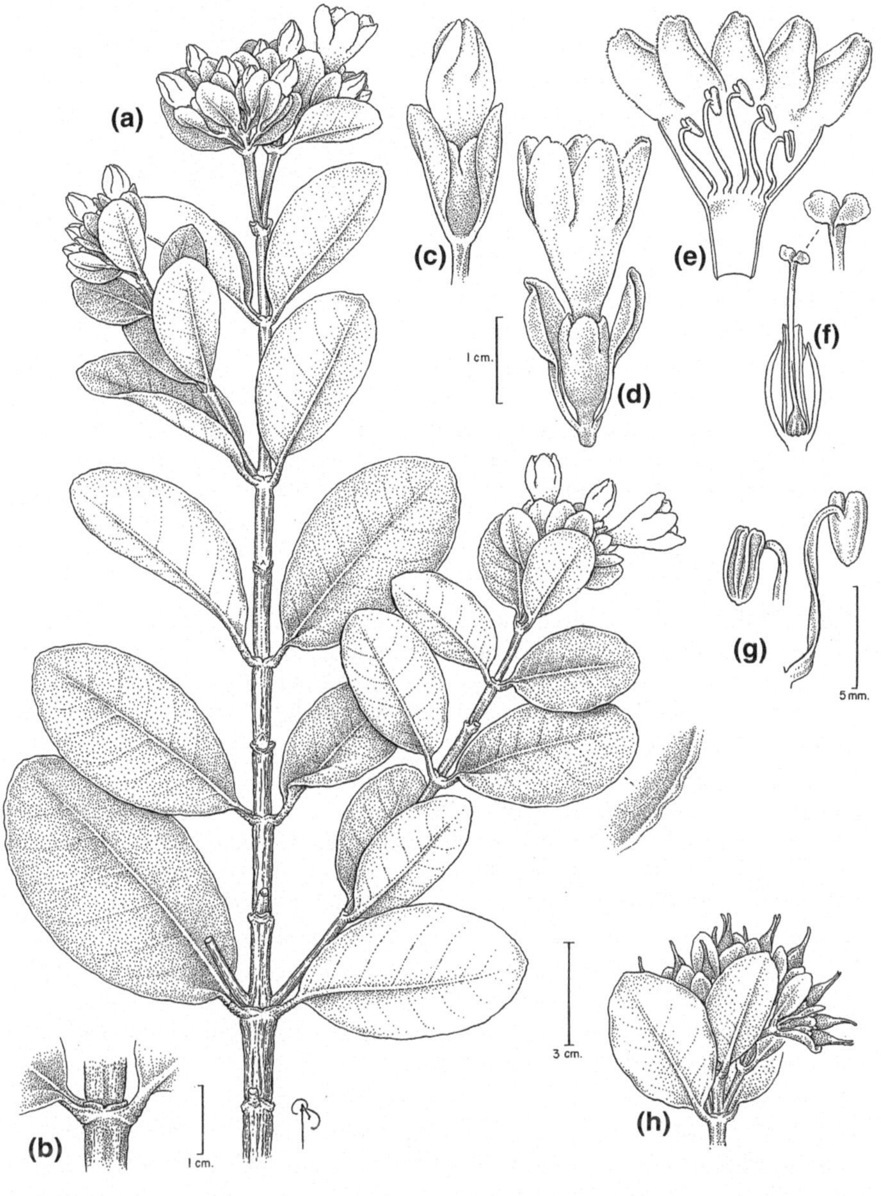

Fig. 3.55 *Macrocarpaea subsessilis* Weaver & J.R. Grant. **a** habit of flowering stem; **b** interpetiolar ridge; **c** flower in bud with distinctive bracteoles; **d** flower; **e** internal cross-section of corolla; **f** gynoecium; **g** anthers, **h** habit of fruiting stem. **a–c** drawn from Luteyn & Clements 7978 (NY); **d–g** drawn from Balslev 1288 (NY); **h** drawn from Steyermark, J. 54436 (NY)

Etymology. Named for its nearly sessile inflorescences.

Specimens examined: ECUADOR. **Loja:** Yangana-Zumba road, km 15–20, N slopes of Cordillera de Sabanilla, 2,550 m, 31 December 1980, *Balslev 1288* (AAU, NY); Río Catamayo Drainage, ridge between Quebrada Amarilla and Quebrada Agua Corra, eastern slope of Cordillera de Santa Rosa, Hacienda Guaycopamba, 21 km S of Vilcabamba, open brushy rocky ridge, 04° 27′ S, 079° 13′ W, low 16 February 1945, *Fosberg & Giler 23097* (NO, US); Nudo de Sabanilla, 21.5 km S of Yangana, 04° 25′ S, 079° 09′ W, 10 February 2001, *Grant, J.R & L. Struwe 3982* (LOJA, NEU, QCNE); Km 8.9 on road from Yangana to Cerro Toledo, 04° 23′ S, 079° 09′ W, 14 February 2001, *Grant, J.R & L. Struwe 4020* (G, LOJA, MO, NEU, NY, QCA, QCNE, S, SBBG, US [2 sheets]); Km 9.5 on road from Yangana to Cerro Toledo, 04° 22′ S, 079° 09′ W, 14 February 2001, *Grant, J.R & L. Struwe 4025* (LOJA, NEU, NY QCNE); Km 8.9 on road from Yangana to Cerro Toledo, 04° 23′ S, 079° 09′ W, 2 November 2002, *Grant, J.R., A. Roguenant & A. Raynal-Roques 02-4267* (MO, NEU, NY, US); Loja Km 8.9 on road from Yangana-Cerro Toledo, 2,450 m, 19-Feb-2006, *Grant, J.R., M. Cheung, F. F. Luisier, & N. Villard 4343* (G, LOJA, MO, NY); Around km 9 on road from Yangana up Cerro Toledo, 04° 22′ 58.7″ S, 079° 09′ 01.6″ W, 2,507 m, 3 February 2011, *Grant, J.R. & J. Vieu 11-4661* (NY); Cerro Bangala, ca. 10 km east of Yangana, 2,500–2,700 m, 18 October 1988, *Harling, G. 25318* (GB, MO); Nudo de Sabanilla, western slope on road to Yangana, 2,600 m, 6 February 1985, *Harling, G. & Andersson 21731* (GB, MO, QCA); Nudo de Sabanilla, western slope ca. 10 km above Yangana on road to Valladolid, 2,500 m, 3 April 1985, *Harling, G. & Andersson 23593* (GB, QCA); Cerro Toledo, road to "La Torre," ca. 7 km southeast of Yangana, 2,500 m, 7 April 1985, *Harling, G. & Andersson 23848* (GB, QCA); Yangana-Nudo de Sabanilla, dry and low mountain forest, 2,500–2,600 m, 16 October 1988, *Harling, G. & E. Madsen 25261* (AAU, MO, GB); Yangana-Nudo de Sabanilla, dense shrub forest and open scrub, 2,200–2,400 m, 13 February 1993, *Harling, G. & Ståhl 26373* (GB, MO, QCA, S); Yangana-Cerro Toledo, burned area, pastures and shrubby vegetation, 04° 23′ S, 079° 08′ W, 2,750–2,850 m, 28 December 1988, *Jørgensen & J. Madsen 65728* (AAU, QCA, QCNE); Yangana-Cerro Toledo, burned area with secondary vegetation, 04° 21′ S, 079° 07′ W, 2,510 m, 14 November 1990, *Jørgensen* et al. *92722* (AAU, QCA, QCNE); Road S out of Loja ca. 14 km S of Yangana, lower slopes of Nudo de Sabanilla, 2,460 m, 16 January 1981, *Luteyn & Clements 7978* (NY); Cordillera de Sabanilla, ca. 15 km S of Yangana, wet montane forest, 2,480 m, 31 December 1980, *Madison & Besse 7497* (SEL); Beside the road from Yangana to 'Cerro Toledo', 2,500 m, January 1998, *Matt, F. 12* (ER); Beside the road from Yangana to 'Cerro Toledo', 2,500 m, January 1998, *Matt, F. 13* (ER); Banderillas, 04° 13′ 52.5″ S, 079° 08′ 34.2″ W, 2,900 m, 20 March 2002, *Merino, B.* et al. *E-1066* (LOJA); Límite del P.N. Podocarpus, carretera Yangana–Cerro Toledo, 2,450 m, January 1995, *Palacios, W. 13016* (MO, QCNE); Between Nudo de Sabanilla and Río Cachiyacu at Tambo Cachiyacu, in dry sotobosque, 3,000–3,500 m, 17 October 1943, *Steyermark, J. 54758* (F); western slopes of Cordillera de Condor and northwest slopes of Nudo de Sabanilla, around Tambo

Cachiyacu, along Río Cachiyacu, about 2 leagues southeast of Yangana, 2,000–3,000 m, 19 October 1943, *Steyermark, J. 54805* (F); Sabanilla, 7 February 1986, *Vivar C., F. 2300* (LOJA). **Zamora-Chinchipe:** Yangana Cantón, ascending road from Yangana to Numbala drainage and antennas on summit of Filo Toledo, 2,400–2,950 m, 27 March 2005, *Clark, J.L. 8891* (QCNE, US).

26. *Macrocarpaea voluptuosa* J.R. Grant

Harvard Pap. Bot. 9(2): 336. 2005 TYPE: ECUADOR. Carchi: Cresta del Cerro Golondrinas Hembra, bosque húmedo montaño bajo, vegetación arbustiva en la cresta de la montaña, 00° 51′ N, 078° 08′ W, 3,000 m, 21 August 1994, *W. Palacios 12515* (Holotype MO; Isotypes QCNE, U) (Fig. 3.30).

Shrub, 1–3 m, hyaline hispid to spiculate with short simple hairs on stems, petioles, leaves, inflorescences, bracts and calyces, cm in diameter. Stems terete to slightly quadrangular, solid to hollow, 9–15 mm in diameter just below the inflorescence. Leaves ovate to subdeltoid, petiolate, 9.5.0–16.5(–25 according to *Boyle* et al. *3394*) cm long. Petioles 5–10 mm long, robust with strong open vagination one half the length of the petiole; interpetiolar ridge 2–3 mm high. Blades 9.0–17.5 × 9.5–11.0 cm, entire, not revolute, dark green, with slightly impressed veins above, and slightly raised veins below, hyaline hispid to spiculate on veins on lower surface, papery thin, to thin coriaceous; base aequilateral to oblique, cuneate to rounded; apex acute, obtuse to rounded. Inflorescence a much branched open thyrse, 36+ cm long; branches 6–36 cm long; 3–15 flowered per branch. Bracts ovate, elliptic, to lanceolate, petiolate, 16–78 × 8–52 mm; base aequilateral to oblique, cuneate; apex acute to acuminate; bract petioles 1–6 mm long. Flowers pedicellate, erect to spreading; pedicels 5–21 mm long; bracteoles ovate, lanceolate to oblanceolate, 6–16 × 2–8 mm. Calyx campanulate, 9–13 × 8–10 mm, glabrous, smooth, green, ecarinate; calyx lobes ovate, suborbicular to reniform, 2–3 × 3–5 mm, apex rounded to obtuse. Corolla funnel-shaped, 41–55 mm long, 18–26 mm wide at the apex of the tube, white, yellow, smooth; corolla lobes ovate to elliptic, 9–17 × 9–14 mm, obtuse to rounded. Stamens 30–35 mm long; filaments 24–28 mm long, filiform, flattened; anthers linear to linear-elliptic, 6–7 × 2–3 mm, sagittate, versatile; pollen *Glabra*-type. Pistil 40–51 mm long; ovary 8–9 × 2–4 mm; style 36–38 × 1.0–1.5 mm; stigma lobes spathulate, 3–4 × 1.5–2.0 mm. Capsules and seeds unknown.

Macrocarpaea voluptuosa belongs to section *Macrocarpaea*, and occurs in the northern Andes, from 3,000 to 3,060 mon the Cordillera Occidental, the Chocó of Ecuador. It is related to *M. duquei* Gilg-Ben. and *M. callejasii* J.R. Grant of the Cordillera Occidental Colombia, and *M. densiflora* (Benth.) Ewan and *M. pachyphylla* Gilg of the Cordillera Central Colombia.

Etymology. From the Latin *voluptuosus*, pleasant or delightful.

Specimens examined: ECUADOR. **Carchi:** Cerro Golondrinas, on crest of N ridge, just below peak, low elfin forest—páramo ecotone scrub, 00° 51′ 15″ N, 78° 08′ 21″ W, 3,000–3,060 m, 24 July 1994, *Boyle* et al. *3394* (MO, QCNE, U).

27. *Macrocarpaea xerantifulva* J.R. Grant

Harvard Pap. Bot. 9(2): 341. 2005. TYPE: PERU. Cajamarca: San Ignacio, La Coipa, vista Florida—La Laguna, bosque primario, 0526′ 00″ S, 7856′ 30″ W, 2,000–2,100 m, 11 June 1997, *J. Campos & Z. García 3960* (Holotype: MO; Isotype: U) (Fig. 3.56).

Shrub or small tree, 1.5–3.0 m, glabrous throughout. Stems terete to slightly quadrangular, hollow, 5–8 mm in diameter just below the inflorescence. Leaves ovate to ovate-elliptic, petiolate, 7.5–32.0 cm long. Petioles 5–40 mm long, robust with strong open vagination 1/2 the length of petiole; interpetiolar ridge 1–4 mm high. Blades 7–28 × 5–16 cm, entire, not revolute, green, with slightly impressed veins above and slightly raised veins below, glabrous above and below, papery thin; base equilateral to attenuate and decurrent on the petiole; apex acuminate to apiculate. Inflorescence a much-branched open thyrse, 18–31+ cm long; branches 11–31 cm long; 5–15 flowered per branch. Bracts ovate to elliptic, sessile to short-petiolate, 17–62 × 8–32 mm; base equilateral to oblique, cuneate to rounded; apex acuminate; bract petioles 0–5 mm long. Flowers pedicellate, erect; pedicels 13–26 mm long; bracteoles inconspicuous and scabrous, triangular, linear to ovate, 0.5–12.0 × 0.5–3.0 mm. Calyx campanulate, 5–9 × 5–10 mm, typically glabrous to hispid (Sánchez Vega 73–80), ecarinate; calyx lobes ovate to reniform, 2.5–3.5 × 3–4 mm, apex rounded to obtuse. Corolla funnel-shaped, 23–31 mm long, 15–20 mm wide at the apex of the tube, green/yellow (Dostert 98/114), greenish yellow (Sánchez Vega 73–80), greenish (Friedberg 696), smooth; corolla lobes ovate, 6–9 × 7–9 mm, obtuse to rounded. Stamens 12–20 mm long; filaments 9–15 mm long, filaments filiform/flattened; anthers ovate-elliptic, 4–5 × 2.00–2.25 mm, sagittate, versatile; pollen Glabra type. Pistil 20–27 mm long; ovary 4–8 × 2–3 mm; style 14.0–16.5 × 1.0 mm; stigma lobes spathulate, 2.0–2.5 × 1.0–1.5 mm. Capsules ellipsoidal to fusiform, 25–35 × 5–7 mm, smooth varying to faintly striated, rugose or ribbed, greenish brown, tan to light brown, erect to slightly nodding; style remnant 2–7 mm long. Seeds "perimetrically winged type," 0.8–1.0 × 0.4–1.0 mm, bicolored, testa tannish, wings straw-colored, testa reticulate, wings ribbed.

Macrocarpaea xerantifulva belongs to section *Choriophylla* and occurs from 1,500 to 2,250 m in the Amotape–Huancabamba region of southern Ecuador and northern Peru. It is a common roadside species from Valladolid south to the border with Peru, and also in northern Peru. This species belongs to a closely related group of species restricted to the Amotape–Huancabamba zone including *M. catherineae, M. claireae, M. dies-viridis, M. illuminata, M. lenae, M. quizhpei,* and *M. xerantifulva*.

Etymology: From the Greek xeros, dry, and the Latin fulvus, yellowish, for the stems and inflorescence that dry rather yellowish-gold in color.

Specimens examined: ECUADOR. **Zamora-Chinchipe:** Valladolid Cantón, road Valladolid-Tapala-El Porvenir del Carmen, 1–2 km E of Valladolid, 04° 33′ 27″ S, 079° 07′ 50″ W, 1,600–1,700 m, 1 April 2005, *Clark, J.L.* et al. *8983* (NY, QCNE, US); 2 km N of Valladolid, 04° 32′ 787″ S, 079° 07′ 838″ W, 1,662 m, 29

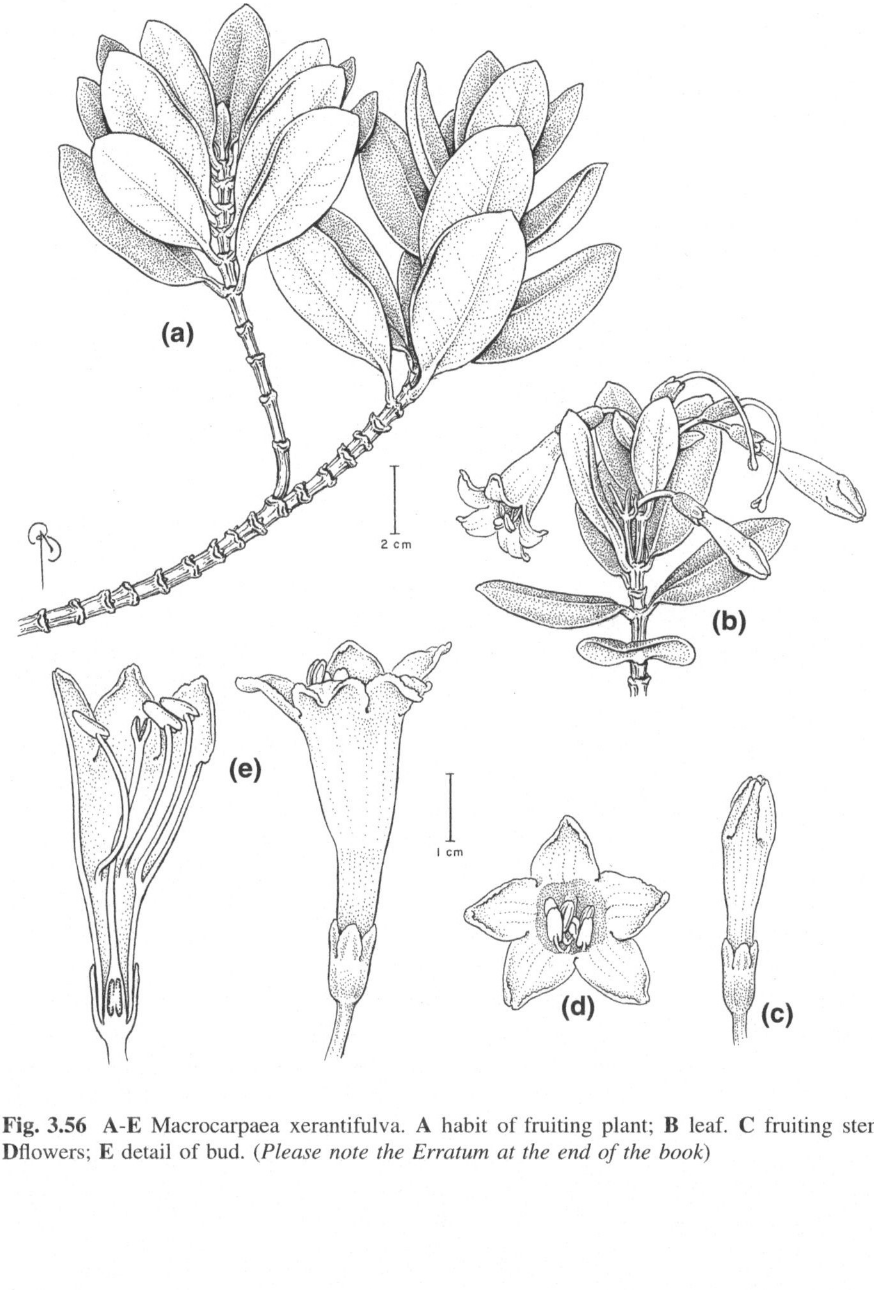

Fig. 3.56 A-E Macrocarpaea xerantifulva. **A** habit of fruiting plant; **B** leaf. **C** fruiting stem; **D**flowers; **E** detail of bud. (*Please note the Erratum at the end of the book*)

January 2009, *Grant, J.R., B. Angell, W. Grant & V. Trunz 09-4569* (NY); 7 km E from Valladolid toward Tapala, 04° 33′ 276″ S, 079° 06′ 655″ W, 1,500 m, 29 January 2009, *Grant, J.R., B. Angell, W. Grant & V. Trunz 09-4572* (NY); 7 km E from Valladolid toward Tapala, 1,500 m, 4 January 2008, *Grant, J.R., C. Agier, C. Arnold & M.L. Cheung 08-4522* (NY); 1 km south of Valladolid, 04° 33′ 32.3″ S, 079° 08′ 13.7″ W, 1,571 m, 6 February 2011, *Grant, J.R. & J. Vieu 11-4667* (NY).

PERU. Cajamarca: San Ignacio, La Coipa, vista Florida-La Laguna, 2,000–2,100 m, 11 June 1997, *Campos, J. & García 3960* (MO, U); Jaén, Colosay, above the village, 1,800–2,300 m, 7 March 1998, *Dostert 98/114* (F, MSB); Cutervo, San Andrés, km 1.5 noriente de San Andrés, 2,250 m, 18 June 1980, *Sánchez Vega 73-80* (MO); San Ignacio, mountain slope above San José de Lourdes, 1,300–1,500 m, 10 June 1998, *Weigend, M.* et al. *98/519* (F, M). Piura: Huancabamba, Cordillera de San José, 1,700 m, 23 September 1961, *Friedberg 696* (NY).

28. ***Macrocarpaea ypsilocaule*** **J.R. Grant**

Harvard Pap. Bot. 9(2): 331. 2005. TYPE: COLOMBIA. Putumayo: Páramo de San Francisco (la Depresión), en el camino carretera de San Francisco a Mocoa, 3,000–3,100 m, January 1942, *P. Fray Miguel 57* (Holotype: F) (Fig. 3.57).

Shrub, 2 m, glabrous throughout except hispid along veins on the lower surface, cm in diameter. Stems terete to slightly quadrangular, solid, 2–5 mm in diameter just below the inflorescence. Leaves elliptic, oval to ovate, petiolate, 5.5–15.5 cm long. Petioles 4–10 mm long, robust with strong open vagination one quarter to one-third the length of the petiole; interpetiolar ridge 1–4 mm high. Blades 5.0–14.5 × 2.5–6.0 cm, entire, not revolute, dark green above, lighter below, with slightly impressed veins above, and strongly raised veins below, glabrous above, hyaline hispid on all veins below, thin to leathery coriaceous; base aequilateral to oblique, cuneate; apex acuminate to acute. Inflorescence a few-branched open thyrse, 6–22 cm long; branches 5–15 cm long; 1(–3) flowered per branch. Bracts ovate to lanceolate, sessile to short-petiolate, 20–70 × 7–15 mm; base aequilateral to oblique, cuneate, to short-attenuate; apex acuminate; bract petioles 0–5 mm long. Flowers pedicellate, spreading; pedicels 8–45 mm long; bracteoles linear to lanceolate, 6–14 × 2–3 mm. Calyx campanulate, 15–22 × 8–18 mm, glabrous, faintly rugose to pusticulate, ecarinate; calyx lobes ovate to elliptic, 9–13 × 8–11 mm, apex rounded to obtuse. Corolla funnel-shaped, 42–45 mm long, 15–20 mm wide at the apex of the tube, green, smooth; corolla lobes ovate, 10–12 × 9–11 mm, obtuse to rounded. Stamens 24–30 mm long; filaments 20–24‘ mm long, filiform, flattened; anthers elliptic to oblong, 4–6 × 2–3 mm, sagittate, versatile; pollen *Glabra*-type. Pistil 40-42 mm long; ovary 8–10 × 2–4 mm; style 29–30 × 0.75–1.0 mm; stigma lobes spathulate, 2–3 × 1.5–2.0 mm. Capsules ovoid, 14–17 × 11–13 mm, smooth, tan, erect to nodding; style remnant 1–3 mm long. Seeds “Rimmed type,” 0.8–0.9 × 0.4–0.5 mm, straw-colored, testa rugose-reticulate.

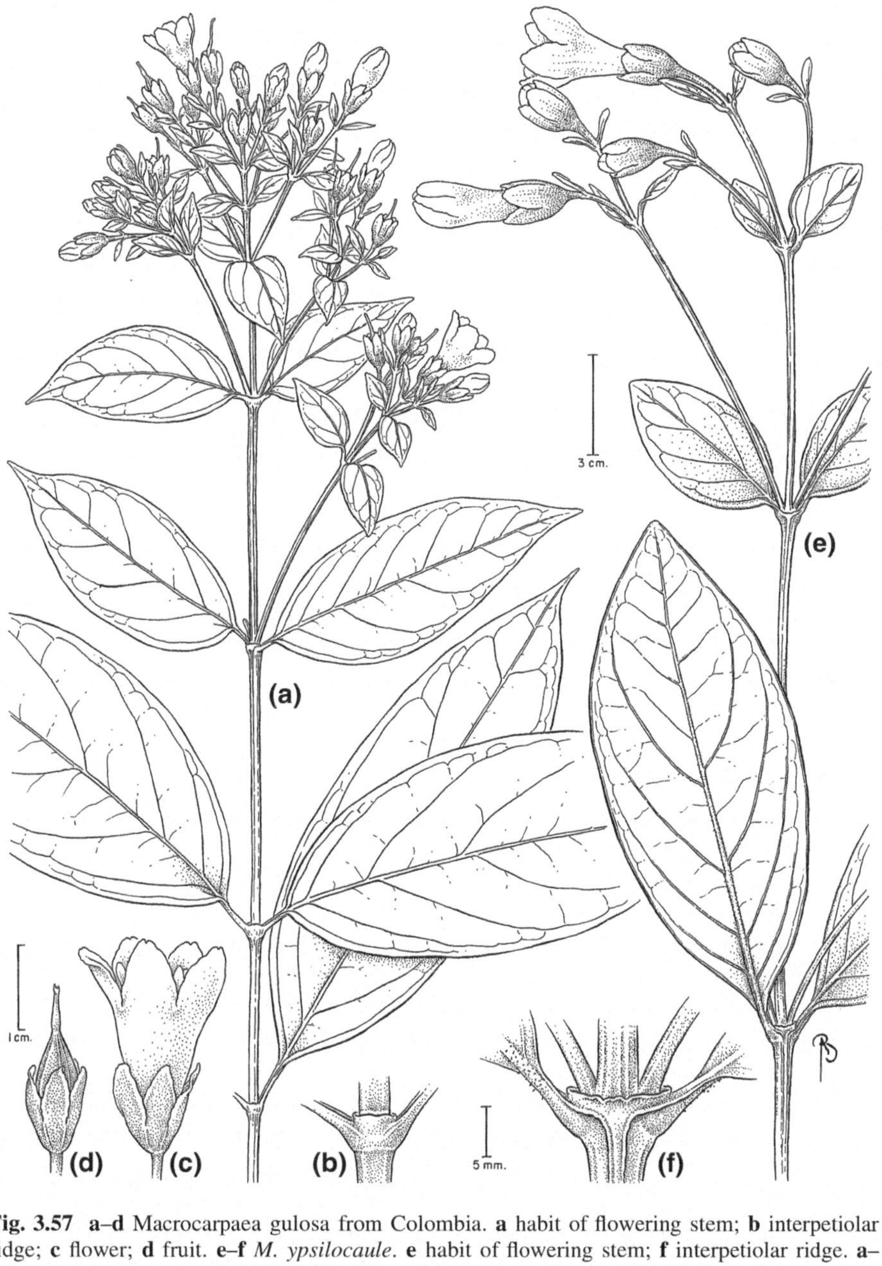

Fig. 3.57 **a–d** Macrocarpaea gulosa from Colombia. **a** habit of flowering stem; **b** interpetiolar ridge; **c** flower; **d** fruit. **e–f** *M. ypsilocaule*. **e** habit of flowering stem; **f** interpetiolar ridge. **a–c** drawn from Forero et al. 3158 (MO, COL), **d** drawn from Forero 3226 (COL), **e–f** drawn from Miguel 57 (F)

Macrocarpaea ypsilocaule belongs to section *Macrocarpaea*, and occurs in the northern Andes, from 2,000 to 2,750 m on the Cordillera Central of Colombia. It is related to *M. biremis* J.R. Grant of the Cordillera Occidental, the Chocó of Colombia on the basis of few-flowered inflorescences, and flowers with large calyces. Often the basal branches of the inflorescence of *Macrocarpaea ypsilocaule* only have a single flower.

Etymology. From the Greek *ypsilon* (or upsilon), and *caulis* (stem) for the "Y-shaped" bifurcated branching pattern of the inflorescence.

Specimens examined: ECUADOR. **Carchi:** Cerro Golondrinas area, access via Chamorro property above El Carmen, which is above Hualchán, flat hilltop before steep ridge crest approach to peak 1840, stunted upper montane forest, heavily festooned with moss, 00° 50′ N, 78° 12′ W, 2,690 m, shrub, in fruit only, 24 July 1993, *Boyle & Hibbs 2333* (MO, QCNE).

COLOMBIA. Huila: Vía a la quebrada Cascajal del Conudo, P.N. Natural Cueva de Los Guácharos, 01° 35′ N, 76° 00′ W, 30 August 1993, *Barbosa 15567* (FMB [2 sheets]); Vía a la quebrada Cascajal del Conudo, P.N. Natural Cueva de Los Guácharos, 01° 35′ N, 76° 00′ W, 30 August 1993, *Barbosa 15592* (FMB); De la Quebrada Negra al primer cerro vía a Cerro Punta, P.N. Natural Cueva de Los Guácharos, 01° 35′ N, 76° 00′ W, 2 September 1993, *Barbosa 15625* (FMB). Putumayo: Municipio de Mocoa, Carretera entre Sibundoy y Mocoa, localidad el Mirador, 01° 04′ 11 N, 76° 44′ 41″ W, 2,000 m, 7 September 1998, *Mendoza* et al. *6034* (COAH, FMB); Municipio Colon, Reserva Natural "La Rejoya," 2,750 m, shrub 2 m, 12 March 1999, *Stancik 2807* (COL, PSO [not seen]).

29. **sp. nov. A "catherineae"**

This species belongs to a closely related group of species restricted to the Amotape–Huancabamba zone including *M. catherineae, M. claireae, M. diesviridis, M. illuminata, M. lenae, M. quizhpei*, and *M. xerantifulva*. It occurs in the Cordillera del Condor region.

Specimens examined: ECUADOR. **Zamora-Chinchipe**: Road from Chuchumbleza, Quimi and Tundayme toward the Condor Mirador, 16 km from end of road at Military camp 'Condor Mirador', 03° 36′ 32.7″ S, 078° 28′ 18.3″ W, 1,059 m, 15 February 2011, *Grant, J.R. & J. Vieu 11-4695* (NY); Région de la Cordillera del Cóndor, vertiente occidental, Cantón El Pangui, parroquia Tundayme, Valle del Río Quimi, Bosque húmedo premontano, arbustivo de 3 m de alto, flores campanulares color blanco-crema, 03° 35′ 05″ S, 078° 27′ 07″ W, 900 m, 8 July 2007, *Quizhpe, W. 2706* (MO, QCNE); El Pangui, Cordillera del Cóndor, vertiente occidental, cuenca del Río Quimi, zona de la futura mina de cobre de la compañia EcuaCorriente, Formación rocosa arenisca, suelo arenoso, bosque muy húmedo montano bajo y bosque arbustivo. Arbusto de 304 m, hojas carnosas, flores con pecíolos amarillos, 03° 34′ 55″ S, 078° 26′ 10″ W, 1,400–1,670 m, 19 March 2006, *Quizhpe, W. & F. Luisier 1977* (QCNE); Cordillera del Cóndor, vertiente occidental, cuenca del Río Quimi, zona de la futura mina de cobre de la

Fig. 3.58 *Macrocarpaea* sp. nov. "cortinae." *Photo* J. Cortina

compañia EcuaCorriente, 03° 34′ 55″ S, 078° 26′ 10″ W, 1,400–1,670 m, 19 March 2006, *Quizhpe, W. & F. Luisier 1981* (LOJA, QCNE).

30. **sp. nov. B "cortinae"**

This species belongs to a closely related group of species with leaves that are long-decurrent on the petiole including *M. cortinae*, *M. pringleana*, *M. sodiroana*, and *M. umbellata*. It occurs in northern Ecuador on Amazon-facing slopes of the Andes in Sucumbios and Napo (Fig. 3.58).

Specimens examined: ECUADOR. **Napo:** Valle alto del Río Quijos, 5 km al Sur de Cuyuja, Finca Agroecologica Antisana, 00° 28′ S, 078° 03′ W, 2850, 16 June 1998, *Vargas, H.* et al. *1851* (MO, NY, QCNE). **Sucumbios:** Rio Bermejo to Cerro Sur Pax, Cofan community of Alto Bermejo, access from Bermejo oil field of Pozo 1, NW between Lumbaqui and Cascales, vicinity of Vista Camp, 00° 18′ 13.8 N, 077° 24′ 32.0″ W, 100–1,200 m, 29 July 2001, *Aguinda, R.* et al. *1204* (F); Sinangoe Station, Rio Cuccono Chico, affluent of Rio Due, SW of Puerto Libre, NW of Lumbaqui, access trail from Rio Sieguyo, vicinity of Ccuccono beach camp, 00° 07′ 48.5″ N, 077° 33′ 19.9″ W, 940–1,000 m, 8 August 2001, *Aguinda, R.* et al. *1585* (F); Gonzalo Pizarro Canton, Parroquia Reventador, 1,800 m, 23 May 1990, *Cerón, C. & Ayala 9988* (MO, QCNE); 13 km NW of La Bonita on road toward Santa Barbara, 00° 32′ 14.7″ N, 077° 31′ 53.6″ W, 2,016 m, 7 February 13, *Grant, J.R & J. Cortina 13-5114* (NY); 6 km south of La Bonita, 00° 27′ 27.1″ N, 077° 32′ 20.8″ W, 1,903 m, 7 February 13, *Grant, J.R & J. Cortina 13-5115* (NY); 4 km N of Rio

Chigual ridge at km 44 from Lumbaqui, 00° 16′ 46.7″ N, 077° 27′ 39.6″ W, 767 m, 7 February 13, *Grant, J.R & J. Cortina 13-5116* (NY).

31. **sp. nov. C "illuminata"**

This species belongs to a closely related group of species restricted to the Amotape–Huancabamba zone including *M. catherineae, M. claireae, M. dies-viridis, M. illuminata, M. lenae, M. quizhpei*, and *M. xerantifulva.*

Specimens examined: ECUADOR. **Zamora-Chinchipe:** Cordilllera del Condor, 9.3 km E of Paquisha (measured from the church in the central square), 1,394 m, 7 January 2008, *Grant, J.R., C. Agier, C. Arnold & M.L. Cheung 08-4529* (LOJA, MO, NY); Cordillera del Condor region, from the road from Zumbi to Paquisha, 8 km south on the road toward Guayzimi, (just W of Paquisha), 03° 58′ 055″ S, 078° 41′ 133″ W, 1,022 m, 25 January 2009, *Grant, J.R., B. Angell, W. Grant & V. Trunz 09-4560* (NY); Cordillera del Condor region, from the road from Zumbi to Paquisha, 8 km south on the road toward Guayzimi, (just W of Paquisha), 03° 58′ 02.1″ S, 079° 41′ 09.9″ W, 910 m, 11-Feb-2011, *Grant, J.R. & J. Vieu 11-4687* (NY).

32. **sp. nov. D "lilliputiana"**

This is an easily identifiable species with small rounded leaves a very short inflorescence, similar to that of *M. subsessilis*. It occurs in the Cordillera del Condor region. This species is part of a group of species with smooth leathery leaves where the secondary leaf veins absent or scarcely visible including *M. arborescens, M. ericii, M. harlingii, M. lilliputiana, M. subsessilis,* and *M. zumbae*.

Specimens examined: ECUADOR. **Zamora-Chinchipe:** Paquisha Cantón, Cordillera del Cóndor, the Machinaza plateau, one of the highest-elevation Hollín sandstone plateaus in the Condór region, about 500 m west of the Ecuador–Peru international border, near end of trail from Paquisha Alto military post. Bare sandstone substrate, or quartzite sand derived therefrom, very nutrient poor. Dwarf scrub vegetation, dominated by shrubs to about 1.5 m tall, with occasional small tree to 4 m tall. Vegetation recovering slowly from an extensive burn about 20 years previously, with charred woody stems in abundance on ground, 03° 54′ 06″ S, 078° 28′ 57″ W, 2,315 m, 23 June 2009, *Neill, D. & C. Kajekai 16910* (LOJA, MO, QCNE); Machinaza plateau summit area, upper Paquisha military post on Ecuador–Peru border, 03° 53′ 50″ S, 078° 28′ 49″ W, 2,420 m, 15 March 2008, *Neill, D. & W. Quizhpe 16113* (MO, NY).

33. **sp. nov. E "pacifica"**

This species is the only species of *Macrocarpaea* known from El Oro province, and is also the southernmost species on the Pacific side of the Andes in *Macrocarpaea* (Fig. 3.59).

Specimens examined: ECUADOR. **El Oro:** Piñas, Moromoro, Reserva Biologica Buenaventura (Fundación de Convervación Jocotoco), bosque Moscos,

Fig. 3.59 *Macrocarpaea* sp. nov. "pacifica." *Photo* J. Grant

Quebrada obscura, 03° 39′ 17″ S, 079° 44′ 32″ W, 940–1,050 m, 28 May 2011, *Clark, J.L.* et al. *12283* (NY); Piñas, Bosque Protector Buenaventura, Fundación de Conservación Jocotoco, 03° 39′ 02.4″ S, 079° 44′ 53.0″ W, 925 m, 16 February 2011, *Grant, J.R. & J. Vieu 11-4704* (NY); Parroquia El Placer, Reserva Buenaventura, propriedad de la Fundación Jocotoco, recorrido por el sendero desde la estación hasta el bosque Puma, bosque secundario con relictos de bosque primario, bosque humedo pre-montano, 03° 38′ 41″ S, 079° 45′ 46″ W, 1,000 m, 4 April 2005, *Vargas, H.* et al. *5191* (MO); Piñas, Parroquia El Placer, Reserva Ecológica Buenaventura, propriedad de la Fundación Jocotoco, sendero al Jardín Botánico, entrada por la Cripta de la Virgen, bosque primario, bosque humedo pre-montano, 03° 39′ 33″ S, 079° 44′ 25″ W, 1,220 m, 8 April 2005, *Vargas, H.* et al. *5344* (MO, NY).

34. **sp. nov. F "umbellata"**

This species belongs to a closely related group of species with leaves that are long-decurrent on the petiole including *M. cortinae*, *M. pringleana*, *M. sodiroana*, and *M. umbellata*. It occurs on Pacific-facing slopes of southern Colombia and adjacent Carchi, Ecuador (Fig. 3.60).

Fig. 3.60 *Macrocarpaea* sp. nov. "umbellata." *Photo* J. Cortina

Specimens examined: ECUADOR. Between Rio Verde and Rio Blanco, 7 January 1962, *Dodson, C.H. & L.B. Thien 1925* (AAU, MO, NY); *Verleysen s.n.* (QPLS). **Carchi:** Camaña El Corazón y alrededores Carchi, 2,200 m, 11 February 1997, *Anton 320* (QCA, QCNE); Tulcán-Maldonado highway, Chical Trail, 1,500 m, 26 January 1977, *Boeke, J.D. 845* (AAU, MO, NY, QCA); Along road from El Chical to Tulcán, 7.4 km E of El Chical, 1,393 m, 19 February 2004, *Croat, T.B. 94939* (MO, NY, QCNE); Road from La Carolina (between Ibarra and San Lorenzo) northeast toward El Chical, 00° 54′ 56.3″ N, 078° 12′ 04.6″ W, 1,497 m, 6 February 13, *Grant, J.R & J. Cortina 13-5107* (NY); Vic. of Maldonado, 1,600–1,900 m, 13 April 1977, *Madison, M.T. 3956* (NY); Environs of Chical, 12 km below Maldonado, 1,200 m, 30 May 1978, *Madison, M.T.* et al. *4795* (F, QCA, SEL); Peñas Blancas, 20 km below Maldonado on the Río San Juan, 900–1,000 m, 27 May 1978, *Madison, M.T.* et al. *4614* (F, QCA, SEL); Environs of Chical, 12 km below Maldonado, 1,200 m, 30 May 1978, *Madison, M.T.* et al. *4795* (F, QCA, SEL); Tulcan, Res. Indígena Awá, 1,800 m, 15 June 1991, *Rubio* et al. *1636* (MO, QCNE); Reserva Golondrinas, recorrido por el sendero de las mulas hasta la casa de la Fundación Golondrinas en Santa Rosa. Bosque muy húmedo montano, bosque primario, arbusto de 2 m, tallos huecos, flores amarillo-verdosas, 00° 50′ N, 078° 08′ W, 1,790 m, 26 January 2004, *Vargas, H.* et al. *4469* (MO, NY, QCNE).

COLOMBIA. *Mutis 334* (US). Cauca: Mun. Santa Rosa, Corregimiento Descanse, Vereda Génova, en cercanía de la quebrada La Isla, Finca La Isla, 1,100 m, 8 April 1995, *Rubiano, L.J. & Moreno O. 644* (COL); Santa Martha, Bota Caucana, en la via Mocoa-Pitalito, ca. 5 km del puente del Rio Caquetá, 1,100 m, 25 January 1990, *Ramos, J.E.* et al. *2501* (CUVC, MO, U). Nariño: Ricáurte, 1,300 m, 7 April 1941, *von Sneidern 512* (S); Mpio. de Ricaurte, Resgardo Indígena Nulpe Medio, Orillas del río Nulpe, 715 m, 6 January 1996, *González & Ramírez P. 1554* (QCA); Mpio. de Ricaurte, Trayecto La Planada-San Isidro, 1,500–2,700 m, 16 November 1991, *Ramírez, B.R. 4270* (MO); Altaquer, 1,000 m, 14 June 1986, *Maas, P.J.M.* et al. *6528* (U); Res. Nat. La Planada, 7 km de Chucunes, 1,800 m, 15 November 1987, *Benavides, O. de 8953* (MO); Trayecto

Fig. 3.61 *Macrocarpaea* sp. nov. "zumbae." *Photo* J. Grant

Pialapi-La Planada, 1,300–1,700 m, 23 July 1988, *Benavides, O. de 10122* (MO); 2–8 km E of Junin on Tumaco-Tuquirres rd, 1,100 m, 26 July 1986, *Gentry, A.* et al. *55232* (MO, U). Valle: Hoya del Río Digua, Piedra de Moler, 900–1,180 m, 19 August 1943, *Cuatrecasas, J. 14986* (F, US, VALLE); Río Anchicaya near CVC hydroelectric plant, 400–500 m, 15 December 1981, *Gentry, A. 35666* (COL, MO); Cali-Buenaventura hwy, 28 km beyond the junction with the new hwy, 1,400–1,420 m, 28 August 1976, *Croat, T.B. 38605* (MO).

35. **sp. nov. G "zumbae"**

This new species ressembles *M. harlingii*, but is actually more closely related to *M. pajonalis* and *M. kuelap* of Peru. It occurs just in the border region of Ecuador with Peru (Fig. 3.61). This species is part of a group of species with smooth leathery leaves where the secondary leaf veins absent or scarcely visible including *M. arborescens, M. ericii, M. harlingii, M. lilliputiana, M. subsessilis*, and *M. zumbae*.

Specimens examined: ECUADOR. **Zamora-Chinchipe:** Along road between Zumba and Amaluza that parallels the Ecuador–Peru border, 04° 47′ 39.6″ S, 079° 18′ 39.6″ W, 1,989 m, 7 February 2011, *Grant, J.R. & J. Vieu 11-4669* (NY); Approx. 50 km S Amaluza on road to Zumba that nearly parallels the Ecuador–Peru border, 2,224 m, 1 January 2008, *Grant, J.R., C. Agier, C. Arnold & M.L. Cheung 08-4518* (NY); *08-4519* (NY).

3.4 Conclusions

This monograph is the result of ten years of study including eight trips to Ecuador for both fieldwork and herbarium studies on *Macrocarpaea*. Examination of over 4,000 herbarium specimens has resulted in the description of numerous new species and the major modification of some species concepts. This also represents the first major monographic revision of *Macrocarpaea* in the Andes, while treatments for both Colombia and Peru under current preparation. While additional new species may be found in remote regions in Ecuador, this monograph may serve as a point of reference for future research on this spectacular example of radiation in the Amotape–Huancabamba zone.

Acknowledgments This chapter was prepared partially for the degree of *Privat Docent* at the Université de Neuchâtel, Switzerland. Financial support is acknowledged from the Fondation Dr. Joachim de Giacomi, the Académie Suisse des Sciences naturelles, and the Université de Neuchâtel. Bobbi Angell prepared the outstanding scientific illustrations. Permission was granted by Harvard University to reproduce illustrations that were originally published in *Harvard Papers in Botany*, from the Missouri Botanical Garden for the *Annals of the Missouri Botanical Garden*, and from the New York Botanical Garden for *Brittonia*.

In Ecuador, I acknowledge assistance and cooperation from colleagues Zhofre Aguirre, Carlos Cerón, Susana Léon, José M. Manzanares, Bolivar Merino, David Neil, Wilson Quizhpe, and Joseph-Marie Torres. The following herbaria (acronyms according to Thiers 2008) are acknowledged for the loan of material, photocopies of specimens, data on their collections, and/or for hospitality extended during visits* to examine material of *Macrocarpaea*: AAU, AFP, ALA*, B, BM, BP, BR*, BRIT, BSB, C, CAS, CAUP, CEPEC*, CESJ, CHOCO, CHRB*, COAH*, COL*, CONN, CR, CUVC*, CUZ*, DAV, DUKE, E, EHH, F*, FAUC, FI, FLAS, FMB, FR*, FTG, G*, GB, GH*, GOET, HAC, HAL, HAM, HAO*, HB*, HUA*, HUCP, HUEFS*, HUQ, HUT*, IAN*, INB, INPA*, JAUM, JBSD, JE, K*, L, LD, LINN, LOJA*, LPB, LS, m, MA*, MANCH, MARY*, MBM*, MBML*, MEDEL*, MER*, MG*, MICH, MIN, MO*, MOL*, MSB, MU, MY, NA, NEU*, NO, NSW*, NY*, OXF, P*, PH, PORT*, PR, PRC, Q*, QAP*, QCA*, QCNE*, QPLS*, QUSF*, R*, RB*, RNG*, S*, SBBG*, SEL*, SP*, SPF*, TEX, U, UC, UCWI, UDBC, UPCB, UPS, UPTC, US*, USM*, VALLE*, VEN*, W*, WIS, WU*, YU, and Z*.

References

Carlquist S, Grant JR (2005) Wood anatomy of Gentianaceae, tribe Helieae: diversification in relation to ecology, habit, and systematics; the effect of sample diameter. Brittonia 57:276–291

Ewan J (1948) A revision of *Macrocarpaea*, a neotropical genus of shrubby gentians. Contr US Natl Herb 29:209–250

Grant JR (2003) De Macrocarpaeae Grisebach (ex Gentianaceis) speciebus novis II: typification of the Ruiz & Pavon names. Harvard Pap Bot 7(2):423–436

Grant JR (2004) De Macrocarpaeae Grisebach (ex Gentianaceis) speciebus novis V: twenty-three new species largely from Peru, and typification of all species in the genus. Harvard Pap Bot 9:11–49

Grant JR (2005) De Macrocarpaeae Grisebach (ex Gentianaceis) speciebus novis VI: seed morphology, palynology, an infrageneric classification, and another twenty-three new species, largely from Colombia. Harvard Pap Bot 9:305–342

Grant JR (2007) De Macrocarpaeae Grisebach (ex Gentianaceis) speciebus novis VII: four new species and two natural hybrids. Harvard Pap Bot 11:129–139

Grant JR (2008) De Macrocarpaeae Grisebach (ex Gentianaceis) speciebus novis VIII: two new species from Ecuador. Harvard Pap Bot 13:253–259

Grant JR (2011) De Macrocarpaeae Grisebach (ex Gentianaceis) speciebus novis IX: a synopsis of the genus in Bolivia. Harvard Pap Bot 16:389–397

Grant JR, Struwe L (2001) De Macrocarpaeae Grisebach (ex Gentianaceis) speciebus novis I: an introduction to the genus *Macrocarpaea* and three new species from Colombia, Ecuador, and Guyana. Harvard Pap Bot 5:489–498

Grant JR, Struwe L (2003) De Macrocarpaeae Grisebach (ex Gentianaceis) speciebus novis III: six new species of moon-gentians (*Macrocarpaea*, Gentianaceae: Helieae) from P.N. Podocarpus, Ecuador. Harvard Pap Bot 8:61–81

Grant JR, Trunz V (2011) De Macrocarpaeae Grisebach (ex Gentianaceis) speciebus novis X: a synopsis of the genus in montane Atlantic Forests of Brazil. Harvard Pap Bot 16:399–420

Grant JR, Weaver RE (2003) De Macrocarpaeae Grisebach (ex Gentianaceis) speciebus novis IV: twelve new species of *Macrocarpaea* (Gentianaceae: Helieae) from Central and South America, and the first report of the presence of a stipule in the family. Harvard Pap Bot 8:83–109

Matt F (2001) Papel ecológico de los murciélagos y orthópteros en el bosque montañoso del PN Podocarpus. Informe parcial parte I: Murcielagos y relaciones con plantas: Propagación de las semillas y polinización. http://www.natureandculture.org/GermanResearchGroups/papers/AG%20Helversen/1%20Murci%E9lagos%20y%20sus%20plantas.pdf

Nilsson S (1968) Pollen morphology in the genus *Macrocarpaea* (Gentianaceae) and its taxonomical significance. Svensk Bot Tidskr 62:338–364

Nilsson S (2002) Gentianaceae: a review of palynology. In: Struwe L, Albert VA (eds) Gentianaceae: systematics and natural history. Cambridge University Press, Cambridge, pp 377–497

Pringle JS (1995) Gentianaceae. In: Harling G, Andersson L (eds) Flora of Ecuador, vol 53. Göteborg university, Sweden, pp 1–132

Struwe LS, Haag S, Heiberg E, Grant JR (2009a) Andean speciation and vicariance in Neotropical *Macrocarpaea* (Gentianaceae-Helieae). Ann Missouri Bot Gard 96:450–469

Struwe L, Albert VA, Calió MF, Frasier C, Lepis KB, Mathews KG, Grant JR (2009b) Evolutionary patterns in neotropical tribe Helieae (Gentianaceae): evidence from morphology, chloroplast and nuclear DNA sequences. Taxon 58:479–499

Thiers B (2008+) [continuously updated]: Index herbariorum: a global directory of public herbaria and associated staff. New York Botanical Garden: published at http://sweetgum.nybg.org/ih/

Thiv M, Grant JR (2002) *Macrocarpaea* (Gentianaceae). In: Flora de la República de Cuba, Fasc Koeltz Scientific Books, Konigstein, Germany,vol 6. pp 23–26

Weigend M (2002) Observations on the biogeography of the Amotape–Huancabamba zone in northern Peru. Bot Rev 68:38–54

Weigend M (2004) Additional observations on the biogeography of the Amotape–Huancabamba zone in northern Peru: defining the south-eastern limits. Rev Peru Biol 11:127–134

Young KR, Reynel C (1997) Huancabamba region: Peru and Ecuador. In: Davis SD, Heywood VH, Herrera-MacBryde O, Villa-Lobos J, Hamilton A (eds) Centres of plant diversity: a guide and strategy for their conservation, vol 3. The Americas IUCN Publications Unit, Cambridge, pp 465–469

Chapter 4
A Review of the Taxonomy and Distribution of the Gentianaceae in the Ukraine

Natalia M. Shiyan

Abstract This chapter represents the author's revision of the family Gentianaceae in the flora of the Ukraine. The review includes information on 21 species with 11 subspecies, and 3 varieties belonging to 7 genera. Two new nomenclatural combinations [*Schenkia spicata* (L.) Mans. var. *tamanica* (Artemczuk) Shiyan, *S. spicata* var. *ramossisima* (Artemczuk) Shiyan] are validated. The checklist of Ukrainian Gentianaceae is provided. Detailed nomenclature citations, synonyms, protologue data, information about the plant type and its location, total distribution in the world, distribution in the Ukraine, and notes are given for each species, subspecies, and variety.

4.1 Introduction

The Ukraine is the second largest country in Europe, following the European part of the Russian Federation. Although occupying less than 6 % of Europe (total surface area is 603,550 km^2), the Ukraine has about 35 % of Europe's biodiversity, in terms of taxonomic groups. This is due to the fact that the Ukraine is located at the crossroads of several natural areas and migration routes of many species of flora and fauna. The Ukraine is home to nearly 70,000 species, including 27,000 species of plants of which 5,100 species are vascular plants and fungi, and more than 45,000 species of animals. There are two major centers of biodiversity and endemism in the Ukraine, namely the Carpathian Mountains and the Crimea. The majority of vascular plant species are concentrated in these hotspots, about 2,000 and 2,250 species, respectively (Mosyakin and Fedoronchuk 1999; Didukh 2009). Whereas the whole of the Crimea is located in the Ukraine, only 11 % of

N. M. Shiyan (✉)
M.G. Kholodny Institute of Botany, National Academy of Sciences of Ukraine, 2 Tereshchenkivska Street, Kyiv-1 01601, Ukraine
e-mail: herbarium_kw@ukr.net

J. J. Rybczyński et al. (eds.), *The Gentianaceae - Volume 1: Characterization and Ecology*, DOI: 10.1007/978-3-642-54010-3_4,

the Carpathian Mountains are in the Ukraine; 49.9 % of the total Carpathian flora (1997 species) and 39.1 % of the flora of the Ukraine is located in this region (Mosyakin and Fedoronchuk 1999).

The basic information about the species of Gentianaceae within the present-day Ukrainian territory can be found in some floristic monographs of the end of the eighteenth and the nineteenth centuries. There are works by P.S. Pallas, F.A. Marschall von Bieberstein, Ch. Steven, W.S.J.G. Besser, K.F. Ledebour, A.L. Andrzejowski, F. Herbich, F. Schur, I.F. Schmalhausen, J.K. Paczoski, and others. Numerous specimens of Gentianaceae collected by these botanists are present in herbaria of the Ukraine and some other countries. The oldest known specimens of Gentianaceae from the Ukraine were collected by V.M. Czernjaev in 1819 (*Gentiana cruciata* L. from Kharkov Region, KW 00038093) and by F. Herbich in 1823 [*Centaurium pulchellum* (Sw.) Druce from Galicia, LW 030723]. In the nineteenth century, botanists usually reported only a few genera and species of the family in the territory of the Ukraine, such as *Erythraea* Rich. (*E. centaurium* Pers., *E. pulchella* Pers., *E. linarifolia* Pers, *E. spicata* Pers.), four species of *Gentiana* L. (*G. pneumonanthe* L., *G. cruciata, G. ciliata* L., *G. amarella* L.), and one species of *Swertia* L. (*S. perennis* L.). However, some species were reported erroneously for the Ukraine (*G. douglasiana* Bong., *G. septemfida* Pall.) and some taxa were described as new ones, for example, *G. bucoviensis* Herb., *G. carpatica* Kit., *G. carpatica* Wettst., *G. laciniata* Kit. ex Kanitz, *G. lancifolia* Besser, *G. wagiana* Andrz. ex Trautv and others (Fedoronchuk et al. 2004).

The first comprehensive taxonomic revision of the Gentianaceae of the Ukraine was made by Wissjulina (1957) in the *Flora of the Ukrainian SSR*, which followed the monographic overview of the family by Grossheim (1952) of the *Flora of the URSS*. In these publications, 26 species of 4 genera were reported (*Blackstonia* Huds., *Centaurium* Hill., *Gentiana*, and *Swertia*) in the flora of the Ukraine. The genus *Centaurium* was subdivided into two sections, namely *Spicaria* Griseb. and *Euerythraea* Griseb.; two subgenera were recognized in the genus *Gentiana* L., subgen. *Eugentiana* Kusn. (with Sections *Coelanthe* DC., *Pneumonanthe* Neck., *Frigida* Kusn., *Aptera* Kusn., *Chondrophylla* Bunge, *Thylacites* Griseb., *Cyclostigma* Griseb.) and the subgen. *Gentianella* (Moench) Kusn. (with sections *Crossopetalum* Froel. and *Endotricha* Froel.). The revision of A.A. Grossheim had a strong taxonomic emphasis (a detailed system of Gentianaceae was proposed), while the work of O.D. Wissjulina focused on details of species distribution in the Ukraine and the development of a new handy key. In the following years, both works were considered the starting points for subsequent Field Guides and local Floras (Privalova 1957; Zerow 1965; Rubtsov 1972; Fodor 1974; Chopyk 1977).

Discussion in *Flora Europea* (Tutin 1972) used the data of main publications dealing with the territory of the Ukraine. As a result, 21 species and 3 subspecies of 5 genera were reported for this region. Some of them, like *Gentianella tenella* (Rottb.) Börner, were mentioned for the first time for the Ukraine; others, such as *Centaurium uliginosum* (Waldst. et Kit.) Fritsch, were interpreted as subspecies, while others were mentioned only in notes. The revision of the Gentianaceae in

Flora Europaea by T.G. Tutin and co-authors was the main base for Tzvelev's (1978) treatment of the family for *Flora of the European Part of the URSS*. Until quite recently, this work was the latest cumulative source of taxonomic information about Gentianaceae in the flora of the Ukraine. In comparison with other treatments, the study by N.N. Tzvelev contained a critical revision of the systematics of the family, with new nomenclatural combinations, some taxonomic changes and with a bias to recognize subspecies and hybrids. As a result, N.N. Tzvelev listed in the flora of the Ukraine 27 species and 8 subspecies belonging to 7 genera.

During the past 20 years, a wealth of information accumulated about the Ukrainian flora (Mosyakin and Fedoronchuk 1999; Shiyan and Dzhus 2005; Ostapko et al. 2010; Ena 2012), and 8 species of Gentianaceae were included in the third edition of the *Red Data Book of Ukraine* (Didukh 2009). The new phylogenetic studies, combining molecular and traditional morphological data, resulted in changes in the classification schemes for the family and some of its genera (Struwe and Albert 2002). The current revision is aimed at the species composition, and the distribution of Gentianaceae in the Ukraine, based on new taxonomic sources. For the taxa that are considered threatened and of conservation concern, data are provided on their conservation status according to the *Red Data Book of Ukraine* (Didukh 2009).

4.2 Procedures Employed in Revising the Taxonomy of the Gentianaceae of the Ukraine

In order to generalize the data on taxonomy and biogeography of the Gentianaceae of the flora of the Ukraine, herbarium materials were explored from CAI, CHER, CSAU, CWU, DNZ, KRA, KRAM, KW, LE, LW, LWKS, LWS, MELIT, MSUD, PHEO, UU, YALT, and the Herbarium of Mykola Gogol State University of Nizhyn. For studies of the current taxonomic composition of the Gentianaceae of the flora of the Ukraine and the distribution of taxa, the BRAHMS database *Gentianaceae of the Flora of Ukraine* was created, with data of 3011 specimens. Records of the distribution of the taxa investigated in the Ukraine, and the following abbreviations for 27 administrative regions of the area namely Cherkasy (Chk), Chernihiv (Chr), Chernivtsi (Chn), Dnipropetrovsk (Dp), Donetsk (Dn), Ivano-Frankivsk (Iv), Kharkiv (Khr), Kherson (Khs), Khmelnytskyi (Khm), Kyiv (Kv), Kirovohrad (Kr), Luhansk (Lh), Lviv (Lv), Mykolaiv (Mk), Odessa (Od), Poltava (Pl), Rivne (Rv), Sumy (Sm), Ternopil (Tr), Vinnytsia (Vn), Volyn (Vl), Zakarpattia (Zk), Zaporizhia (Zp), Zhytomy (Zht), the Autonomous Republic of Crimea (Cr), and cities with the special status such as Kyiv (KvC) and Sevastopol (SvC), are provided in this review.

4.3 Checklist of the Gentianaceae of the Flora of Ukraine

4.3.1 Genus Blackstonia Huds.

Fl. Angl.: 146. 1762.—*Chlora* L., Mant. Pl.: 10. 1767.—*Seguiera* Manetti, Virid. Fl.: 81. 1751, ex Kuntze, Rev. Gen. Pl.: 430. 1891.

Blackstonia perfoliata (L.) Huds., Fl. Angl.: 146. 1762; Grossh., Fl. SSSR, 18: 536. 1952 [in Russ.]; Wissjul., Fl. URSR, 8: 226. 1957 [in Ukr.]; Ţopa, Fl. Roman. 8: 436. 1961; Schiman-Czeika in Rechinger, Fl. Iranica, 41: 75. 1967; Privalova in Wulf, Fl. Taurica, 3, 1: 41. 1957 [in Russ.]; Tutin, Fl. Europ. 3: 56. 1972; Edmondson in Davis, Fl. Turkey, 6: 177, 1978; Tzvelev, Fl. Europ. p. SSSR, 3: 64. 1978 [in Russ.]; Kožuharov et Petrova, Fl. Narodna Republ. Bulg. 8: 395. 1982 [in Bulg.]; Z. Díaz, Fl. Iber. 11: 94. 2012; *Gentiana perfoliata* L., Sp. Pl.: 232. 1753. *Chlora perfoliata* (L.) Willd., Sp. Pl. 2, 1: 340. 1799; Ledeb., Fl. Ross. 3: 52. 1847–1849; Schmalh., Fl. Centr. & S. Russia 2: 216. 1897; *Seguiera perfoliata* (L.) Kuntze, Revis. Gent. Pl. 2: 430. 1891.

Ind. col.: "Habitat in Gallia, Anglia, Hispania."
Type material: Type not designated (probably in BM or LINN?).
Total distribution: Atlantic, Mediterranean, C. and W. Europe, N. Africa, Asia Minor, Iraq. Introducted into Argentiana and Australia.
Distribution in the Ukraine: Cr.
Notes. *B. perfoliata* was reported in the Ukraine from the Crimea in the nineteenth century as growing in wet areas near streams and on the Southern Coast of Crimea between Alupka and Alushta (Ledebour 1847–1849; Privalova 1957; Tzvelev 1978). Botanists collected the species regularly between 1824 and 1953 and even in sufficient quantity to be published in *Herbarium Florae Rossicae* (No. 375, K. Golde, 1898). *B. perfoliata* has not been found in the Crimea for a long time, and only in August 2013 it was found in Baydar valley (Fateryga et al. 2013).

In the literature (Fodor 1974; Tzvelev 1978; Mosyakin and Fedoronchuk 1999), there is mention that *B. acuminata* (W.D.J. Koch et Ziz) Domin also occurs in the Ukrainian territory. The source of it was the monograph *Flora Trancarpathica* (Fodor 1974), where the author reported the species from the neighborhoods of the Wilok (in Romanian, Tiszaújlak), Transcarpathian Region, based on the publication of Domin (1933). However, no specific records indicating unambiguously the species from the Ukraine are present in the cited publication of Domin (1933). Specimens are not known from the Ukrainian territory in herbaria. Therefore, this record of *B. acuminata* from the Ukraine is probably erroneous.

4.3.2 Genus Centaurium Hill

Brit. Herb.: 62. 1756, emend.; *Erythraea* Borkh., in Roem., Arch. Bot. 1, 1: 30. 1796.

Centaurium erythraea Rafn, Danm. Holst. Fl. 2: 75. 1800; Melderis, Fl. Europ. 3: 57. 1972; Jacobsen in Davis, Fl. Turkey, 6: 178. 1978; Tzvelev, Fl. Europ. p. SSSR, 3: 62. 1978 [in Russ.]; Avetisian, Fl. Armenia, 7: 32. 1980 [in Russ.]; *Gentiana centaurium* L., Sp. Pl.: 299. 1753, p.p.; *Centaurium umbellatum* Gilib., Fl. Lithuan. 1: 135. 1781, nom. invalid.; Grossh., Fl. SSSR, 18: 533. 1952 [in Russ.]; Szafer et al., Rośliny Polskie: 599. 1953; Privalova in Wulf, Fl. Taurica, 3, 1: 38. 1957 [in Russ.]; Ţopa, Fl. Roman. 8: 441. 1961; Kožuharov et Petrova, Fl. Narodna Republ. Bulg. 8: 390. 1982; Z. Díaz, Fl. Iber. 11: 55. 2012; *Chironia centaurium* Curt., F1. Lond., 2, 4: t. 22. 1782, p.p.; Moench, Meth. Pl.: 449. 1794; Wissjul., Fl. URSR, 8: 223. 1957 [in Ukr.]; *Chironia centaurium* (L.) F.W. Schmidt, Fl. Bohem. 2: 31. 1794; *Erythraea centaurium* (L.) Borkh. in Roem., Arch. Bot. 1, 1: 30. 1796; *E. centaurium* auct. non Pers. in Ledeb., Fl. Ross. 3, 1: 49. 1847–1849; Herbich, Fl. v. Bucovina: 217. 1859; Schmalh., Fl. S.-W. Russia: 394. 1886; Druce, Fl. Berks.: 342. 1898, p.p. nom. inval.

Ind. col.: "Jeg har fundet den paa Bornholm ved Boberne, og, im jeg minded ret, ved Rønne, paa høje Steder i tør Jord" (Denmark).
Type material: Type not designated (probably in BM or LINN?).

(a) **Subsp.** ***erythraea***

Total distribution: Europe (excluding northen parts of Scandinavia and Russia), Mediterranean (including islands), northwest Africa and Libya (?), Caucasus, W. Asia, Persia, Afghanistan (?), W. Siberia (Altai). Naturalized in North America (Canada, USA, Costa Rica, Jamaica) and South America (Ecuador, Peru, Argentina), Australia, and New Zealand, and in some islands: Azores (Portugal), Canaries (Spain), Hawaii, and Midway (USA).
Distribution in the Ukraine: All regions.
Notes. Grows commonly in all regions of the Ukraine. In the Carpathian Mountains, the species occurs up to the lower forest zone.

(b) **Subsp.** ***turcicum*** (Velen.) Melderis. Bot. J. Linn. Soc. 65, 2: 232. 1972; id. Fl. Europ. 3: 57. 1972; Jacobsen in Davis, Fl. Turkey, 6: 178. 1978; Tzvelev, Fl. Europ. p. SSSR, 3: 62. 1978 [in Russ.]; *Erythraea turcica* Velen. Fl. Bulg.: 384. 1891; *Centaurium turcicum* (Velen.) Bornm. Verh. K.K. Zool.-Bot. Ges. Wien 60: 150. 1910, nom. nud.; *C. turcicum* (Velen.) Ronniger, Mitt. Naturw. Ver. Steierm. 52: 320. 1916; Druce, Rep. Bot. Exch. Club Brit. Isles, 1916: 614. 1917; Hayek, Prodr. Fl. Penins. Balc. 2: 415. 1930; Kožuharov et Petrova, Fl. Narodna Republ. Bulg. 8: 391. 1982; *C. umbellatum* Gilib. subsp. *austriacum* Ronniger f. *turcicum* (Velen.) Egey, Ind. Horti. Bot. Budapest: 16. 1932; Ţopa, Fl. Roman. 8: 442. 1961; *C. minus* Moenh subsp. *turcicum* (Velen.) Soó, in Soó et Jav., Magyar Növ. Kéz. 1: 480. 1951; *C. umbellatum* Gilib. var. *turcicum* (Velen.) Stoj. Stef. et Kitan., Fl. Bulg. 2: 845. 1967.

Ind. col.: "In graminosis prope Varna, Sliven, Gorna Banja" [Bulgaria].
Type material: Type not designated (probably in PRC?).
Total distribution: Balkans and East Aegean Islands, Black Sea region (Ukraine, southern part; Russian Federation, Krasnodar Krai; N-E. Turkey), Asia Minor, Lebanon, Syria (?), N. Iraq, N. and E. Iran, Caucasus (Georgia, Armenia, Azerbaijan), Afghanistan, Pakistan (?).
Distribution in the Ukraine: Cr (including SvC), Dn, Khs, Mk, Od, Zp.
Notes. The species occurs occasionally in southern regions of continental Ukraine and in the Crimea. Tzvelev considered *Erythraea turcica* Velen. as a later synonym of *E. anatolica* Koch (Linnaea, 19: 27. 1846) and used the first name as the basionym of *Centaurium anatolicum* (Koch) Tzvelev (Tzvelev 1993). However, in the original description, Koch considered *E. anatolica* as a species closely related to *E. caspica* Fisch.: "*E. caspica* Fisch. unterscheidet sich durch die dem Kelche gleiche Kronröhre". In *Beiträge zu einer Flora des Orientes* (1850), Koch provided the second and more detailed description of the species and pointed out that the cyme and calyx are "puberulo-canescens." In the article, the author first indicated the type of species, specifically, "Von Dr. Thirke aus der Umgebung von Brussa erhalten" (Anatolia, Turkey). Consequently, the names *E. turcica* and *E. anatolica* are most probably not conspecific, and the plants occurring in the Southern part of Ukraine are not identical with *Centaurium anatolicum*. Probably the (holo) type and other authentic materials of Koch (collections of 1836–1844) are incorporated in the Herbarium Berolinense (B). Grossheim (1952), Schiman-Czeika (1967), Jacobsen, (1978) and others treated *C. anatolica* and *C. caspica* as synonyms of *C. pulchellum* (Sw.) Druce.

Centaurium littorale (D. Turner) Gilmour subsp. *uliginosum* (Waldst. et Kit.) Rothm. ex Melderis. Bot. J. Linn. Soc. 65, 2: 241. 1972; Rothm., Exkursionsfl. Deutschl., Krit. Ergänz.: 250. 1963, comb. invalid.; Melderis, Fl. Europ. 3: 58. 1972; Tzvelev, Fl. Europ. p. SSSR, 3: 60. 1978 [in Russ.]; *Chironia uliginosa* Waldst. et Kit., Descr. Icon. Pl. Rar. Hung. 3: 287, tab. 259. 1809; *Erythraea uliginosa* (Waldst. et Kit.) Roemer et Schultes in L., Syst. Veg., ed. 2, 4: 168. 1819; *E. linarifolia* auct. non Pers. in Ledeb., Fl. Ross. 3, 1: 50. 1847–1849; Schmalh., Fl. S.-W. Russia: 394. 1886; *Centaurium uliginosum* (Waldst. et Kit.) G.Beck ex Ronniger, Mitt. Naturw. Ver. Steierm. 52: 320. 1916; Grossh., Fl. SSSR, 18: 534. 1952 [in Russ.]; Wissjul., Fl. URSR, 8: 235. 1957 [in Ukr.]; Ţopa, Fl. Roman. 8: 440. 1961; *C. vulgare* Raf. subsp. *uliginosum* (Waldst. et Kit.) Soó, in Soó et Jav., Magyar Növ. Kéz. 1: 480. 1951; Fodor, Fl. Transcarpath.: 105. 1974.

Ind. col.: "...in pratis et pascuis humidis, uliginosis" [Hungary].
Type material: Type not designated.
Total distribution: Central and Western Europe, Ciscaucasia.
Distribution in the Ukraine: Chk, Chr, Khr, Kv, KvS, Pl, Sm.
Notes. The subsp. *littorale* is little-known in the Ukraine as compared to the subsp. *uliginosum*, which grows in most parts of the country and is well defined by papillose-scabrid stems, leaves, and calyx. Some specimens identified as

C. littorale from YALT were collected a longtime ago and their identity is doubtful. Perhaps this taxon grew on sandy areas of the coast of the Black Sea of Crimea, where it is now absent. Four specimens listed in the *Flora of The Ukrainian SSR* (1957) are considered as *C. vulgare* Raf. and belong to the subsp. *uliginosum*. This subspecies grows occasionally in wet alkaline meadows, forest glades, edges of swamps, and sometimes on sands all over the northeastern part of the Ukraine.

Centaurium pulchellum (Sw.) Druce, Fl. Berks.: 342. 1898; Grossh., Fl. SSSR, 18: 528. 1952 [in Russ.]; Wissjul., Fl. URSR, 8: 230. 1957 [in Ukr.]; Szafer et al., Rośliny Polskie: 599. 1953; Privalova in Wulf, Fl. Taurica, 3, 1: 36. 1957 [in Russ.]; Ţopa, Fl. Roman. 8: 440. 1961; Schiman-Czeika in Rechinger, Fl. Iranica, 41: 5. 1967; Melderis, Fl. Europ. 3: 59. 1972; Jacobsen in Davis, Fl. Turkey, 6: 180. 1978; Tzvelev, Fl. Europ. p. SSSR, 3: 62. 1978 [in Russ.]; Avetisian, Fl. Armenia, 7: 30. 1980 [in Russ.]; Kožuharov et Petrova, Fl. Narodna Republ. Bulg. 8: 392. 1982; Z. Díaz Fl. Iber. 11: 75. 2012; *Gentiana centaurium* L., Sp. Pl.: 229, quoad var. *β* et *γ*. 1753; *G. pulchella* Sw. in Kongl. Vetensk. Akad. Nya Handl., 4: 85, t. 3. ff. 8, 9. 1783; *G. ramosissima* Vill. Fl. Delphinalis: 23. 1785; *Erythraea ramosissima* (Vill.) Pers., Syn. Pl., 1: 283. 1805; *E. pulchella* (Sw.) Hornem., Fl. Dan. 10: tab. 1637. 1809; *E. pulchella* (Sw.) Fr., Nov. Fl. Suec.: 30. 1814; Ledeb., Fl. Ross. 3, 1: 50. 1847–1849; Schmalh., Fl. S.-W. Russia: 394. 1886; Herbich, Fl. v. Bucovina: 217.1859; *Centaurium pulchellum* subsp. *ramosissimum* (Vill.). P. Fourn., Quatre Fl. Fr.: 855. 1938.

Ind. col.: Described from Sweden.
Type material: Type not designated (probably in S?).

(a) **Subsp. *pulchellum***

Total distribution: Europe (including islands), North Africa, Western, and Central Asia.
Distribution in the Ukraine: Chk, Chr, Chn, Cr, Dp, Dn, Iv, Khr, Khs, Khm, Kv included KvS, Kr, Lh, Lv, Mk, Od, Pl, Rv, Sm, Vn, Vl, Zk, Zp, Zht.
Notes. Subsp. *pulchellum* occurs occasionally in open wet saline meadows, salt marshes, stony ground and rocks, on shores of rivers, streams and seas, sometimes within settled areas all over the Ukraine, except the Carpathian Mountains.

(b) **Subsp. *meyeri* (Bunge) Tzvelev**, Fl. Europ. p. SSSR, 3: 63. 1978 [in Russ.]; *Erythraea meyeri* Bunge in Ledeb. Fl. Alt. 1: 220. 1829; *E. ramosissima* Pers. *γ*. *altaica* Griseb. in DC., Prodr. 9: 57. 1845; *E. pulchellum β. albiflora* Ledeb., Fl. Ross. 3, 1: 51. 1847–1849; *E. ramosissima* Pers. var. *albiflora* Boiss., Fl. Or. 4: 67. 1879; *Centaurium meyeri* (Bunge) Druce, Rep. Bot. Exch. Cl. Brit. Isles, 1916: 613. 1917; Grossh., Fl. SSSR, 18: 529. 1952 [in Russ.]; Wissjul., Fl. URSR, 8: 231. 1957 [in Ukr.]; Schiman-Czeika in Rechinger, Fl. Iranica, 41: 6. 1967; Jacobsen in Davis, Fl. Turkey, 6: 178. 1978, in nota; *C. pulchellum* var. *altaicum* (Griseb.) Kitag. Et Hare, 1937, Journ. Jap. Bot. 13: 26. 1937; *C. pulchellum* var. *albiflora* (Ledeb.) Privalova in Wulf, Fl.

Taurica, 3, 1: 37. 1957; *C. pulchellum* f. *meyeri* (Bunge) Em. Ţopa, Fl. Roman. 8: 441. 1961; *C. pulchellum* var. *meyeri* (Bunge) Omer, Fl. Pakistan 197: 10. 1995.

Ind. col.: "Hab. In pratis humidiusculis ad fl. Irtysch prope Ablaikit et alibi (M.)" [Kazakhstan].
Type material: Type not designated (probably in LE in the collections of Meyer from Middle Asia).
Total distribution: C. and E. Europe (Romania, Moldova, Ukraine, Russia), N. Africa (Egypt), Anatolia, Caucasus, W. and C. Asia, Iran, Afghanistan, W. Siberia (Altai), E. Siberia (south).
Distribution in the Ukraine: Cr, Dp, Dn, Khr, Khs, Kr, Lh, Mk, Zp, Zht.
Notes. The taxon occurs quite rarely in the southeastern part of the Ukraine, and in the Crimea, on open saline meadows and the banks of water.

Centaurium tenuiflorum (Hoffmanns. et Link) Fritsch in Mitt. Naturw. Ver. Wien. 5: 97. 1907; Grossh., Fl. SSSR, 18: 530. 1952 [in Russ.]; Wissjul., Fl. URSR, 8: 232. 1957 [in Ukr.]; Schiman-Czeika in Rechinger, Fl. Iranica, 41: 6. 1967; Jacobsen in Davis, Fl. Turkey, 6: 181. 1978; Tzvelev, Fl. Europ. p. SSSR, 3: 62. 1978 [in Russ.]; Privalova in Wulf, Fl. Taurica, 3, 1: 38. 1957 [in Russ.]; Melderis, Fl. Europ. 3: 59. 1972; *Erythraea tenuiflora* Hoffmanns. et Link, 1813–1820, Fl. Portug. 1: 354, tab. 67. 1813–1820; *E. pulchella* subsp. *tenuiflora* (Hoffmanns. et Link) Rohlena in *Sitz.-Ber. Bohm. Ges. Wzss.,* 1912: 88. 1912; *Centaurium pulchellum* subsp. *tenuiflora* (Hoffmanns. et Link), Maire, 1928, Mém. Soc. Sci. Nat. Maroc. 17: 41.

Ind. col.: "sur les bonds de in mer, pris d' A. Costa" [Portugal].
Type material: Type not designated.
Total distribution: N–W. Africa, Atlantic Europe, Mediterranean, Anatolia, Crimea, W. Asia.
Distribution in the Ukraine: Cr.
Notes. *C. tenuiflorum* grows occasionally on wet coastal sands of the Black Sea in the Crimea. K. Jacobsen in the *Flora of Turkey* (l.c., 1978) reported for the Crimea, *C. tenuiflorum* subsp. *acutiflorum* (Shott) Zeltner. This taxon has not been found among the specimens investigated. Probably only the typical subspecies occurs in the Ukraine.

4.3.3 Genus Schenkia Griseb.

Bonplandia 1: 226. 1853; G. Mans., Taxon 53 (3): 724. 2004; *Erythraea* section *Spicaria* Griseb., Genera and species Gentianearum: 147. 1839; *Centaurium* section *Spicaria* (Griseb.) Ronniger, Mitt. Naturw. Ver. Steierm. 52: 312–321. 1916.

Schenkia spicata (L.) G. Mans., Taxon, 53, 3: 726. 2004; *Gentiana spicata* L., Sp. Pl.: 230. 1753, *Hippion spicatum* (L.) F. W. Schmidt, Roem. Arch. 1: 11. 1796;

Chironia spicata (L.) Willd., Sp. Pl. 1. 1: 1069. 1797; *Erythraea spicata* (L.) Pers. 1805, Syn. Pl. 1: 283; Ledeb., Fl. Ross. 3, 1: 51. 1847–1949; *E. babylonica* Griseb. in DC, Prodr. 9: 60. 1845; *Centaurium spicatum* (L.) Fritsch., Mitt. Naturw. Ver. Wien, 5: 97. 1907; Grossh. Fl. SSSR 18: 535. 1952, [in Russ.]; Vissijul. Fl. URSR, 8: 229. 1957 [in Ukr.]; Privalova in Wulf, Fl. Taurica, 3, 1: 40. 1957 [in Russ.]; Ţopa, Fl. Roman. 8: 439. 1961; Schiman-Czeika in Rechinger, Fl. Iranica, 41: 7. 1967; Melderis, Fl. Europ. 3: 59. 1972; Jacobsen in Davis, Fl. Turkey, 6: 182. 1978; Tzvelev, Fl. Europ. p. SSSR, 3: 64. 1978 [in Russ.]; Avetisian, Fl. Armenia, 7: 30. 1980 [in Russ.]; Kožuharov et Petrova, Fl. Narodna Republ. Bulg. 8: 394. 1982.

Ind. col.: "in solo arenoso humidoque stratorum rupium altarum orae maritimae, loco dicto Praia das Maças" [Portugal].
Holotype: "Portugal, "in solo arenoso humidoque stratorum rupium altarum orae maritimae, loco dicto Praia das Maças", 2 Sep 1961, *J. Matos, A. Marques, M. Alves 8284*" (COI) (Mansion 2004).

1. **Var.** ***spicata***

Total distribution: Mediterranean (from N. Portugal to Greece and Bulgaria; from S. Morocco to Libya), including the islands, S. and E. Europe (south of Ukraine and Russia), Caucasus, W. Asia (Turkey, Syria, Jordan), Iran, Afghanistan. Naturalized in North America and Australia.
Distribution in the Ukraine: Cr, Dn, Khs, Mk, Od, Zp.
Notes: *S. spicata* grows rarely on wetlands, on the banks of saline lakes and estuaries, and salt marshes in the coastal regions of the Ukraine and the Crimea.

2. **Var.** ***tamanica*** (Artemczuk) Shiyan, comb. nov. = *Erythraea spicata* (L.) Pers. var. *tamanica* Artemczuk, J. Bot. Acad. Sci. Ukraine, 1, 1: 136. 1940.

This variety differs from the typical one in having white flowers in dense almost one-sided spiciform cymes.

Ind. col.: "Habitat in salsis et salsuginosis humidi peninsulae Tamanicae [Tuzla Spit]" [Ukraine].
Holotype: "Taman Peninsula, Tuzla Spit, 21.VII 1936 *I.V. Artemchuk*" (KW 001 02880!; isotypes: KW 00102879!, KW 00102878!)
Total distribution: Coasts of the Sea of Azov and the Black Sea in the Ukraine (including East Crimea) and the Russian Federation (near Anapa, Krasnodar Kraj).
Distribution in the Ukraine: Cr, Dn.
Notes. In the Ukrainian part of the protologue, I.V. Artemczuk specified the locality: "Tuzla Spit, humid marine saline soils and salt marsh, 21.VII 1936 *I.V. Artemchuk*". Now Tuzla is a sandy island in the middle of the Strait of Kerch between the Kerch Peninsula (Ukraine) in the west and the Taman Peninsula (Russian Federation) in the east. On the label of the specimen KW 00102880, I. Artemchuk indicated in Ukrainian: "typus." The taxon occurs rarely in humid marine saline soils and salt marshes along the coasts of the Sea of Azov and Kerch Peninsula (Crimea).

3. **Var.** ***ramosissima*** (Artemczuk) Shiyan, comb. nov. = *Erythraea spicata* (L.) Pers. var. *ramosissima* Artemczuk, J. Bot. Acad. Sci. Ukraine, 1, 1: 136. 1940. This variety differs from the typical one in being very ramose from the base of the stems to 40 cm with subelongated spiciform cymes, with no numerous flowers placed a noticeable distance from each other.

Ind. col.: "Habitat in ora boreali Ponti et Maeotidis" [Ukraine].
Lectotype (N. Shiyan, designated here): "Kherson Reg[ion], Kinburnska Spit, littoral area. 11.X 1925 *A. Prianishnikov* [In Ukr.]" (Lectotype: KW 00096993!; isolectotype: KW 00099283!, paratypes: KW 00096994, KW 00096995, KW 00096998, KW 00097000–KW 00097002!)
Total distribution: On the coast of the Sea of Azov and the Black Sea in the Ukraine (including E. Crimea) and the Russian Federation (Anapa, Krasnodar Kraj).
Distribution in the Ukraine: Cr, Dn, Khs, Mk, Zp.
Notes. In the Ukrainian part of the description, Artemchuk cited the type (In Ukr.: "Kinburnska Spit. Littoral Area. 11.X 1925 *A. Prianishnikov*") and four paratypes. The lectotype designated here is one of the two very similar duplicates of the 'type' with detailed information about the taxon on the label. This variety occurs rarely on humid marine saline soils and salt marshes on the sea coast of the Sea of Azov and the Black Sea, and in N. Crimea.

4.3.4 Genus Gentiana L.

Sp. pl.: 227. 1753; id. Gen. Pl., ed. 5: 107. 1754.

Gentiana lutea L., Sp. Pl.: 227. 1753; Kusn., Subgen. Eugent. gen. Gent.: 9. 1894 [in Russ.]; Grossh., Fl. SSSR, 18: 540. 1952 [in Russ.]; Szafer et al., Rośliny Polskie: 599. 1953; Wissjul., Fl. URSR, 8: 237. 1957 [in Ukr.]; Ţopa, Fl. Roman. 8: 451. 1961; Tutin, Fl. Europ. 3: 60. 1972; Tzvelev, Fl. Europ. p. SSSR, 3: 68. 1978 [in Russ.]; Kožuharov et Petrova, Fl. Narodna Republ. Bulg. 8: 397. 1982; Z. Díaz, Fl. Iber. 11: 10. 2012; *Asterias lutea* (L.) Borkh. in Roem., Arch. Bot. 1, 1: 25. 1796.

Ind. col.: "Habitat in alpibus Norvegicis, Helveticis, Apenninis, Pyrenaeis, Tridentinis" [Europe].
Lectotype: BM: "Herb. Clifford: 80, Gentiana 1." (BM-000558179)[1].
Total distribution: C. and S. Europe, Carpathians, Rhodopes, Asia Minor (Turkey).

---

[1] Image of lectotype see: http://www.nhm.ac.uk/research-curation/research/projects/linnaean-typification/database/detail.dsml?ID=389800&listPageURL=list%2edsml%3fVarqtype%3dstarts%2bwith%26CVarqtype%3dstarts%2bwith%26CGenusqtype%3dstarts%2bwith%26CSpeciesqtype%3dstarts%2bwith%26sort%3dGenus%252cSpecies%26Speciesqtype%3dstarts%2bwith%26Genus%3dGentiana%26Genusqtype%3dstarts%2bwith%26CSspqtype%3dstarts%2bwith.

Distribution in the Ukraine: Iv, Zk.
Conservation status in the Ukraine: Vulnerable.
Notes. The Ukrainian part of the Carpathians is the northeastern limit of the range of *G. lutea*. Here, the species grows in alpine and subalpine zones at altitudes 750–2,000 m above sea level (a.s.l.) mainly on slopes with a southern aspect. In some previously reported localities in the Ukrainian Carpathians, *G. lutea* is now extinct, for example, in the Zeleny Zholob area (Chornohora Range) and Nidea Mt. (Gorgany Range). At present this species is confirmed on the Chornohora and the Svydovets ranges, the Marmarosh Alps and the Polonyna Borzhava.

A hybrid taxon, *G. dorfleri* Ronn. (=*G. lutea* x *G. punctata*), was reported for the territory of the Ukraine (Tzvelev 1978; Mosyakin and Fedoronchuk 1999). The field research and herbarium materials investigated cannot prove the occurrence of this hybrid.

Gentiana punctata L., Sp. Pl.: 227. 1753; Kusn., Subgen. Eugent. gen. Gent.: 16. 1894 [in Russ.]; Grossh., Fl. SSSR, 18: 540. 1952 [in Russ.]; Szafer et al., Rośliny Polskie: 600. 1953; Wissjul., Fl. URSR, 8: 239. 1957 [in Ukr.]; Ţopa, Fl. Roman. 8: 459. 1961; Tutin, Fl. Europ. 3: 60. 1972; Tzvelev, Fl. Europ. p. SSSR, 3: 68. 1978 [in Russ.]; Kožuharov et Petrova, Fl. Narodna Republ. Bulg. 8: 398. 1982.

Ind. col.: “Habitat in Helvetia, Sibiria”.
Type material: Type not designated. (probably in LINN or BM?).
Total distribution: C. Europe, W. Europe (Carpathians), Balkany.
Distribution in the Ukraine: Chn, Iv, Zk.
Conservation status in the Ukraine: Vulnerable.
Notes. The Ukrainian Carpathians are the northeastern limit of the range of *G. punctataa*. Here, the species grows frequently on open alpine and subalpine meadows (*polonyny*) at altitudes of 1,000–2,060 m a.s.l., where it is more frequent than the previous species. *G. punctata* grows in sunny, stony, or clay habitats with acidic or slightly acidic soils and thick snow cover in winter.

Gentiana pneumonanthe L., Sp. Pl.: 228. 1753; Ledeb., Fl. Ross. 3, 1: 66. 1847; Herbich, Fl. v. Bucovina: 214. 1859; Kusn., Subgen. Eugent. gen. Gent.: 82. 1894 [in Russ.]; Grossh., Fl. SSSR, 18: 547. 1952 [in Russ.]; Szafer et al., Rośliny Polskie: 601. 1953; Wissjul., Fl. URSR, 8: 243. 1957 [in Ukr.]; Ţopa, Fl. Roman. 8: 464. 1961; Tutin, Fl. Europ. 3: 61. 1972; Tzvelev, Fl. Europ. p. SSSR, 3: 70. 1978 [in Russ.]; Kožuharov et Petrova, Fl. Narodna Republ. Bulg. 8: 400. 1982; Gagnidze, Fl. Georg., 10: 127. 1985 [In Georg.]; Z. Díaz, Fl. Iber. 11: 33. 2012; *Gentianusa pneumonanthe* Pohl, Tent. Fl. Bohem. 1: 251. 1810; *Pneumonanthe vulgaris* F.W.Schmidt, Arch. Bot. [Leipzig], 1(1): 10. 1796; *Dasystephana pneumonanthe* (L.) Soják, Čas. Nár. Muz. Praze, Rada Přír. 148(3–4): 200. 1980; *D. pneumonanthe* (L.) Zuev, Bot. Zhurn. (Moscow et Leningrad) 75(9): 1300. 1990.

Ind. col.: “Habitat in Europae pascuis humidiusculis.”
Type material: Type not designated (probably in LINN or BM?).

Total distribution: Europe (excluding N. Scandinavia and S. Europe), Caucasus, W. Siberia to Sayan Mts.
Distribution in the Ukraine: Chk, Chr, Chn, Dp, Iv, Khr, Khs, Kv, KvS, Lh, Lv, Od, Pl, Rv, Sm, Tr, Vn, Vl, Zk, Zp, Zht.
Notes. The taxon occurs in meadows, forest clearings, along marshes, rivers and lakes, mostly in the plain part of the Ukraine, except the southern and southeastern regions and the Crimea. Three forms of *G. pneumonanthe* can be distinguished in the Ukraine, based on morphological characteristics, in particular, *G. pneumonanthe* f. *diffusa* Griseb., *G. pneumonanthe* f. *latifolia* Scholler, and *G. pneumonanthe* f. *minor* Brot. Plants of the first form have a stem to 60–70 cm and are very branched above the middle of the plant; those of the second form have a strong stem to 60–65 cm tall and elliptic-lanceolate, 3-5-veined leaves, up to 3–4 cm wide. Both forms are found most commonly in Polissya on humus-rich soils. Forma minor has thin stems 4–15 (20) cm tall, and the inflorescence is usually 1(3)-flowered. This form grows in marshy meadows and in the submountains of the Carpathians to 400 m a.s.l.

Gentiana asclepiadea L. Sp. Pl.: 227. 1753; Ledeb., Fl. Ross. 3, 1: 67. 1847; Herbich, Fl. v. Bucovina: 214.1859; Kusn., Subgen. Eugent. gen. Gent.: 70. 1894 [in Russ.]; Grossh., Fl. SSSR, 18: 543. 1952 [in Russ.]; Szafer et al., Rośliny Polskie: 601. 1953; Wissjul., Fl. URSR, 8: 241. 1957 [in Ukr.]; Ţopa, Fl. Roman. 8: 463. 1961; Tutin, Fl. Europ. 3: 60. 1972; Pritchard in Davis, Fl. Turkey, 6: 184. 1978; Tzvelev, Fl. Europ. p. SSSR, 3: 69. 1978 [in Russ.]; Kožuharov et Petrova, Fl. Narodna Republ. Bulg. 8: 400. 1982; *Dasystephana asclepiadea* (L.) Borkh., Arch. Bot. [Leipzig], 1(1): 26. 1796; *Pneumonanthe asclepiadea* F.W.Schmidt, Arch. Bot. [Leipzig], 1(1): 10. 1796; *P. plicata* F.W.Schmidt, Arch. Bot. [Leipzig], 1(1): 10, 13. 1796; *Coilantha asclepiadea* G.Don, Gen. Hist. 4: 186. 1837.

Ind. col.: "Habitat in alpibus Helvetiae, Pannoniae, Maurtaniae."
Type material: Type not designated (probably in BM, LINN or S-LINN?).
Total distribution: Mountain parts of C. and E. Europe, Balkans, Turkey.
Distribution in the Ukraine: Chn, Iv, Lv, Zk.
Notes. The species is in the submountain region, forests, and subalpine zones of the Ukrainian Carpathians at altitudes of 230–2,050 m a.s.l. In the forest zone of the mountains, it grows along roads and trails in meadows and among oak, hornbeam, linden, and spruce trees.

Gentiana cruciata L. Sp. Pl.: 231. 1753; Ledeb., Fl. Ross. 3, 1: 69. 1847; Herbich, Fl. v. Bucovina: 215. 1859; Kusn., Subgen. Eugent. gen. Gent.: 184. 1894 [in Russ.]; Grossh., Fl. SSSR, 18: 568. 1952 [in Russ.]; Szafer et al., Rośliny Polskie: 601. 1953; Wissjul., Fl. URSR, 8: 245. 1957 [in Ukr.]; Privalova in Wulf, Fl. Taurica, 3, 1: 41. 1957 [in Russ.]; Ţopa, Fl. Roman. 8: 452. 1961; Schiman-Czeika in Rechinger, Fl. Iranica, 41: 12. 1967; Tutin, Fl. Europ. 3: 61. 1972; Pritchard in Davis, Fl. Turkey, 6: 185. 1978; Tzvelev, Fl. Europ. p. SSSR, 3: 69. 1978 [in Russ.]; Avetisian, Fl. Armenia, 7: 37. 1980 [in Russ.]; Kožuharov et Petrova, Fl. Narodna Republ. Bulg. 8: 402. 1982; Gagnidze, Fl. Georg., 10: 134.

1985 [in Georg.]; Z. Díaz, Fl. Iber. 11: 31. 2012; *Hippion cruciatum* F.W.Schmidt, Fl. Boem, 2: 14. 1793; *Ericoila cruciata* Borkh., Arch. Bot. [Leipzig], 1(1): 27. 1796; *Tretorrhiza cruciata* Delarbre, Fl. Auvergne (Delarbre), ed. 2: 24. 1800; *Gentianusa cruciata* Pohl, Tent. Fl. Bohem. 1: 250. 1810; *Ericala cruciata* G.Don, Gen. Hist. 4: 191. 1837; *Pneumonanthe cruciata* (L.) Zuev, Bot. Zhurn. (Moscow et Leningrad), 70, 7: 921. 1985.

Ind. col.: "Habitat in Pannoniae, Apenninorum, Helvetiae montosis apricis."
Type material: Type not designated (probably in BM, LINN or S-LINN?).
Total distribution: C., S., and E. Europe, Balkans, Asia Minor, Caucasus, N. Iran, Talysh, and W. Siberia.
Distribution in the Ukraine: Chk, Chr, Chn, Cr, Dp, Dn, Iv, Khr, Khm, Kv, Lh, Lv, Pl, Sm, Tr, Vn, Zk, Zp, Zht.
Notes. *G. cruciata* occurs sporadically or in groups in forest clearings, among shrubs, in open steppe areas, pastures, river valleys, ravines, and on limestone outcrops almost throughout the Ukraine.

Gentiana acaulis L. Sp. Pl.: 228. 1753; Kusn., Subgen. Eugent. gen. Gent.: 287. 1894 [in Russ.]; Tutin, Fl. Europ. 3: 62. 1972; Tzvelev, Fl. Europ. p. SSSR, 3: 72. 1978 [in Russ.]; Kožuharov et Petrova, Fl. Narodna Republ. Bulg. 8: 404. 1982; Z. Díaz, Fl. Iber. 11: 16. 2012; *Ciminalis acaulis* Borkh., Arch. Bot. [Leipzig], 1, 1: 26. 1796; *Gentiana excisa* C.Presl, Flora (Regensb.), 11: 263. 1828; Grossh., Fl. SSSR, 18: 568. 1952 [in Russ.]; Wissjul., Fl. URSR, 8: 247. 1957 [in Ukr.]; Ţopa, Fl. Roman. 8: 460. 1961; *G. kochiana* Perr. et Song., Bull. Soc. Hist. Nat. Sovoie, 1853: 33. 1855; Szafer et al., Rośliny Polskie: 602. 1953.

Ind. col.: "Habitat in alpibus Helveticis, Austriacis, Pyrenaicis".
Type material: Type not designated (probably in BM or LINN?).
Total distribution: Europe (Pyrenees, Jura, Alps, Apennines, Carpathians, Rhodopes).
Distribution in the Ukraine: Iv, Zk.
Conservation status in the Ukraine: Rare.
Notes. In the Ukrainian Carpathians, the species occurs sometimes in alpine and subalpine zones at altitudes of (600) 900–2,060 m a.s.l. on limestone soils among stones and in rock crevices.

Gentiana laciniata Kit. ex Kanitz, Verh. Zool.-Bot. Ges. Wien, 12: 572. 1862; Grossh., Fl. SSSR, 18: 572. 1952 [in Russ.]; Wissjul., Fl. URSR, 8: 246. 1957 [in Ukr.]; Ţopa, Fl. Roman. 8: 467. 1961; Tzvelev, Fl. Europ. p. SSSR, 3: 70. 1978 [in Russ.]; *G. vagneriana* Janka, Öestr. Bot. Zeitschr., 35: 109. 1885, nom. nud.; *G. pyrenaica* var. *laciniata* (Kit. ex Kanitz) Jav. Shed. Fl. Hung. Exs. 8: № 786. 1927; *G. pyrenaica* auct. non L.: Tutin, Fl. Europ. 3: 61. 1972, pp.

Ind. col.: "In sumniitate alpium Bereghiensium."
Type material: Type not designated (probably in W?).
Total distribution: Easter (Ukrainian) Carpathian Mountains.
Distribution in the Ukraine: Iv, Zk.

Conservation status in the Ukraine: Rare.
Notes. *G. laciniata* is an endemic taxon of the Carpathians. Sometimes it grows in the alpine and subalpine zones at altitudes of 950–2,060 m a.s.l. only in the Chornohora, Svydovets, and Polonina Borzhava Ranges.

Gentiana nivalis L. Sp. Pl.: 228. 1753; Knapp, Pflanz. Galiz. u. Bukov.: 189. 1872; Kusn., Subgen. Eugent. gen. Gent.: 338. 1894 [in Russ.]; Grossh., Fl. SSSR, 18: 588. 1952 [in Russ.]; Szafer et al., Rośliny Polskie: 603. 1953; Wissjul., Fl. URSR, 8: 250. 1957 [in Ukr.]; Ţopa, Fl. Roman. 8: 468. 1961; Tutin, Fl. Europ. 3: 63. 1972; Pritchard in Davis, Fl. Turkey, 6: 190. 1978; Tzvelev, Fl. Europ. p. SSSR, 3: 74. 1978 [in Russ.]; Gagnidze, Fl. Georg., 10: 140. 1985 [In Georg.]; Kožuharov et Petrova, Fl. Narodna Republ. Bulg. 8: 406. 1982; Z. Díaz, Fl. Iber. 11: 22. 2012; *Hippion nivale* (L.) F.W. Schmidt, Fl. Boem. 2: 10, 18. 1794; A. et D. Löve, Acta Hort. Gotoburg. 20, 4: 147. 1956; *Calathiana nivalis* (L.) Delarb., Fl. Auv., ed. 2, 1: 28. 1800.

Ind. col.: "Habitat in Lapponiae, Helvetiae, summis, alpibus."
Type material: Type not designated (probably in BM or LINN?).
Total distribution: North America (Labrador, Greenland), Arctic Europe including Iceland, mountains of C., S., and E. Europe, the Balkans, Asia Minor, W. Caucasus.
Distribution in the Ukraine: Zk.
Conservation status in the Ukraine: Endangered.
Notes. The presence of this species in the flora of the Ukraine has not been confirmed for a long time by herbarium specimens. In 1991, Kardash re-discovered *G. nivalis* in the Ukrainian Carpathians. Now the species is known in the Ukraine only from the Svydovets Range (Shiyan and Dzhus 2005).

Gentiana utriculosa L. Sp. Pl.: 229. 1753; Kusn., Subgen. Eugent. gen. Gent.: 342. 1894 [in Russ.]; Grossh., Fl. SSSR, 18: 589. 1952 [in Russ.]; Wissjul., Fl. URSR, 8: 250. 1957 [in Ukr.]; Ţopa, Fl. Roman. 8: 471. 1961; Tutin, Fl. Europ. 3: 63. 1972; Tzvelev, Fl. Europ. p. SSSR, 3: 74. 1978 [in Russ.]; Kožuharov et Petrova, Fl. Narodna Republ. Bulg. 8: 408. 1982; Z. Díaz, Fl. Iber. 11: 22. 2012; *Hippion utriculosum* F.W.Schmidt Fl. Boem. 2: 22. 1794; *Ericoila utriculosa* Borkh., Arch. Bot. [Leipzig], 1, 1: 27. 1796; *Gentiana bucoviensis* Herbich, Select. Pl. Rar. Galic. Bucov.: 10. 1836.

Ind. col.: "Habitat in Helvetiae, Italiea".
Type material: Type not designated (probably in BM or LINN?).
Total distribution: C., S. and E. Europe, Balkans.
Distribution in the Ukraine: Chn.
Conservation status in the Ukraine: Endangered.
Notes. For a long time, herbarium specimens of *G. utriculosa* were not known from the Ukraine; the species was reported only in the literature. In 2002–2003, it was found in Bukovyna in the vicinity of the village of Sarata in the Putyla District (Chernivtsi Region). Two known populations of the species are protected (Didukh 2009).

Gentiana verna L. Sp. Pl.: 228. 1753; Kusn., Subgen. Eugent. gen. Gent.: 313. 1894 [in Russ.]; Grossh., Fl. SSSR, 18: 587. 1952 [in Russ.]; Szafer et al., Rośliny Polskie: 603. 1953; Wissjul., Fl. URSR, 8: 249. 1957 [in Ukr.]; Ţopa, Fl. Roman. 8: 472. 1961; Tutin, Fl. Europ. 3: 62. 1972; Pritchard in Davis, Fl. Turkey, 6: 178. 1978; Tzvelev, Fl. Europ. p. SSSR, 3: 72. 1978 [in Russ.]; Avetisian, Fl. Armenia, 7: 40. 1980 [in Russ.]; Kožuharov et Petrova, Fl. Narodna Republ. Bulg. 8: 405. 1982; Z. Díaz, Fl. Iber. 11: 24. 2012; *Gentianusa verna* Pohl, Tent. Fl. Bohem. 1: 253. 1810; *Calathiana verna* (L.) Holub, Folia Geobot. Phytotax. 8, 2: 170. 1973.

Ind. col.: "Habitat in alpibus Helvetiae".
Type material: Type not designated (probably in BM or LINN?).
Total distribution: Scotland, C. Europe, mountains of S. and E Europe (Carpatians, Arctic region, N. Urals, Belarus (invasive), Balkans, Asia Minor).
Distribution in the Ukraine: Zk.
Conservation status in the Ukraine: Endangered.
Notes. Arctic-alpine *G. verna* is known in the Ukraine from one habitat near Yasinia (Zakarpatska Region). Among the herbarium material investigated, there are two specimens collected by W. Dzeduszycki in 1861 (LWS 5914) and by S. Turceji in 1878 (LWS 5913) from the Czerwony Werch Range of the Eastern Beskids (Lviv Region). In subsequent years, specimens from this habitat were not collected. This habitat of the species is probably now lost.

4.3.5 Genus Gentianella Moench

Methodus (Moench): 482. 1794; *Gentiana* sect. *Endotricha* Froel., Gentian.: 86. 1796; *Gentiana* subsp. *Gentianella* Kusn., Trudy Imp. S-Peterburgsk. Bot. Sada, 13,4: 58. 1893.

Gentianella amarella (L.) Börner, Fl. Deut. Volk: 543. 1912; Tutin, Fl. Europ. 3: 65. 1972; Tzvelev, Fl. Europ. p. SSSR, 3: 77. 1978 [in Russ.]; Z. Díaz Fl. Iber. 11: 37. 2012; *Gentiana amarella* L., Sp. Pl. 1: 230. 1753; Herbich, Fl. v. Bucovina: 216.1859; Szafer et al., Rośliny Polskie: 606. 1953; Privalova in Wulf, Fl. Taurica, 3, 1: 42. 1957 [in Russ.]; Kožuharov et Petrova, Fl. Narodna Republ. Bulg. 8: 412. 1982; *Hippion amarella* F.W.Schmidt, Fl. Boem. 2: 28. 1793; *Gentiana lancifolia* Besser, Fl. Galic. 1: 192. 1809; *G. amarella γ axillaris* Ledeb., Fl. Ross. 3, 1: 53. 1847; *Gentianella axillaris* (F.W. Schmidt) Murb., Acta Horti Berg. 2, 3: 20. 1892; Wissjul., Fl. URSR, 8: 255. 1957 [in Ukr.]; Grossh., Fl. SSSR, 18: 606. 1952 [in Russ.]; Ţopa, Fl. Roman. 8: 477. 1961; *G. axillaris* (F.W.Schmidt) Á. Löve et D. Löve, Bot. Not. 114: 42. 1961; *G. axillaris* (F.W. Schmidt) Jovan.-Dunj., Fl. SR Srbije, 5: 430. 1973.

Ind. col.: "Habitat in Europae pratis."
Lectotype (Gillett, 1957): "Herb. Linn. No. 328.32, photo, A" (LINN).

(a) **Subsp.** ***amarella***

Total distribution: Temperate and arctic zones of Europe, Asia, and North America.
Distribution in theUkraine: Chn, Cr, Iv, Khr, Kv, Lh, Lv, Sm, Tr, Zk.
Notes. Subsp. *amalella* occurs in the Ukraine in meadows, forest clearings, in the submountain region of the Carpathians and in the Crimean Mountains on meadows and gravelly outcrops. At present, this taxon occurs in the Ukraine more rarely than in the past.

(b) **Subsp.** ***lingulata*** (Agardh) Holub, Folia Geobot. Phytotax. (Praha), 2, 1: 118. 1967; *Gentiana lingulata* Agardh, Lunds Physiogr. Sälsk. Årsberättelse: 29. 1825; Grossh., Fl. SSSR, 18: 605. 1952; *G. amarella γ livonica* Ledeb., Fl. Ross. 3, 1: 53. 1847; *Gentianella lingulata* (Agardh) Pritchard, Watsonia, 4, 4: 178. 1959; *Gentiana amarell*a L. var. *lingulata* (C.Agardh) Ţopa, Fl. Roman. 8: 477. 1961; *G. amarella* L. var. *lingulata* (C. Agardh) Zuev, Fl. Sibiri, 14: 80. 2003.

Ind. col.: "Suecia".
Type material: Type not designated (probably in S?).
Total distribution: W., N. and E. Europe, W. Siberia, E. Siberia (south west), N. Kazakhstan.
Distribution in the Ukraine: Chn, Iv, Lv, Zk.
Notes. Subsp. *lingulata* grows in meadows and in forest clearings from the submountain to the subalpine zone of the Ukrainian Carpathians in grass cover, or on rocks up to 1950 m a.s.l.

Gentianella lutescens .(Velen.) Holub, Folia Geobot. Phytotax. (Praha), 2, 1: 117. 1967; Tutin, Fl. Europ. 3: 66. 1972; Tzvelev, Fl. Europ. p. SSSR, 3: 78. 1978 [in Russ.]; *Gentiana lutescens* Velen., Věstn. Král. České Společn. Nauk, 1888: 29. 1889; *G. praeflorens* Wettst. Öesterr. Bot. Z. 42: 234. 1892; *G. praecox* auct. non A.Kern. et J.Kern. Verh. K.K. Zool.-Bot. Ges. Wien, 38: 669. 1889: Grossh., Fl. SSSR, 18: 602. 1952 [in Russ.]; Ţopa, Fl. Roman. 8: 474. 1961; Wissjul., Fl. URSR, 8: 252. 1957 [in Ukr.]; Kožuharov et Petrova, Fl. Narodna Republ. Bulg. 8: 415. 1982.

(a) **Subsp.** ***lutescens***

Ind. col.: "In graminosis excelsioribus m. Stara Planina prope Petrov Han in solo arenoso" [Bulgaria].
Type material: Type not designated.
Total distribution: C. Europe, Carpathians, Balkans.
Distribution in the Ukraine: Chn, Iv, Lv, Zk.
Notes. Subsp. *lutescens* often grows in meadows and forest clearings from the submountain region to subalpine zones of the Ukrainian Carpathians at 110–2,060 m a.s.l.

(b) **Subsp. *carpatica*** (Hayek) Holub, Folia Geobot. Phytotax. (Praha), 2, 1: 119. 1967; Tutin, Fl. Europ. 3: 67. 1972; Tzvelev, Fl. Europ. p. SSSR, 3: 78. 1978 [in Russ.]; *Gentiana germanica* subsp. *carpatica* Hayek, Prodr. Fl. Penins. Balkan. 2: 423. 1930; *Gentiana carpatica* Wettst., Öesterr. Bot. Z. 42: 4. 1892; *Gentiana carpaticola* Borb. in F. Schultz, Herb. Norm. 38: 288. 1923; id. in Prodan, Fl. Roman. 1: 828. 1923; Grossh., Fl. SSSR, 18: 603. 1952 [in Russ.].

Ind. col.: "in pratis et dumetis montanis montium Karpaten nec non in montibus Sudeten, Riesengebirge, Beskiden."
Type material: Type not designated.
Total distribution: W. Europe (Carpathians, Balkans).
Distribution in the Ukraine: Chn, Iv, Lv, Zk.
Notes. Like the previous subspecies, subsp. *carpatica* grows in meadows and forest clearings from the submountain region to subalpine zones of the Ukrainian Carpathians at 280–1,940 m a.s.l. The taxon is an autumn race of *G. lutescens* (flowering in August–October), unlike subsp. *lutescens* that blooms in the summer (June–July).

4.3.6 Genus Gentianopsis (L.) Ma.

Acta Phytotax. Sin. 1, 1: 7. 1951; *Gentiana* sect. *Crossopetalum* DC. in Lam. et DC., Fl. Fr., ed. 3, 3: 659. 1805; *Gentianella* sect. *Crossopetalum* (Froel.) Pritchard, Bot. Journ. Linn. Soc. (London), 65, 2: 260. 1972.

Gentianopsis ciliata (L.) Ma, Acta Phytotax. Sin. 1,1: 19. 1951; Tzvelev, Fl. Europ. p. SSSR, 3: 75. 1978 [in Russ.]; Z.Díaz, Fl. Iber. 11: 44. 2012; *Gentiana ciliata* L. Sp. Pl.: 231. 1753; Grossh., Fl. SSSR, 18: 591. 1952 [in Russ.]; Szafer et al., Rośliny Polskie: 606. 1953; Wissjul., Fl. URSR, 8: 251. 1957 [in Ukr.]; Ţopa, Fl. Roman. 8: 452. 1961; Schiman-Czeika in Rechinger, Fl. Iranica, 41: 18. 1967; Avetisian, Fl. Armenia, 7: 40. 1980 [in Russ.]; Z. Díaz, Fl. Iber. 11: 24. 2012. *Gentianella ciliata* (L.) Borkh. In Roem., Arch. Bot. 1, 1: 29. 1796; Tutin, Fl. Europ. 3: 64. 1972; Pritchard in Davis, Fl. Turkey, 6: 192. 1978; Kožuharov et Petrova, Fl. Narodna Republ. Bulg. 8: 409. 1982.

Ind. col.: "Habitat in Helvetiae, Italiae, Canadae montibus."
Type material: Type not designated (probably in BM or LINN?).
Total distribution: Central, South, and Western Europe, Asia Minor, Caucasus (?), Talysh, Iran
Distribution in the Ukraine: Chn, Iv, Khm, Kr, Kv, Lv, Tr, Vn, Zk.
Notes. *G. ciliata* grows mainly to the west of the Dnieper River on the plain and in the Carpathians up to the alpine zone. This species prefers dry and well-lit places, meadows, and forest glades. In the mountains it grows near trails, on stony slopes, and along streams. For some time, *G. ciliata* has not been collected in the plain of the Ukraine. Probably those habitats have been lost already.

4.3.7 Genus Swertia L.

Sp. Pl.: 226. 1753; id. Gen. Pl. 5: 107, p.p., excl. Helenia.

Swertia perennis L., Sp. Pl.: 226. 1753; Ledeb., Fl. Ross. 3, 1: 74. 1847; Herbich, Fl. v. Bucovina: 216.1859; Grossh., Fl. SSSR, 18: 630. 1952 [in Russ.]; Szafer et al., Rośliny Polskie: 606. 1953; Wissjul., Fl. URSR, 8: 257. 1957 [in Ukr.]; Ţopa, Fl. Roman. 8: 446. 1961; Tutin, Fl. Europ. 3: 67. 1972; Pissjauk., Fl. Europ. p. SSSR, 3: 85. 1978 [in Russ.]; Kožuharov et Petrova, Fl. Narodna Republ. Bulg. 8: 418. 1982; Renobales, Fl. Iber. 11: 85. 2012.

Ind. col.: "Habitat in Helvetia, Bavaria".
Lectotype (P. Hartvig in A. Strid et Kit Tan (ed.), Mountain Fl. Greece 2: 7. 1991): LINN: "*Gmelin s.n.*, Herb. Linn. No. 328.7."

(a) **Subsp. perennis**

Total distribution: Europe (plain and piedmont)
Distribution in the Ukraine: Chn, Iv, Lv, Rv, Tr, Zk.
Conservation status in the Ukraine: Vulnerable.
Notes. *S. perennis* occurs in the Volyno-Podolia Upland and on the submountain and forest zone of the Ukrainian Carpathians along rivers, lowland wet, moss meadows, and marshes.

(b) **Subsp. *alpestris* (Baumg. ex Fuss) Domin et Podp**. in Polivka et al., Klic Uplné Květene: 525. 1928; *Swertia alpestris* Baumg. ex Fuss, Enum. Stirp. Transsilv. 4, Mant. 1: 19. 1846; Wissjul., Fl. URSR, 8: 259. 1959 [in Ukr.]; Tutin, Fl. Europ. 3: 67. 1972, ad not.; Pissjauk., Fl. Europ. p. SSSR, 3: 85. 1978 [in Russ.]; *S. perennis* L. var. *alpestris* (Baumg. ex Fuss) Sag. et Schn., Fl. Carr. Centr.: 396. 1891; Ţopa, Fl. Roman. 8: 446. 1961; *S. perennis* L. var. *alpestris* (Baumg. ex Fuss) Kožuharov et Petrova Fl. Narodna Republ. Bulg. 8: 418. 1982.

Ind. col.: "In alpibus petrosis Rodnensibus prope rivulos saxosus versus montes Györgelo, Pietroszul, Galatz. Jul. Baumg[arten]" [Eastern Carpathian Mountains].
Type material: Type not designated.
Total distribution: C. and E. Europe (mountains)
Distribution in the Ukraine: Chn, Iv, Zk.
Conservation status in the Ukraine: Vulnerable.
Notes. In the Ukraine, the taxon grows in the alpine and subalpine zones of the Carpathian Mountains, mainly along rivers and streams, in swampy meadows and wetlands, and on wet rocks. This subspecies is considered an alpine race of *S. perennis*.

Swertia punctata Baumg. Enum. Stirp. Transsilv. 1: 190. 1816; Szafer et al., Rośliny Polskie: 607. 1953; Wissjul., Fl. URSR, 8: 259. 1957 [in Ukr.]; Ţopa, Fl. Roman. 8: 446. 1961; Tutin, Fl. Europ. 3: 67. 1972, ad not.; Pissjauk., Fl. Europ. p. SSSR, 3: 85. 1978 [in Russ.].

Ind. col.: "In alpibus Fagarasiensibus ad pagum Dreguts pertinentibus, ibidem prope fossiculos rupestres, et in Rodnensibus versus Dscheammeanie".
Type material: Type not designated.
Total distribution: C. and E. Europe (mountains)
Distribution in the Ukraine: Iv, Zk.
Conservation status in the Ukraine: Vulnerable.
Notes. *S. punctata* is a very rare species in the Ukraine. It is known from some herbarium specimens collected at several localities within the alpine and subalpine zones of the Chornohora Range of the Carpathian Mountains, namely the mountain area of Kozi Ulogy (KW 00102206; LW sn, LE sn), Pip Ivan Chornohorski Mt (KBR 04862), Polonina Pohorilets (KBR 00833, KBR 00834), and from the area near Lake Berebenescul (KBR 4862)[2]. Like the previous species, *S. punctata* grows in damp and swampy places.

4.4 Conclusions

The present revision of the Gentianaceae in the flora of the Ukraine has demonstrated that, at present, the family consists of 21 species with 11 subspecies and 3 varieties belonging to 7 genera. The center of species richness of the Gentianaceae in the Ukraine is the Carpathian Mountains, where there are 18 species with 3 subspecies belonging to 5 genera. Some of them (*Gentiana lutea*, *G. punctata*, *G. nivalis*, *Swertia punctata*, and others) have the Northeastern limit of their range in the Ukrainian Carpathians. Climate change and the influence of man's activities are the main causes of reduction of the range for *Gentiana cruciata*, *Gentianopsis ciliata*, *Swertia spicata* and other species of the Gentianaceae in the Ukraine. Various literature sources on the flora of the Ukraine reported *Blackstonia acuminata*, *Centaurium tenuiflorum* subsp. *acutiflorum*, *Gentiana clusii* Perrier et Songeon, *G.* x *dorfleri* and *G. frigida* Haenke (Wisjulina 1957; Fodor 1974; Tzvelev 1978; Mosyakin and Fedoronchuk 1999), but specimens of these taxa were not found among the herbarium specimens investigated.

References

Chopyk VI (1977) Manual of plants of the Ukrainian Carpathians. Naukova Dumka Press, Kyiv (in Ukrainian)
Didukh YP (2009) Red data book of Ukraine. Plants. Global Consulting Press, Kyiv (in Ukrainian)
Domin K (1933) Rod Blackstonia Huds. se zvláštnim zřenim k Československu. Třídy Českě Akademie 43:1–10

[2] KBR = Herbarium of the Carpathian Biosphere Reserve.

Ena AV (2012) Spontaneous flora of the crimean Peninsula. N. Orianda, Simferopol (in Russian)
Fateryga AV, Svirin SA, Fateryga VV, Shiyan NM (2013) New finding of *Blackstonia perfoliata* (Gentianaceae) in the Crimea. Opt Prot Ecosyst 8:26–29 (in Russian)
Fedoronchuk MM, Shevera MV, Krytzka LI, Shiyan NM, Tsarenko OM (2004) The species of vascular plants described from Ukraine: Families Valerianaceae, Dipsacaceae, Apocynaceae, Campanulaceae, Gentianaceae. Ukrainian Bot J 61:44–54 (in Ukrainian)
Fodor SS (1974) Flora of Transcarpathia. Vyshcha Shkola Press, Lviv (in Ukrainian)
Gillett TM (1957) A revision of the North American species of *Gentianella* Moench. Ann Missouri Bot Gard 44:249
Grossheim AA (1952) Gentianaceae. In: Shyshkin BK, Bobrov EG (eds). Flora URSS, vol 18. Academy of Science of URSS Press, Moskow and Leningrad (in Russian)
Jacobsen K (1978) Centaurium. In: Davis PH (ed) Flora of Turkey, vol 6. Edinburgh at the University Press, Edinburgh, pp 178–183
Koch K (1850) Beiträge zu einer Flora des Orientes. Linnaea 23:581
Mansion G (2004) A new classification of polyphyletic genus *Centaurium* Hili (Chironiinae, Gentianaceae): description of New World endemic Zeltneria, and reinstatement of *Gyrandra* (by Griseb. and Schenkia Griseb.). Taxon 53:719–740
Mosyakin SL, Fedoronchuk NM (1999) Vascular plants of Ukraine. A nomenclatural checklist. M.G. Kholodny Institute of Botany, Kyiv
Ostapko VM, Boiko GV, Mosyakin SL (2010) Vascular plants of the Southeast of Ukraine. Knowledge, Donetsk
Privalova LA (1957) Gentianaceae. In: Wulff EW (ed) Flora Taurica, vol 3(1). Agricultural Literature Press, Yalta and Moscow, pp 33–44 (in Russian)
Rubtsov NI (ed) (1972) Manual of vascular plants of crimea. Nauka Press, Leningrad (in Russian)
Schiman-Czeika H (1967) Gentianaceae. In: Rechinger KH, Flora Iranica, vol 41, pp 1–28
Shiyan NM, Dzhus MM (2005) *Gentiana nivalis* L. (Gentianaceae) in the flora of the Ukrainian Carpathian Mountains. Ukrain Bot J 62:22–28
Struwe L, Albert VA (2002) Gentianaceae: systematics and natural history. Cambridge University Press, Cambridge and New York
Tutin TG (1972) Flora Europea, vol 3. Cambridge University Press, Cambridge
Tzvelev NN (1978) Gentianaceae. In: Fedorov AA (ed) Flora of the European parts of the URSS, vol 3. Nauka Press, Leningrad, pp 57–82 (in Russian)
Tzvelev NN (1993) Synopsis of the species of the family Gentianaceae in the flora of Caucasus. Bot J 78:131–138 (in Russian)
Wisjulina OD (1957) Gentianaceae. In: Kotov MI, Barbarych AI (eds) Flora of the Ukrainian SSR, vol 8. Academy of Science of Ukrainian SSR Press, Kyiv, pp 221–260
Zerow DK (ed) (1965) Manual of plants of the Ukraine. Urozhai, Kyiv (in Ukrainian)

Chapter 5
Distribution, Ecology, and Some Taxonomical Notes of the Genera *Gentiana* L. and *Gentianella* Moench (Gentianaceae) in the Balkans

Vladimir Stevanović and Ksenija Jakovljević

Abstract To date, 17 taxa at species and subspecies rank of the genus *Gentiana* and 13 species of the genus *Gentianella* have been recorded in the Balkan Peninsula. The majority of species of both these genera are distributed in mountainous areas of the Balkans, i.e., in the Dinaric Alps in South West Croatia, Montenegro, and Northern Albania, in the northern parts of the Scardo-Pindhic mountains in South West Serbia, Northern Macedonia, and North East Albania, as well as in the Balkan and the Rhodope mountain ranges in Bulgaria. A noticeable feature is a relatively small number of endemic taxa of both genera in the Balkan mountains in comparison to their presence in the Alps and Carpathians. Certain problems persist in the taxonomy of *Gentiana* and *Gentianella* in the Balkans, arising from the complex history and evolution of the populations of different species during the Ice Age, i.e., in the course of the succession of glacial and interglacial periods.

5.1 Introduction

The exceptional floristic richness of the Balkan Peninsula is brought about by historical and evolutionary changes, and by the changing conditions of the habitats in this part of Europe, from the Tertiary, through the Ice Age, to the present day. In consequence, it harbors an outstanding number of highly diverse plant taxa, and among them many with unresolved, even controversial, taxonomic position. Examples of this kind are the genera *Gentiana* and *Gentianella*.

V. Stevanović (✉) · K. Jakovljević
The Institute of Botany and Botanical Garden "Jevremovac", Faculty of Biology, University of Belgrade, Takovska 43 11000 Belgrade, Serbia
e-mail: vstev@bio.bg.ac.rs

K. Jakovljević
e-mail: kjakovljevic@bio.bg.ac.rs

J. J. Rybczyński et al. (eds.), *The Gentianaceae - Volume 1: Characterization and Ecology*, DOI: 10.1007/978-3-642-54010-3_5,

The genus *Gentiana* L. comprises more than 360 species distributed mainly in mountainous regions of the Northern Hemisphere, notably in Eurasia, spanning the mountain massifs of Central Europe, the Alps in particular, of Asia Minor and the Caucuses, through the mountains of Iran and Central Asia (Mts Pamir, Altai), and reaching China and Japan in the east. A smaller part of this genus range (with a small number of species) encompasses North and Central America, South America (Chile), Noth West Africa (Mt Atlas in Morocco), and Eastern Australia. The center of the genus with the largest number of species is in the mountains of China, where 250 species have been recorded (Ho and Pringle 1995). Heading westwards, the number of species gradually decreases on Mts Pamir and Altai (Grossheim 1952), so much so that only 21 species have been recorded in the mountains of Iran (Schiman-Czeika 1967), 18 species in the Caucasus (Doluhanov 1967) and 12 species in the mountains of Asia Minor (Pritchard 1978a). In Europe, including all mountain massifs, about 40 taxa are present ranked as species and subspecies (Tutin 1972), i.e., 8.5 % out of the total number of species of this genus in the world. The genus is taxonomically very complex and is divided into 15 sections, out of which 12 are present in the mountains of China (Ho and Lui 1990; Yuan et al. 1996), with a large number of taxa, the status of which has been discussed by different authors, from the rank of species to that of variety. A large number of synonyms within the genus *Gentiana* illustrate taxonomic complexity of this genus.

In addition, some authors include a closely related genus *Gentianella* Moench in the genus *Gentiana* (Turrill 1929; Hayek 1930; Topa 1961; Hess et al. 1972; Greuter et al. 1986; Aeshiman et al. 2004), whereas others regard it as a subgenus of the genus *Gentiana* (Grossheim 1952; Doluhanov 1967). Currently, most botanists agree that *Gentianella* is separate but closely related to the genus *Gentiana*. Thus, the genus *Gentianella* itself encompasses about 300 species distributed mainly in mountainous regions of the Northern and Southern Hemispheres. This genus is also taxonomically complex, and comprises a large number of taxa, the status of which is not clear. In Europe, this refers particularly to an aggregate *Gentianella germanica*, which should be the subject of broader studies on the molecular phylogeny of populations in the mountains of Central Europe (Mts Alps, Tatras, Carpathians), as well as in the Balkan Peninsula.

5.2 Mapping and Abbreviations

Data on the distribution were obtained through a survey of the abundant literature of regional floras and floristic papers, as well as of the revisions and the checking of rich herbarium material, derived from Serbia, Macedonia, Montenegro, Bosnia and Herzegovina, Greece and Bulgaria, deposited in the collections of the University of Belgrade Herbarium (BEOU) and Herbarium of the Natural Museum in Belgrade (BEO) (Thiers 2013) and including the authors' own field investigations.

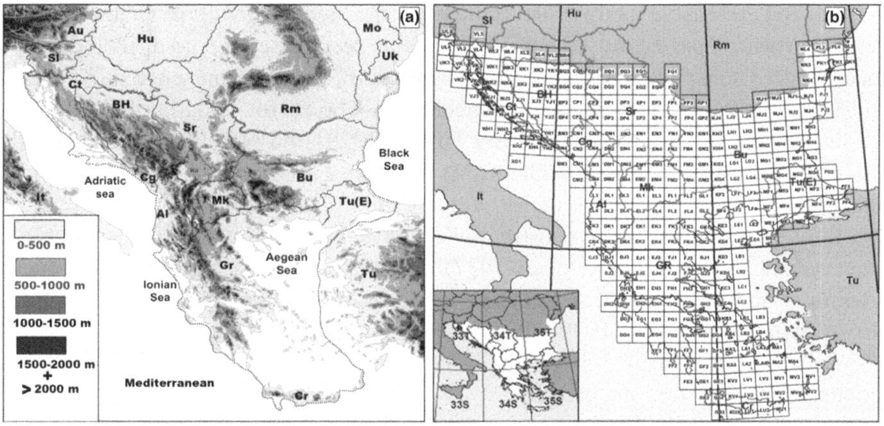

Fig. 5.1 Map of the Balkans with phytogeographical borders of the Peninsula, after Hayek (1927) and Turrill (1929)—*dotted line* (**a**), Map of the Balkans in MGRS, grid cells 50 × 50 km, as used in the mapping of *Gentiana* and *Gentianella* taxa (**b**)

Distribution mapping of the taxa of the genus *Gentiana* in the Balkan Peninsula was carried out on a military grid reference system (MGRS) using 10 × 10 km grid cells, and converted to MGRS 50 × 50 km grid cells, as is used in mapping for the *Atlas Florae Europaeae* (Kurtto et al. 2010). The Balkan Peninsula is encompassed by MGRS zones 33T, 34T, 35T, 34S, and 35S (Fig. 5.1b). Taxa mapping was done within the boundaries of the Balkan Peninsula as outlined by Hayek (1927) and Turrill (1929) (Fig. 5.1a). Acronyms of the countries on the maps are given according to the Euro+Med-database (2011).

5.3 Review and Distribution of Species of the Genus Gentiana and Gentianella in the Balkan Peninsula

5.3.1 Gentiana L.

In Europe, this genus is represented by about 40 taxa ranked as species and subspecies which are classified into 7 sections, namely *Gentiana*, *Pneumonanthae* (Gled.) Link., *Frigida* Kusn., *Cruciata* Gaudin, *Ciminalis* (Adans.) Dum., *Chondrophyllae* Bunge, and *Calathianae* Froelich (Tutin 1972). Marhold (2011) also included two species, *Gentianella albanica* (Jav.) Holub and *G. laevicalyx* Rohlena, belonging to the genus *Gentianella* in the genus *Gentiana*, although by morphological characters, those taxa undoubtedly belong to the genus *Gentianella*. The Sections *Gentiana, Ciminalis* and *Calathianae* characterize plants from the European mountains, notably the Alps, which are the centers of the diversity of these plants and probably the origin of these sections.

The first extensive taxonomic review of the genus *Gentiana* in the Balkan Peninsula was reported by Hayek (1930), who recognized 17 taxa, including 14 species and 3 subspecies. However, among these 17 taxa, the presence of *Gentiana pannonica* in the Balkan Peninsula, checked by Hayek for Croatia, is disputable, but this record was later taken over by Turrill (1929) and Nikolić (1997). Hayek's old information refers to Southern Mt Velebit (Visočica). However, despite intensive subsequent surveys of Mt Velebit, this plant has never been found, so that the report of this plant is justifiably doubtful and seems that it was *Gentiana lutea* subsp. *symphiandra* (Degen 1937). In other respects, a concise phytogeographic review of the species of the genus *Gentiana* in the Balkan Peninsula was also reported by Turrill (1929). It should be stressed that the number of species known in the Peninsula, since the time when Hayek reported his data, has remained more or less unchanged.

In summary, on the basis of previous and contemporary taxonomic and floristic data, 17 taxa, ranked as species and subspecies, classified into six sections, grow in the Balkan Peninsula, i.e., about 45 % of the total number of species of this genus inhabiting Europe. Those are in the Section *Gentiana*, three taxa within two species (*G. lutea* L. subsp. *lutea*, *G. lutea* subsp. *symphiandra* (Murb.) Hayek and *Gentiana punctata* L.), 3 taxa from *Pneumonanthae* (*Gentiana asclepiadea* L., *Gentiana pneumonanthe* L. subsp. *pneumonanthe, G. pneumonanthe* subsp. *nopscae* Jav.), 1 species from *Frigida* (*Gentiana frigida* Haenke), 1 species from *Cruciata* (*Gentiana cruciata* L.), 1 species from the Section *Chondrophyllae* (*Gentiana pyrenaica* L.), 3 species from *Ciminalis* (Adans.) Dum (*Gentiana clusii* Perr., *Gentiana acaulis* L. and *Gentiana dinarica* G. Beck), and 5 taxa from the Section *Calathianae* (*Gentiana verna* L. subsp. *balcanica* Pritch., *G. verna* L. subsp. *verna*, *G. verna* subsp. *tergestina* (G. Beck) Hayek, *Gentiana nivalis* L., and *Gentiana utriculosa* L.).

5.3.1.1 Section *Gentiana* L.

This Section comprises 11 taxa, i.e., 5 species and 6 subspecies, which are distributed in the Pyrenees, Alps, Apennines, in the Balkan Mountains, Tatras and Carpathians, extending to the east up to Northern Anatolia. The distribution center of this section is in the Alps. Only 2 species are present in the Balkan Peninsula.

Gentiana lutea L. (Fig. 5.2a)
According to Marhold (2011) and Rossi (2011), *G. lutea* consists of 4 subspecies, the type subsp. *lutea* growing in the mountains of the Iberian Peninsula, including the Cantabrian Mountains and the Pyrenees, the Massif Central and the Jura mountains, as well as the Western and Central Alps. The subspecies *monserratii* inhabits the Central Pyrenees, subsp. *vardjanii* Wraber is in the South East Alps, with subsp. *symphiandra* (Murb.) Hayek spreading into the Eastern Alps, the Balkan Peninsula, and the Carpathian Mountains, as well as in the mountains of North West Asia Minor. However, Tutin (1972) recognized only two subspecies of

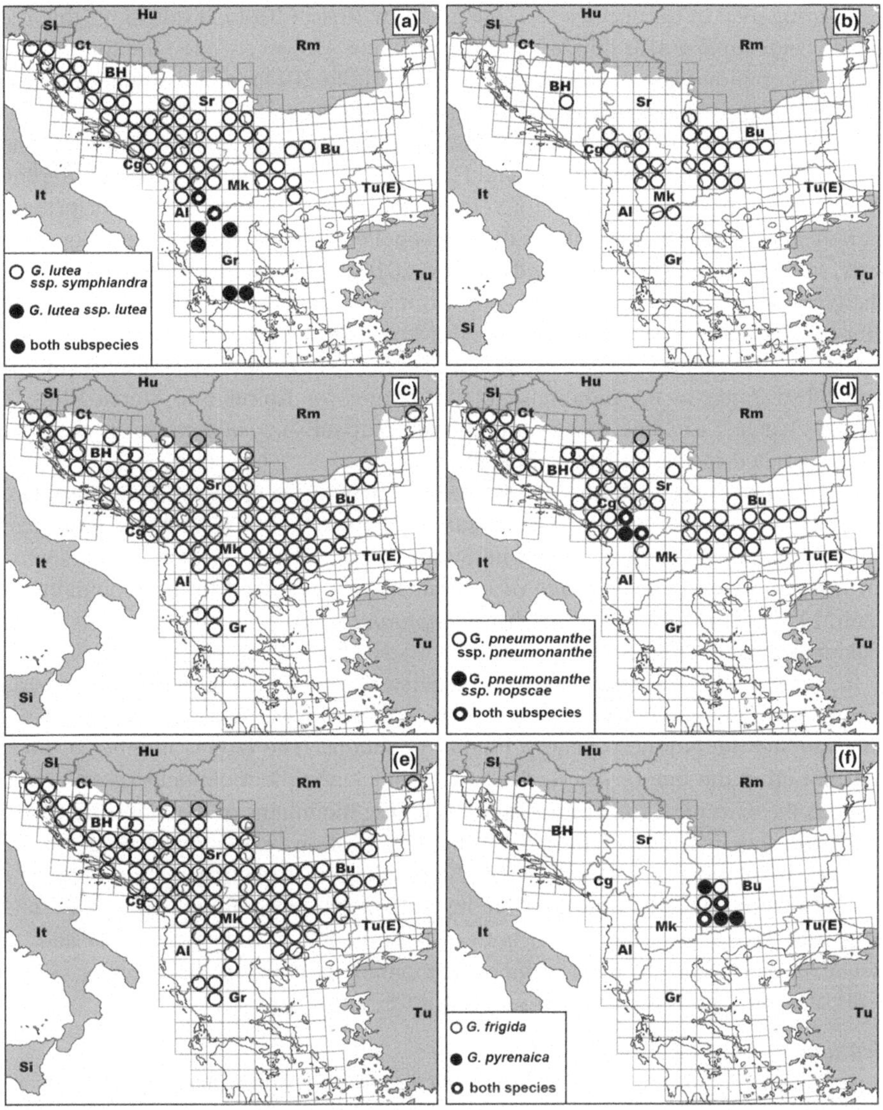

Fig. 5.2 Distribution of *Gentiana lutea* L. subsp. *symphiandra* (Murb.) Hayek and *G. lutea* subsp. *lutea* (**a**), *G. punctata* L. (**b**), *G. asclepiadea* L. (**c**), *G. pneumonanthe* L. subsp. *pneumonanthe* and subsp. *nopscae* (Jav.) Wraber (**d**), *G. cruciata* L. (**e**), *G. frigida* Haenke and *G. pyrenaica* L. (**f**) in the Balkans given in MGRS, grid cells 50 × 50 km

G. lutea, whereby the distribution area of the type subsp. *lutea* encompasses the mountains of Central and Southern Europe except for the Balkan Peninsula. In contrast, subsp. *symphiandra* stretches from the South East Alps, through the mountains of the Balkan Peninsula up to North West Anatolia. According to the distribution map of the species *G. lutea* given by Rossi (2011), it transpires that the subsp. *symphiandra* also inhabits the Carpathians. However, most authors do not share this opinion (Topa 1961; Tutin 1972; Marhold 2011), and maintain that it is the type subsp. *lutea* which grows in the Carpathians.

The subsp. *symphiandra* was recorded in the greatest number of localities in the Balkan Peninsula (Croatia, Bosnia and Herzegovina, Montenegro, Albania, Serbia, Bulgaria, North East Greece). Conversely, the type subsp. *lutea* is only present sporadically in the mountains of the Balkan Peninsula, notably in its Southern parts, in Macedonia (Matevski and Wraber 2010a) and in Greece (Hartvig 1991). The type subsp. *lutea* was reported once from Serbia (Jovanović-Dunjić 1973), but after our detailed revision of the material from this part of the Peninsula (deposited in BEOU, BEO), only subsp. *symphiandra* was confirmed. The presence of the type subsp. *lutea* in the mountains of the regions of Epirus and Sterea Ellas in Greece changes, to some extent, the opinion that subsp. *symphiandra* is the only one distributed in the Balkan Peninsula.

It should be noted that there is certain regularity in the distribution of these two subspecies. Namely, the type subsp. *lutea* inhabits South West Macedonia and Greece, and may also be found in South West Albania, which is somewhat unexpected, given that the range of this subspecies is linked to the mountains of Central Europe. In contrast, the subsp. *symphiandra* grows in all other parts of the Balkan Peninsula (Fig. 5.2).

In general, the distribution of the species *G. lutea* in the Balkan Peninsula encompasses montane and sub-alpine belts of the Dinaric Alps, as well as of the Moesian and the Northern-Scardo-Pindic mountains. The largest number of finds does not cross the line of the Northern Mt Pindus and the mountains of Thessaly, whereas the extreme Southern boundary is on the mountains of Aetolia (Mts Giona and Vardusia) in Greece (Hartvig 1991). The Eastern border of its range in the Balkans is in the Central part of Mt Stara Planina in Bulgaria (Fig. 5.2). The species inhabits mostly limestone, though it may be also found on silicate and serpentine bedrocks. It grows on both deep soils and in limestone clefts and, in fact, it is there that the most numerous populations develop. Its habitats on the Peninsula are within the altitude range from 800 to 2,100 m.

Gentiana punctata L. (Fig. 5.2b)
G. punctata is a typical acidophilic species of sub-alpine and alpine belts of siliceous mountains of the Central part of the Balkan Peninsula. It is seldom found on siliceous Dinaric Alps such as Mt Vranica in Central Bosnia, whereas it is somewhat more common in the South Eastern Dinaric Alps from Mt Bjelasica to Mt Prokletije in Montenegro, on the border with Albania. In Macedonia, this plant inhabits the entire mountain chain of the Šarplanina and Korab extending southwards to Mts Jablanica, Pelister, and Kajmakčalan (Matevski and Wraber 2010a), which are the Southern boundary of its distribution range in the Balkan Peninsula. In Eastern Serbia and in Bulgaria, it occupies the Western and Central parts of Mt Stara Planina. It also occurs on Mts Vitoša, Rila, Pirin, and on the Western and

Central parts of Mt Rhodopes in Bulgaria (Evstatieva 2011). The habitats of this species in the Peninsula are located at on altitude of 1,500–2,400 m.

5.3.1.2 Section *Pneumonathe* Gled. (Gaudin)

The section comprises about 40 species distributed in Europe, in Western, Northern and Central Asia, as well as in North and Central America. This section is represented by three species in Europe, and also in the Balkan Peninsula.

Gentiana asclepiadea L. (Fig. 5.2c)
This species is distributed in mountain and sub-alpine zones of the Central European mountains from the Mts of Jura, Ligurian Apennines, Alps, Tatras, Sudetes, Carpathians, all across the Balkans mountains and extending eastwards via the mountains of Northern Anatolia, all the way to the Caucasus. In the Balkan Peninsula, it is frequent on clearings and fringes of mountainous beech forests, of mixed and coniferous forests, especially spruce and spruce-fir forests between latitudes 41° and 45° N. The Southern border of its range in the Peninsula is mountain Iti located in the Central part of Greece, in Aetolia (Hartvig 1991), whereas most part of its area ends at the line of the mountains of Thessaly (Mts Olympus, Pelion) and on the Northern Pindus (Mt Smolikas). The Eastern border of its range on the Peninsula is in the Eastern part of the Mt Stara planina and the Central Rhodopes in Bulgaria, as well as on the slopes of Mt Strandža in the European part of Turkey (Pritchard 1978a).

Gentiana pneumonanthe L. (Fig. 5.2d)
The species is widespread in Europe and stretches as far as Central Asia, occupying temperate and subboreal zones between latitudes 37° and 65° N and longitudes 9° W and 84° E. *G. pneumonanthe* extends into Western and Central parts of the Balkan Peninsula, but nowhere is it frequent. The species inhabits predominantly mountainous wet meadows and peat bogs, as the characteristic member of plant communities of the *Molinietum* type. It is rarely found in lowland regions of the Southern parts of the Panonnian Plain, in Vojvodina and Slavonia, where it is usually an indicator of relict wet habitats. The Southern border of its range is on the line of the mountains beginning in North West Macedonia (Mt Bistra) and extending to Mts Pirin-Rhodopes in Bulgaria; its Eastern border is on the Eastern part of Mt Stara planina in Bulgaria.

Gentiana pneumonathe L. subsp. *nopscae* (Jav.) Wraber (Fig. 5.2d)
This taxon has been described from the surroundings of Tropoja in North East Albania (Javorka 1921), but it was also found in wet serpentine habitats in Kosovo province and Northern Albania (Markgraf 1931; Wraber 1990). However, this taxon was regarded by some authors to be a synonym of *Gentiana pneumonanthe* (Hayek 1930; Greuter et al. 1986; IOPI 1996–2003; Marhold 2011), whereas the others consider it to be a separate species (Javorka 1921; The Plant List 2010), or a subspecies (Wraber 1990; Vangeli et al. 1994; Tomović 1999).

5.3.1.3 Section *Cruciata* Gaudin

This section comprises 20 species distributed throughout Europe and Asia, with the largest number of species in China. There are 2 species in Europe, but only 1 in the Balkans.

Gentiana cruciata L. (Fig. 5.2e)
This species is distributed widely in the temperate zone of Europe and Asia between latitudes 37°–60° N and longitudes 2° W and 88° E. It is noteworthy that it is this species of the genus *Gentiana* which has the largest range in the Balkan Peninsula. The Eastern border of its range in the Peninsula is in South East Romania (in the hilly region of Dobrudja), while the Southern border is in the Central part of Mount Pindus in Greece. The plant inhabits the montane meadows and extends to subalpine ones, while its area penetrates even into the steppe zone of the Vlaška Plain. *G. cruciata* grows in different plant communities, from mesophilous to xerophilous grasslands, mainly on calcareous bedrock. It is also present in clearings of beech and coniferous forests, in mountain and subalpine zones, all along an altitude of 200–2,000 m.

5.3.1.4 Section *Frigida* Kusn

The section comprises 18 species distributed in the mountains of Eurasia and North America. The majority of species (17) grow in the mountains of China. In Europe, this section is represented by 2 species, one of which is also present in the Balkans.

Gentiana frigida Haenke (Fig. 5.2f).
This species is distributed in the Eastern Alps, Mts Tatras, and Carpathians, as well as in the mountains of West and South West Bulgaria. *G. frigida* is most closely related to the Eastern alpine species, *Gentiana froelichii* Jan ex Reichenb., and to the Central Asian species, *Gentiana algida* Pall. In the Balkan Peninsula, this relict orophyte was recorded only in several localities in Bulgaria, on Mts Rila, and Pirin, as well as in the Western part of Mt Stara planina (Kožuharov and Petrova 1982), where it is presently absent (Dimitrov 2011). This species most commonly inhabits alpine siliceous rocky ground communities within the alliance *Caricion curvulae* at altitudes from 2,600 to 2,900 m.

5.3.1.5 Section *Chondrophyllae* Bunge

This is the largest section of the genus *Gentiana*, given that it encompasses 158 species. The latter is distributed in Eurasia, North Africa, North and Central America, and Australia. This section is represented in Europe by 4 species, of which only one, *G. pyrenaica*, is present in the Balkan Peninsula.

Gentiana pyrenaica L. (Fig. 5.2f)
The species is disjunctly distributed in Eurasia, i.e., in the Pyrenees, Carpathians, and the mountains of Bulgaria, as well as in the mountains of North East Anatolia and on Mt Caucasus. It is most closely related to *Gentiana dshimilensis* C. Koch from Mt Caucasus, which is considered by some authors to be a synonym of *G. pyrenaica* (Pritchard 1978a; The Plant List 2010). In the Balkan Peninsula, this species inhabits the mountains of SW Bulgaria (Mts Vitoša, Rila and Pirin), growing in wet peat meadows of subalpine and alpine areas from 1,500 m and above.

5.3.1.6 Section *Ciminalis* (Adans.) Dum. (*Megalanthae* Gaudin, *Thylacites* Griseb.)

Representatives of this section are distributed in the mountains of Central Europe. The section consists of 7 species of which three, *G. clusii, G. acaulis*, and *G. dinarica*, grow in the Balkan Peninsula. The largest number of species is found in the Alps which is the center of distribution of plants in this section. Studies in molecular phylogeny reveal the existence of a complex of related vicarious species inhabiting various substrata (Hungerer and Kadereit 1998).

Gentiana clusii Perr. (Fig. 5.3a)
Distribution of this species encompasses the Alps, Tatras, Carpathians, and the mountains of the Westernmost part of the Balkan Peninsula. In the Peninsula, this species grows on the limestone mountains of the Western Dinaric Alps, such as Mts Risnjak, Kapela, Velebit, and Plješevica, as well as on the mountainous region Gorski Kotar (Degen 1937; Gaži-Baškova 1994). The species inhabits calcareous stony grounds and grasslands of the upper montane and subalpine zones, and is a member of the plant communities within the alliance *Onobrychi-Seslerion*. It is known that *G. clusii* is vicarious with the related species *G. acaulis*, which grows on siliceous shallow stony grounds, or on deep soils from which lime has been removed by washing. In Croatia, the status of those two species appears somewhat inconclusive. As early as 1909, Hirc expressed doubt on the existence of *G. acaulis* in this region. In his Prodromus, Hayek (1930) registered only the species *G. clusii* for Croatia. Degen (1937), for his part, reported only the species *G. clusii* also for the mountains of Croatia (Mts Risnjak, Velebit, Kapela, Plješevica). Similarly, Meusel et al. (1978) provided a map in which those two species in the Balkan Peninsula are separated chronologically, whereby the range of *G. clusii* encompasses the extreme western parts of the Peninsula, whereas *G. acaulis* reaches the west only up to the Central Dinaric Alps. However, Pavletić (1994) reported both *G. clusii* and *G. acaulis* for the same localities in Croatia. At present, it is not clear whether there exists the edaphic vicariance of those two species in these same localities, or if the discrepancy in data is the result of erroneous determination.

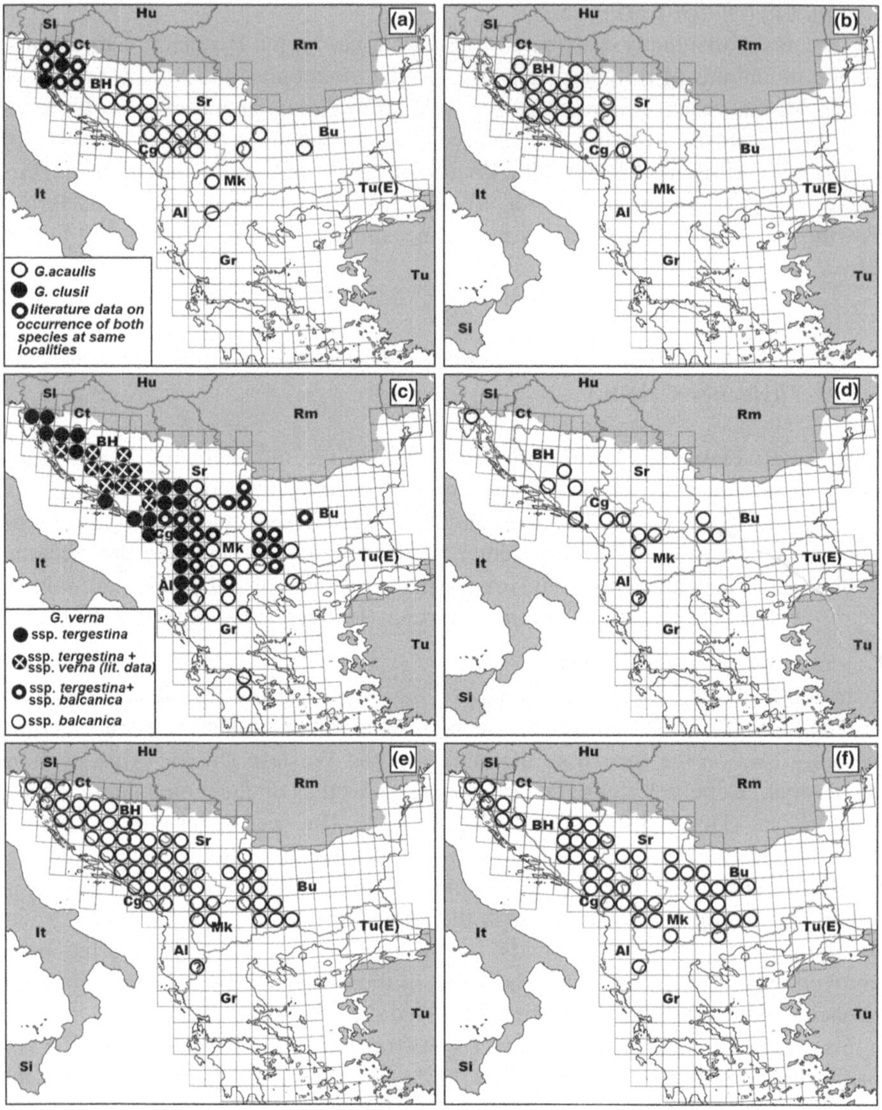

Fig. 5.3 Distribution of *Gentiana clusii* Perr. and *G. acaulis* L. (**a**), *G. dinarica* G. Beck (**b**), *G. verna* aggr. (**c**), *G. nivalis* L. (**d**), *G. utriculosa* L. (**e**) and *Gentianella ciliata* (L.) Borkh. (**f**) in the Balkans given in MGRS, grid cells 50 × 50 km

Gentiana acaulis L. (Fig. 5.3a)

This is the species with the greatest distribution range in the section *Ciminalis*. It is spread throughout in the Pyrenees, Alps, Tatras, Carpathians, and in the mountains of the Balkan Peninsula. In the Peninsula, its range extends from Central Bosnia Eastwards via Northern and South Eastern Montenegro (Mts Sinjavina, Bjelasica,

Komovi, Prokletije) and Serbia (Mts Prokletije, Kopaonik, Stara planina) to Bulgaria (Central parts of Mt Stara planina), and Southwards to Mt Pelister on the boundary between Macedonia and Greece (Matevski and Wraber 2010a). In the Westernmost part of the Peninsula, in Croatia, it is replaced by the vicarious species *G. clusii*. As already emphasized, the chorological relation between those two vicarious species, has not been elucidated entirely since both species have been reported for the same localities in Croatia (Gaži-Baškova 1994; Pavletić 1994; Nikolić 2012). In general, *G. acaulis* grows on mountainous and subalpine rocky grounds on silicate bedrock.

Gentiana dinarica G. Beck (Fig. 5.3b)

The species is distributed on the calcareous Dinaric Alps in Croatia (Southern Mt Velebit, Mt Plješevica), in Bosnia and Herzegovina (Mts Osječenica, Klekovača, Jadovnik, Dinara, Cincar, Prenj, Čabulja, Čvrsnica, Bjelašnica, Treskavica, Hranisava, Lelija, Volujak, Maglić), in Montenegro, and in the Central Apennines (Mišić 1965). The Western border of the range in the Balkans is on Visočica peak of the Southern Mt Velebit in Croatia (Mišić 1965), while the Eastern border is in Mt Koritnik in Kosovo and Metohia province.

Unlike the two previous species of the section, which mainly inhabit mountainous rocky grounds on limestone (*G. clusii*) or on silicate (*G. acaulis*), *G. dinarica* is most often found on the steep ledges of stony calcareous or dolomite rocks. Generally, in comparison to the other two mentioned species, *G. dinarica* grows in more stony habitats.

Taking into account the available literature data on all the three described species of the Sect. *Ciminalis*, it should be borne in mind that their chorological and ecological relationship in the Western part of the Balkan Peninsula is not entirely clear.

5.3.1.7 Section *Calathianae*

The section comprises 10–15 species with a greater number of subspecies, distributed in Europe, Northern and Western Asia, North East North America, and in North West Africa. Eight species with nine subspecies are present in Europe, the subspecies being regarded by some authors as species. Three species of this section are distributed in the Balkan Peninsula, namely, *G. verna* aggr., *G. nivalis*, and *G. utriculosa*. *G. verna* aggr. belongs to the subsection *Vernae* which brings together perennial species, while two other species belong to the subsect. *Calathianae*. The latter encompasses annual representatives of the genus *Gentiana*.

Subsect. *Vernae*

Gentiana verna L. aggr. (Fig. 5.3c)

This species as an aggregate comprises a number of taxa, which are distributed in the mountains of Central and Eastern Europe, in Asia Minor, on Mt Caucasus, and in Central Asia, as well as in the Arctic region of Western Siberia. A number of taxa ranked as species and subspecies were included within this aggregate. In his Prodromus, Hayek (1930) reported the occurrence of 3 subspecies of *G. verna* in the Balkan Peninsula, these being, type subsp. *verna*, subsp. *tergestina* (Beck) Hayek and subsp. *pontica* (Soltok.) Litard. & Maire. In addition, Mayer and Bjelčić (1974a) gave the status of species to subsp. *tergestina.*

This opinion, that in the Balkan Peninsula there are present three subspecies, prevailed for quite a longtime in the standard Floras of Europe (Tutin 1972) and in Floras of the Balkan countries (Jovanović-Dunjić 1973; Kožuharov and Petrova 1982). Thus, type subsp. *verna* should be widespread in all mountainous regions of the West and Central Balkans, subsp. *tergestina* mostly in the Western part of the Peninsula, but also in Greece and Bulgaria, whereas subsp. *pontica* should be found mainly in Central and Eastern parts of the Peninsula (Hayek 1930). Consequently, it may be inferred that the ranges of all three subspecies overlap and that among them there is not a clear chorological interdifferentiation. Pritchard (1977) produced a new concept on the taxonomy of the aggregate *G. verna,* describing a new subspecies, *balcanica,* which he separated from the subspecies *pontica* s.l. In Pritchard's opinion (1978a), the distribution range of subsp. *pontica* is confined exclusively to Asia Minor.

The presence of subsp. *pontica* in the Balkan Peninsula remains controversial. This subspecies has been reported for Serbia, Bulgaria, Macedonia, and Greece (Hayek 1930; Jovanović-Dunjić 1973; Kožuharov and Petrova 1982; Greuter et al. 1986; Marhold 2011). However, there exists the belief that all the statements about the presence of subsp. *pontica* in the Balkans in fact relate to subsp. *balcanica* (Hartvig 1991; Matevski and Wraber 2010a). This is also confirmed by our own review and revision of abundant herbarium material from Macedonia and Serbia, available in BEO and BEOU. Although the chorological delimitation of those three subspecies in the Balkan Peninsula is not settled completely, from our point of view the range boundaries of the three subspecies are differentiated, being in the Western parts of the Peninsula between subsp. *verna* and subsp. *tergestina*, and in the Central part between subsp. *tergestina* and subsp. *balcanica.*

Adding to the still barely settled matters are the most recent studies in molecular phylogeny and taxonomy of the Sect. *Calathianae* in Europe, which hold that the type subsp. *verna*, and even the species *G. verna* does not grow at all on the Balkan Peninsula (Hammerli 2007). According to these data, there is in the Balkan Peninsula a newly positioned species, *G. tergestina*, comprising two subspecies, i.e., subsp. *tergestina* and subsp. *balcanica*, whereas the third subspecies *pontica* remains confined to Asia Minor. Such a taxonomic concept is contrary to all the established views about the aggregate *G. verna* in the Balkan Peninsula. This new opinion is based on the results of molecular analyses, according to which two clades were clearly separated; *G. verna* on one side, and *G. tergestina* with subsp. *balcanica* and subsp. *pontica,* on the other side. However, some doubt still remains regarding this conclusion, since molecular investigations were carried out on rather scarce plant material derived from the Balkan Peninsula. Currently, the presence of *G. verna*, i.e., *G. verna* subsp. *verna* in the Balkans appears open for debate, quite contrary to the view of connoisseurs of this genus (Mayer and Bjelčić 1974a), who were convinced of its existence in the Balkans. It is highly probable that *G. verna* is distributed in the Dinaric Alps, and its presence also on the high mountains of Bulgaria cannot be excluded. In any case, migrations of the species *G. verna* from the Alps and Carpathians to the mountains of the Balkan Peninsula during the Ice Age was not only possible, but almost certainly occurred. More extensive studies are required in order to support this statement further. Usually, taxa of aggregate *G. verna* occupy different types of grasslands and stony grounds within the altitude range from 1,300 to 2,800 m, on various geological bedrocks.

Subsect. *Calathianae*

Gentiana nivalis L. (Fig. 5.3d).
The species is a glacial Amphi-Atlantic-Arctic-Alpine relict which is rather rare in the Balkan Peninsula and, until now, was recorded only in some mountains of Croatia, Bosnia and Herzegovina, Montenegro, South West Serbia, and Western Bulgaria. Of particular interest are the findings on the relatively low maritime Dinaric Alps such as Učka, Biokovo, and Orjen (Kušan 1969). It grows in these mountains in subalpine cold sinkholes where snow remains for a longtime. These habitats certainly represent glacial refugia for this species. In other mountains of the Peninsula, such as Mts Prokletije, Šarplanina, Rila, and Pirin, it also inhabits very cold sites of alpine regions above 1,800 m.

Gentiana utriculosa L. (Fig. 5.3e)
The species is distributed widely in the mountains of Central Europe, from the Alps, Ligurian Alps, Apennines, and Carpathians, to all the mountains of the Balkan Peninsula. It is common in the Peninsula in the Dinaric mountains, extending eastward to the line from the Western side of Mt Stara planina and across Mts Vitoša, Rila and Pirin to Western Rhodopes. The Southern border in the Peninsula is in the mountains of North West Macedonia (Mts Korab, Šarplanina, Bistra). There is one unconfirmed report of the occurrence of this species on Mt Gramos in Epirus (Hartvig 1991). Commonly, this species inhabits different types of grasslands on all types of bedrocks, within an altitudinal range from 570 to 2,200 m. The most favorable habitats are situated in the mountain, subalpine zone.

5.3.2 Distribution of the Genus Gentiana in the Balkan Peninsula: An Overview

The general survey showing the distribution of all 13 species (including subspecies of *G. verna* and *G. lutea*) of the genus *Gentiana* in the Balkan Peninsula, presented in UTM grid cells 50x50 km, makes it evident that the greatest number of species was recorded in mountainous areas (Fig. 5.4). The majority of species, i.e., 9–10, were found on Mt Vranica in Central Bosnia, and on Mt Prokletije, a mountain massif bordering Montenegro, Albania, and Kosovo Province, as well as on the Western part of Mt Stara planina and on Mt Rila in Bulgaria. Mountains with a

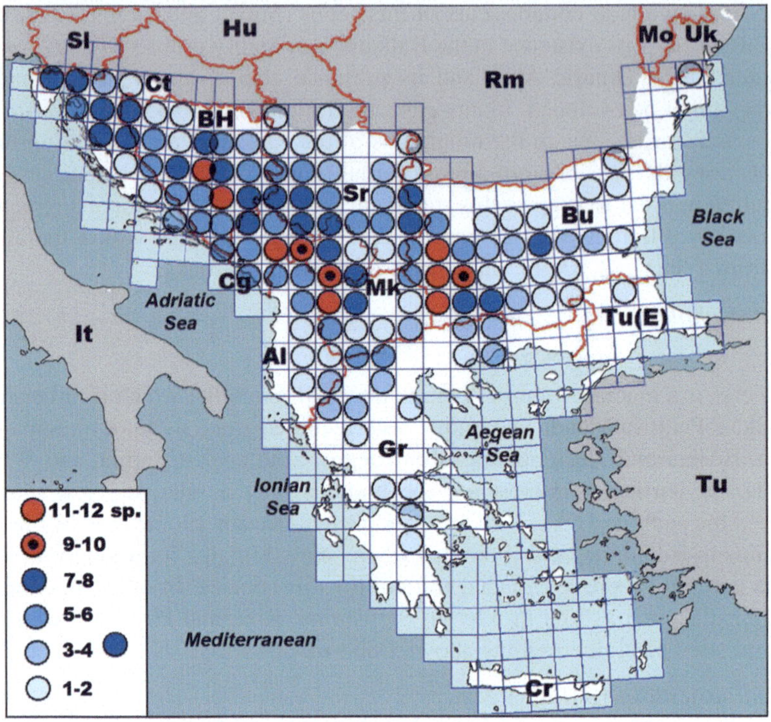

Fig. 5.4 General distribution of the genus *Gentiana* L. in the Balkans presented as number of taxa of different ranks (species and subspecies) per MGRS, grid cell 50 × 50 km

slightly smaller number of species (7–8) include Mts Velebit, Čvrsnica, Čabulja, and Prenj in Herzegovina, Maglić, Volujak, Durmitor, Bjelasica, and Komovi in Montenegro, mountain massif Šarplanina and Korab in Macedonia, Western parts of Mt Stara planina in Serbia and Bulgaria, as well as the Central part of Mt Stara planina, Mts Vitoša and Pirin, and the Western part of Rhodopes in Bulgaria.

The range of the two species, *G. clusiii* and *G. dinarica*, does not cross the Dinaric Alps, whereas the distribution range of the largest number of species, such as *G. pneumonathe, G. pneumonathe* subsp. *nopscae, G. frigida. G. pyrenaica, G. nivalis* and *G. utriculosa*, ends on the line of the mountains Šarplanina-Pirin-Rhodopes. The Southernmost distribution of the species *G. punctata* and *G. acaulis* reaches Mt Pelister, situated on the boundary between Macedonia and Greece. In the Peninsula, the range of the most distributed species, *G. cruciata*, extends somewhat more to the south, to Epirus in Greece. The ranges of *G. lutea*, *G. asclepiadea* and *G. verna* extend further to the south, to Mts Olympus, Pelion and to the mountains of Aetolia, while *G. verna* reaches as far south as the Northern Peloponnesus. This is also the Southernmost point of distribution of this genus in the Balkan Peninsula.

5.3.3 Gentianella Moench

The genus *Gentianella* was included originally in the genus *Gentiana* (Hayek 1930; Greuter et al. 1986; Aeshiman et al. 2004), which in its broadest sense, covered most genera of the subtribe *Gentianinae* of the family Gentianaceae (Yuan and Küpfer 1995). As it has been pointed out earlier, some taxonomists considered *Gentianella* as a subgenus of the genus *Gentiana* (Grossheim 1952; Doluhanov 1967). The genus has a somewhat contorted history of taxonomic classification being several times segregated, and then included *de novo* within the much larger genus *Gentiana*. Currently, it is generally accepted that *Gentianella* is a distinct genus.

The genus consists of annual, biennial, and more seldom perennial plants which differ from the species of the genus *Gentiana* by having corolla throat or lobes which are usually ciliate, whereas appendages between corolla lobes are absent. About 250 species have been recorded worldwide, distributed mainly in the mountainous areas from temperate to boreal Holarctic ecozones, and also in South America, New Zealand, and Australia (von Hagen and Kadereit 2001). In Europe, according to Pritchard and Tutin (1972), the genus *Gentianella* is represented by 22 species as well as by a number of subspecies and by taxa, whose status has not been resolved satisfactorily. The genus is characterized by unusually high variability that is manifested in the size and branching of its habit, the occurrence of aestival and autumnal ecotypes, and the type of inflorescence, as well as size, shape and type of indumentum of sepals and petals. Fairly poor genetic differentiation within this genus enables hybridization and introgression among the species. The numerous taxa described to date point unambiguously to the complex taxonomy of this genus.

In Europe, this genus consists of several sections, as claimed by Pritchard and Tutin (1972). These sections are *Comastoma* (Wettst.) Pritchard, with two species, *Crossopetalae* (Froelich) Pritchard, comprised of three species, *Gentianella* represented by 17 species and the largest number of taxa of undetermined status, and *Arctophila* (Griseb.) J. Holub, with only 1 species (Pritchard and Tutin 1972). However, according to current taxonomic assessments, several distinct genera that were separated from the genus *Gentianella* could be related in certain aspects to some of the above-mentioned sections, namely, the genus *Gentianopsis* Ma (Sect. *Crossopetalae* (Froelich) Pritchard) and the genus *Comastoma* (Wettst.) Toyok (Sect. *Comastoma* (Wettst.) Pritchard).

In this article, the genus *Gentianella* of the Balkan Peninsula is presented following the classification of Pritchard and Tutin (1972).

Based on all available literature sources, the Balkan Peninsula harbors 13 species from two sections of the genus *Gentianella*: one species, *G. ciliata* (L.) Borkh. from Sect. *Crossopetalae* and 12 species from Sect. *Gentianella*, such as *G. amarella* (L.) Börner, *G. laevicalyx* (Rohlena) Rohlena, *G. anisodonta* (Borbás) Á. & D. Löve, *G. liburnica* (Mayer and Kunz) Holub, *G. albanica*, *G. engadinensis* (Wettst.) J. Holub, *G. austriaca* (A. and J. Kerner) J. Holub, *G. germanica*

(Willd.) E. F. Warburg, *G. lutescens* (Velen.) J. Holub, *G. bulgarica* (Velen.) J. Holub, *G. crispata* (Vis.) J. Holub and *G. pevalekii* Mayer and Bjelčić.

The presence of the species *Gentianella nana* (Wulfen) N. M. Pritch. (*Comastoma nanum* (Wulfen) Toyok) from the Sect. *Comastoma* in the Balkan Peninsula is controversial, although it is reported for Croatia (Pritchard and Tutin 1972; Nikolić 1997). The reason is that the species in question grows in high mountainous siliceous stony grounds from the alliance *Androsacion alpinae*, in alpine and nival zones in the range from 2,200 to 2,800 m (Pignatti 1982; Aeshiman et al. 2004). The preference of the species for this type of habitat renders its existence in Croatia rather improbable, except possibly in some glacial refugial habitats in subalpine zones in this region of the Balkan Peninsula.

5.3.3.1 Section *Crossopetalae* (Froelich) Pritchard (*Gentianopsis* Ma)

This section comprises about 25 species distributed in temperate areas of Eurasia and North America, of which 11 grow in China. In Europe, it is represented by 3 species, of which only one, *G. ciliata*, grows in the Balkan Peninsula.

Gentianella ciliata (L.) Borkh. (Syn.: *Gentianopsis ciliata* (L.) Ma) (Fig. 5.3f)
This species is widely distributed in the Balkan Peninsula, from the Western parts of Croatia via Bosnia, Montenegro and Serbia, and toward the East to Bulgaria. In the Central part of Mt Stara planina, it reaches the Eastern border of its distribution range in the Peninsula. In the Southern part of the Peninsula, it extends over Western Macedonia and North West Greece (Epirus) to the mountains of the region Sterea Ellas (Mts Giona, Vardousia, Iti), where has its Southernmost distribution border. It is somewhat rare everywhere and is present only in small populations. Not enough is known of its distribution in Albania, as it has been reported only from Mt Maja Hekurave in the Northern part of this territory (Javorka 1926).

5.3.3.2 Section *Gentianella*

This is the type section of this genus and comprises about 125 species distributed mainly in mountainous areas of the temperate zone of Eurasia, North and South America, and New Zealand. In Europe, this section is represented by about 25 taxa, ranked as species and subspecies (Prichard and Tutin 1972). The largest number of species in Europe occupies in the Alps and Carpathians, the Tatra mountains, and also mountains in the Balkan Peninsula, where 12 species have been recorded. In Europe, a number of taxa were described initially within this section, from the rank of species to varieties, but their taxonomical status has been changed. It so happened that the aestival and the autumnal ecotypes, having different habits and appearing in the populations at the same localities but in different seasons, were perceived as distinct species, subspecies, or varieties. In addition, each and every taxon shows variation in its appearance from one population to another, within its distribution range. It should be emphasized that, to further compound the difficulties in determination and differentiation of taxa from this section, there occurs frequent introgression and hybridization between species. This has contributed to the present confusion in the taxonomy of this group.

Nevertheless, it has been possible to classify several reasonably well-differentiated groups within this section, primarily according to the following features; the shape of the sinus among calyx lobes, lobe and calyx tube length ratio, size and uniformity of calyx lobes, corolla to calyx length ratio, gynophore length, type of papillae on calyx lobes, and the type of inflorescence and its branching. According to Pritchard and Tutin (1972), Mayer and Bjelčić (1974b), Greimler et al. (2004) and Greimler (2010), these groups are *G. crispata*, *G. campestris*, *G. amarella*, *G. anisodonta*, *G. germanica*, and *G. austriaca*.

In addition, studies in molecular phylogeny, based on Internal Transcribed Spacer Sequences (ITS), are adding new data to this controversial field. Molecular analyses suggest great closeness among several European representatives such as *G. amarella, G. aspera, G. campestris* and *G. austriaca* (von Hagen and Kadereit 2001). A recent study on molecular phylogeny of the taxa from the Sect. *Gentianella* collected in the Alps, based on cpDNA RFLP (Restriction Fragment Length Polymorphism) data (Greimler et al. 2011), shows a complex reticulate evolution of this genus generated by climatic oscillations during the Pleistocene. It remains an open question as to how these changes during the Ice Age affected the population of various taxa of the genus *Gentianella* in the Balkans. Molecular and morphometric analyses of a large number of samples from as many populations as possible from the Balkan Peninsula will be essential to elucidate the problems of phylogeny and taxonomy of the genus *Gentianella* in this part of Europe.

G. crispata group

Crispate calyx lobe margins are a characteristic feature of this group, by which it is readily distinguished from other representatives of the Sect. *Gentianella*. The distribution range of this group is associated mostly with the Balkan Peninsula.

Gentianella crispata (Vis.) Holub (Syn.: *Gentiana crispata* Vis.) (Fig. 5.5a)

G. crispata is characterized by variability in habit, in the type of branching and in the shape of leaves, so that several taxa ranked as a variety and/or a subspecies have been described within the species. It is mainly distributed in the Western part of the Peninsula, extending from Mt Učka in Croatia via the Dinaric Alps in Bosnia and Herzegovina, Montenegro, and in Serbia, further spreading Eastwards to Mt Slavjanka (Ali Botush, Orvilos) on the Greek—Bulgarian border. This is the Easternmost point of the distribution range of the species. The Southern part of its area stretches across the mountain range of Šarplanina and Mt Korab, reaching its border in Southern Macedonia (Mts Jablanica, Galičica, Nidže). Plants are isolated on Mt Olympus in Thessaly, as is the species in the Southernmost point of its range in the Balkan Peninsula (Hartvig 1991).

Aside from this one in the Balkan Peninsula, an additional isolated part of the range exists in the South of the Apennine Peninsula, in the Calabria district, which

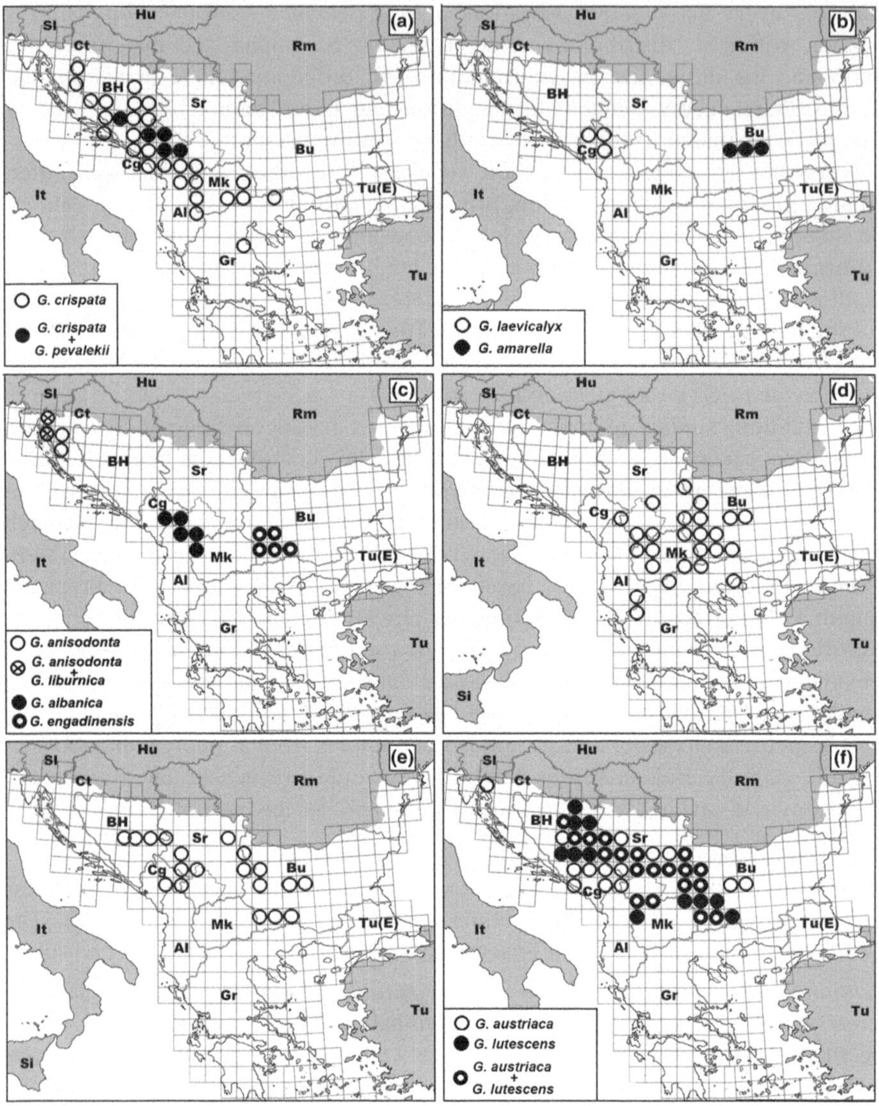

Fig. 5.5 Distribution of *Gentianella crispata* (Vis.) Holub and *G. pevalekii* Bjelčić and Mayer (**a**), *G. laevicalyx* Rohlena and *G. amarella* (L.) Boner (**b**), *G. anisodonta* Borb., *G. liburnica* E. Mayer and Kuntz, *G. albanica* (Jav.) Holub and *G. engadinensis* (Wettst.) Holub (**c**), *G. bulgarica* (Velen.) Holub (**d**), *G. germanica* (Willd.) E. F. Warb. (**e**), *G. austriaca* and *G. lutescens* (Velen.) Holub (**f**) in the Balkans presented in MGRS, grid cells 50 × 50 km

is the Southernmost point of the entire range of this species. The Western border of the range is in Slovenia. The species *G. crispata* inhabits limestone mountains at altitudes from 1,000 to 2,350 m.

Gentianella pevalekii Bjelčić and Mayer (Syn.: *Gentiana crispata* Vis. subsp. *bosnjakii* Pevalek; *Gentianella crispata* (Vis.) Holub subsp. *pevalekii* (Pevalek) Holub.) (Fig. 5.5a)
This species was described from Mt Komovi in South East Montenegro, but was also found later in several surrounding mountains (Durmitor, Sinjavina, Bjelasica, and Prokletije) and on Mt Čvrsnica in Bosnia and Herzegovina (Bjelčić and Mayer 1973). Its range is restricted to the area of the Central and South Eastern Dinaric Alps. *G. pevalekii* is the species most closely related to *G. crispata* and their areas overlap in these mountains.

G. campestris group
The group comprises taxa that are readily distinguished from other representatives of the Sect. *Gentianella* by unequal calyx lobes, whereby two lobes are much wider and larger, covering the other two smaller and more narrow ones. The flowers of the *G. campestris* group are usually 4–5 merous. The range of the species of this group encompasses subarctic and boreal zones from Iceland via Great Britain to Scandinavia, also Central Europe and mountains of this region from the Pyrenees to the Apennines, with an isolated part of the range in the South Eastern Dinaric Alps. The species *G. laevicalyx* is the sole representative of the *G. campestris* group in the Balkan Peninsula.

Gentianella laevicalyx (Rohlena) Stevanović and Jakovljević comb nov. (Syn.: *Gentiana campestris* var. *laevicalyx* Rohlena; *Gentiana laevicalyx* (Rohlena) Rohlena) (Fig. 5.5b)
This species was described by Rohlena (1912) as *G. campestris* L. var. *laevicalyx* found on the highest peak, Savin Kuk, of Mt Durmitor in Montenegro. Subsequently, the same author raised it to the rank of species within the genus *Gentiana*, assigning it the name *G. laevicalyx* (Rohlena) Rohlena (1937–1938, 1942). More recently, the plant has been also recorded in other localities on Mt Durmitor, such as Sedlena Greda, Klještina, Pošćensko Lake and Lokvice (plant material deposited in BEOU), and in neighboring Mt Sinjavina, as well as on the somewhat more distant Mt Komovi (Lakušić and Grgić 1969). This is an endemic plant whose range is limited to several mountains of the South East Dinaric Alps in Montenegro. In contemporary floristic literature, this taxon has been inconsistently positioned as the species of the genus *Gentiana* (Pritchard and Tutin 1972; Meusel et al. 1978; Greuter et al. 1986; Marhold 2011). Pritchard and Tutin (1972) pointed out that this taxon might be similar to the species *Gentianella campestris* or *Gentianella columnae* found in the Apennines. It is important to emphasize that Rohlena in his publications (1912, 1937–1938, 1942) indicated the similarity of *G. laevicalyx* with the species initially named as *Gentiana campestris*, which, in subsequent revisions was included in the genus *Gentianella*. It is obvious that this plant was not easily recognizable because it is rare in herbarium collections throughout Europe, and confusion therefore remains among taxonomists about its taxonomic status.

From our standpoint and after detailed revision of abundant herbarium material, this taxon, by all its features, undoubtedly belongs to the genus *Gentianella* and is most closely related to the ancestral species *G. campestris*. It differs from the species *G. campestris* by a short and delicate habit, 4-merous flowers and smooth margins of the calyx lobes. The evidence that *Gentianella laevicalyx* is the only representative of the *Gentianella campestris* group in the Balkan Peninsula is noticeable in the distribution map of *Gentiana campestris* and its relatives in Europe presented by Meusel et al. (1978), except that there it is marked as *Gentiana laevicalyx*.

The differentiation of *G. laevicalyx* from its ancestor, *G. campestris*, has almost certainly taken place during the migration of orophytes from the Alps toward the Dinaric Alps during the Pleistocene glacial periods.

Gentianella amarella group
The group *G. amarella* is commonly distinguished by small 4 - 5 merous flowers, more or less unequal, delicate calyx lobes and inflorescences on the upper part of the stem. Representatives of the group *G. amarella* have a circum-boreal distribution, and also inhabit the mountains of Central Europe (Pyrenees, Alps, Carpathians, Crimea and Caucasus) (Hultén 1961; Meusel et al. 1978). Therefore, their range in the broader sense is that of the Circum-Arctic-Boreal-Central-European-Caucasian mountain type. Within such a large distribution area, a number of taxa have been described in the rank of species, subspecies, and varieties. This group is most closely related to the *G. germanica* group.

Gentianella amarella (L.) Börner (Fig. 5.5b)
This species is very rare in the Balkan Peninsula and has been found in Bulgaria, in the Central part of Mt Stara planina (Ivanova 2011) and in Serbia (Jovanović-Dunjić 1973). However, after the revision of plant material from Serbia, the present authors have not confirmed the presence of this species in this region of the Balkan Peninsula, since all the specimens collected there and previously determined as *G. amarella* belong, in fact, to taxa from the group *G. germanica*.

Gentianella anisodonta group
Taxa belonging to this group are distinguished by the markedly unequal length and width of their calyx lobes, with revolute margins. The margins are usually with long-conical, occasionally short-conical, or cylindrical papillae (Greimler et al. 2004). These differently positioned taxa are predominantly distributed in the Alps and the Apennines, whereby the majority inhabit the Central and Eastern Alps, as well as the Western Dinaric Alps. The isolated part of the range of certain representatives of this group is in the South East Dinaric Alps and in the mountains of South West Bulgaria. Four representatives of this group are found in the Balkan Peninsula.

Gentianella anisodonta Borb. (Syn.: *Gentiana anisodonta* Borbás, *G. germanica* subsp. *calycina* (Koch) Hayek, *G. calycina* (Koch) Wettst.) (Fig. 5.5c)
The species is distributed in the South Central and South Eastern Alps, Apennines and in the mountains of Slovenia and North West Croatia. Its range in the Balkan Peninsula encompasses only the mountains of the Westernmost parts of the Dinaric Alps in Croatia (Mts Risnjak, Velebit, Kapela, Plješevica). On Mt Velebit, in particular, as stated by Degen (1937), it is one of the most common species of the

genus *Gentianella*. It may be assumed that it also grows in the bordering mountains between Croatia and Bosnia and Herzegovina (Mts Plješevica, Jadovnik, Klekovača). The distribution of this species in the Peninsula points to strong floristic relations between the Alpine and the Western Dinaric flora. The plant inhabits calcareous stony grounds of the montane and subalpine belt of the above-mentioned Balkan mountains in Croatia.

Gentianella liburnica E. Mayer and H. Kunz (Fig. 5.5c)
G. *liburnica* is an endemic species of the Balkan Peninsula distributed on Mt Velebit and in the bordering mountains between Croatia and Slovenia. It differs from *G. anisodonta* by more acute cauline leaves, a narrower corolla and a stipitate capsule. It is noteworthy that in the Northern part of Mt Velebit, the range of *G. anisodonta* and *G. liburnica* overlap. According to Mayer (1969a), *G. liburnica* is most closely related to *G. engadinensis*.

Gentianella engadinensis (Wettst.) J. Holub (Syn.: *Gentiana engadinensis* (Wettst.) Br.-Bl. and Samuelsson) (Fig. 5.5c)
G. engadinensis is a disjunctly distributed orophyte inhabiting the Central and Eastern Alps, on one side, and mountains of South West Bulgaria (Mts Rila, Pirin, the Western Rhodopes), on the other side. However, according to Pritchard and Tutin (1972) this species is distributed only in the Alps, in Italy and Switzerland, but Kožuharov and Petrova (1982) cited this species from the mountains of Bulgaria. Descriptions of the plants from the Bulgarian mountains correspond undeniably to the description of *G. engadinensis* from the Alps. Thus, it is an example of high mountain disjunction between representatives of the flora of the Alps and that of the mountains of Central parts of the Balkan Peninsula. This kind of disjunct distribution seems most likely due to migration of Alpine flora during the Ice Age, so that populations in Bulgaria might be regarded as relicts. It follows then, that comparative morphometric analyses and molecular studies of *G. engadinensis* from the Alpine and Balkan populations would be not only of interest, but even indispensable for the justification of its taxonomic status.

Gentianella albanica (Jav.) Holub (Syn.: *Genetiana anisodonta* Borb. subsp. *albanica* Jav., *Gentiana albanica* (Jav.) A.W. Hill, *G. bulgarica* Velen. var. *albanica* Jav. (Jovanović-Dunjić), *Gentiana germanica* Willd. subsp. *bulgarica* (Velen.) Hayek subvar. *albanica* (Jav.) Hayek) (Fig. 5.5c)

G. albanica is an endemic orophyte reported from Mt. Škelzen in Northern Albania, initially as a subspecies of the Alpine-Western Balkan species, *G. anisodonta* Borb. (Javorka 1921). According to Pritchard and Tutin (1972), its taxonomic status is still not fully resolved and requires further research. However, as Javorka correctly pointed out in the description of this taxon, *G. albanica* is related to *G. anisodonta* from which it differs by having a smaller number of 4–5 merous flowers, and a simple stem, which rarely branches from the base. Additionally, in the species *G. albanica*, the calyx lobe margins have short conical papillae. Revised material, deposited in BEO and BEOU from Mt Prokletije in Montenegro, corresponds thoroughly to the description presented by Javorka (1921) and shows beyond doubt that *G. albanica* belongs to the group *G. anisodonta*. Moreover, detailed analyses of this material provide evidence that this species is also distributed in the southern parts of the Balkan Peninsula, in the mountains Paštrik and Korab.

Some 80 years ago, Hayek (1930) determined erroneously the status of this taxon, ranking it as a variety of the species *G. bulgarica*, i.e., *Gentiana bulgarica* var. *albanica*. This taxonomic status was later accepted by Jovanović-Dunjić (1973), Kožuharov and Petrova (1982), Hartvig (1991), and Matevski and Wraber (2010b). As a result, the distribution of this species is cited for Macedonia, Bulgaria, and Central and Eastern Serbia. In our opinion, it is related to *G. bulgarica.*

Based on the available data on distribution, this is a narrow endemic species of Mt Prokletije and the neighboring mountains of Northern Albania, with a range extension toward border between Albania and Macedonia (Mt Korab), and also between South West Serbia, i.e., Kosovo and Metohia province, and Albania (Mt Paštrik).

Gentianella germanica group (*G. bulgarica*, *G. germanica* aggr., *G. austrica* aggr.) Taxonomically, this is certainly the most complex and the least elucidated group in the genus *Gentianella* and Sect. *Gentianella* in Europe, including the Balkan Peninsula. A large number of taxa from this group were cited for the Balkan Peninsula. Thus, Hayek (1930) included within the species *G. germanica* a number of species and varieties which, according to current views, are either synonymous plant names or distinct species. Mayer (1969b) reported the following species from this group as growing in the territory of former Yugoslavia, these being *G. germanica* with two varieties, *rhaetica* (A. and J. Kerner) E. Mayer, and *solstitialis* (Wettst.) E. Mayer, which are autumnal and aestival ecotypes, respectively. He also cited G. *austriaca* with two varieties, i.e., *austriaca* and *lutescens* (Velen.) E. Mayer, as well as *G. praecox* (A. and J. Kerner) Dostal also with two varieties, i.e., aestival ecotype as var. *praecox* and autumnal ecotype as var. *carpatica* (Wettst.) E. Mayer. In Europe, Pritchard and Tutin (1972) retain the species status only for *G. germanica* and *G. austriaca,* whereas the taxon *G. austriaca* var. *lutescens* has been raised to the species rank, i.e., *G. lutescens*. *G. praecox* was classified, instead, as a synonym of *G. austriaca*. An identical view was shared by Marhold (2011).

It should be emphasized that the taxonomy of this group in the Balkan Peninsula is very far from being resolved in a satisfactory way. When we examined the abundant plant material, collected in the past 150 years from Serbia, Macedonia, Montenegro, Bosnia and Herzegovina, and partly from Albania and Bulgaria, and deposited in BEOU and BEO, we found several specimens with transitional diagnostic characters. Determination and revision of old herbarium material proved difficult and literature data on distribution were not always reliable. However, relying on verified herbarium material and on trustworthy data from the literature, the mapping of this group in the Balkan Peninsula encompasses *G. bulgarica*, *G. germanica* s.l., *G. austriaca* s.l. and *G. lutesens*.

For Bosnia and Herzegovina, Mayer and Bjelčić (1974b) cited the following species from this group, namely *G. germanica*, *G. austriaca* and *G. praecox*, which is consistent with the taxonomic views reported by Mayer (1969b). In Bulgaria, according to Kožuharov and Petrova (1982), grow the three species *G. germanica*, *G. austriaca* (sub nomine *G. praecox*), and *G. lutescens*. For Serbia, the species cited are *G. austriaca* (including *G. praecox*) and *G. lutescens* (sub nomine *G. austriaca* var. *lutescens*) (Jovanović-Dunjić 1973). Matevski and Wraber (2010b) cited for Macedonia only the species *G. lutescens*. The data for Greece are controversial; Greuter et al. (1986) cited only the species *G. lutescens*, whereas Hartvig's (1991) opinion was that there were no other representatives of this group in Greece, except *G. bulgarica* and *G. ciliata*.

Since the present revision did not cover the material from the entire Balkan Peninsula (Bulgaria and Croatia were not included), the full register of all the localities could not be undertaken. It should be borne in mind that the distribution map of species from the group *Germanica* in the Balkan Peninsula is presented here based on our revision and in agreement with the taxonomic concept of Greimler (2010), which for the most part matches that of Pritchard and Tutin (1972).

Gentianella bulgarica (Velen.) Holub (Syn.: *Gentiana bulgarica* Velen., *Gentiana germanica* Willd. subsp. *bulgarica* (Velen.) Hayek) (Fig. 5.5d)
This species is included in the group *G. germanica*, but it differs from all other taxa of this group by having small flowers with recurved calyx lobes which are much longer than tubes, and pedicles much longer than the calyx, also with procumbent or ascending stems, the latter being very branched from the base. The species is mainly distributed in the mountains of the Central part of the Peninsula, in Serbia, Northern Albania, Macedonia, Bulgaria, and Northern Greece. There is an isolated part of the range in the Southern Carpathians. Data exist on the presence of this species in North West Anatolia (Mt Ulu Dagh) (Wettstein 1896; Hartvig 1991), but according to Pritchard (1978b), these findings refer to the species *G. caucasea* (Loddiges ex Sims) Holub. Furthermore, *G. bulgarica* is cited for Mt Prokletije in South West Serbia-Kosovo province and in South West Montenegro (Markišić 2002), but those data have not been checked. The Western border of its range is on the line of the Mts Šarplanina-Korab-Northern Pindus, and Pindus is the Southernmost point of its distribution; the Eastern border is on a line through the Central part of Mt Stara Planina-Central Rhodopes–Karlik Dagh.

The plant inhabits mainly mountainous and high-mountain pastures and stony grounds of siliceous mountains at 1,000–2,500 m.

Gentianella germanica (Willd.) E. F. Warb. aggr. (*Gentianella germanica* subsp. *rhaetica* (A. Kern. and Jos. Kern.) Hayek, *Gentianella germanica* subsp. *solstitialis* (Wettst.) Holub) (Fig. 5.5e)
Verified data for the presence of this species in the Balkan Peninsula exist for Bosnia and Herzegovina (Hayek 1930; Mayer 1969b), for Bulgaria (Kožuharov and Petrova 1982) and also for Croatia (Nikolić 1997). Jovanović-Dunjić (1973)

regarded the presence of this species in Serbia as uncertain, and thought it should be checked. However, by revision of material, the presence of the species *G. germanica* was confirmed in Western and South West Serbia in the localities on Mts Tara and Kopaonik, as well as in the region of Novi Pazar. Through their characters, revised specimens fully match those from Bosnia, initially determined as *G. rhaetica* or *G. solstitialis*.

G. germanica is distributed in the Balkan Peninsula in Central and Eastern Bosnia and Herzegovina, in Western and South West Serbia and in Bulgaria; in Croatia it is probably found near the Slovenian border. The distribution of this species in the western parts of the Peninsula, in Bosnia and Western Serbia, is scattered and often sympatric with *G. austriaca* and/or *G. lutescens*.

Gentianella austriaca aggr. (*G. austriaca* (A. Kerner and Jos. Kerner) Holub, *G. lutescens* (Velen.) Holub, *G. praecox* (A. Kern. and Jos. Kern.) Holub) (Fig. 5.5f)
This taxon is the one most widely distributed in the group *G. germanica* in the Balkan Peninsula. Taxonomically and morphologically it is not differentiated enough. Pritchard and Tutin (1972), Greimler et al. (2004) and Greimler (2010) based their delimitation between the species *G. austriaca* and *G. lutescens* on the calyx lobe length to tube length ratio, pointing out that calyx teeth are longer than the lobe in *G. austriaca*, and are the same length or shorter in *G. lutescens*. However, the plants were found frequently in Western and Central parts of Serbia, collected from the same locality and at the same time, where on one single specimen there exist flowers some of which exhibit ratios characteristic for *G. austriaca*, and the others those of *G. lutescens*. In our opinion the delimitation between the species *G. austriaca* and *G. lutescens* on the basis of those differential characters is not justifiable and we consider that *G. lutescens* should be regarded only as a variety of the species *G. austriaca*, as suggested by Mayer (1969b), Jovanović-Dunjić (1973) and Kožuharov and Petrova (1982). On the distribution map of the *G. austriaca* s.l. are marked the localities on which those two taxa grow in sympatry.

5.3.4 Distribution of the Genus Gentianella in the Balkan Peninsula: An Overview

The distribution of the species of the genus *Gentianella* in the Balkan Peninsula is more or less continual in Western and Central parts of the Peninsula to the line Šarplanina-Pirin-Rhodope in the South (Fig. 5.6). From that line, the number of species declines rapidly so that Southwards in Greece, only *G. ciliate* is found.

Three centers, comprising four–six species, are prominent, these being 1. Mt Prokletije and surrounding mountains of North Montenegro (Mts Komovi, Durmitor), 2. Mts Šarplanina and Korab bordering Montenegro, Albania and Kosovo province and 3. Mt Rila in Bulgaria. On Mt Prokletije six species were

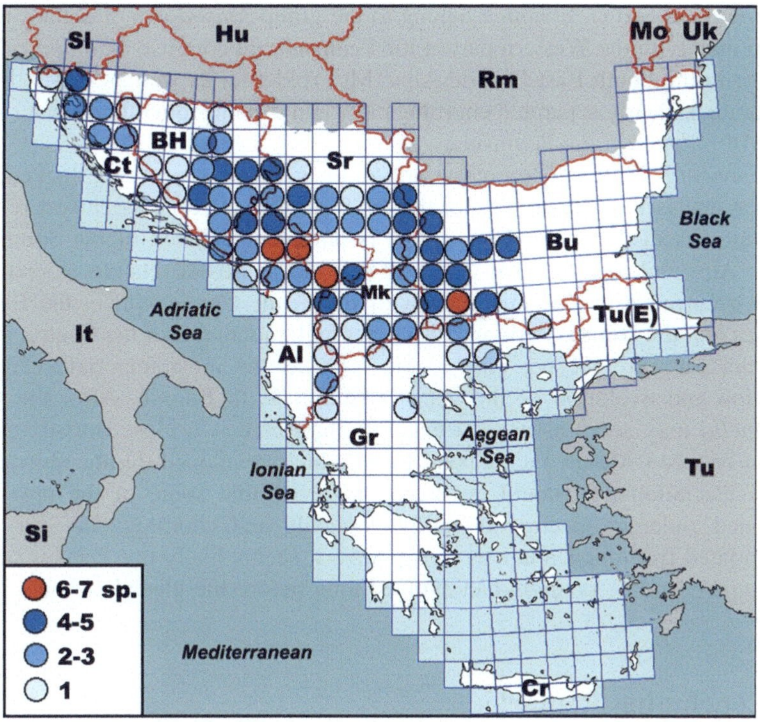

Fig. 5.6 General distribution of the genus *Gentianella* Moench. in the Balkans presented as numbers of species per MGRS, grid cell 50 × 50 km

recorded (*G. ciliata*, *G. austriaca* s.l., *G. crispata*, *G. pevelekii*, *G. albanica*, *G. laevicalyx*), on Mts Šarplanina and Korab five species (*G. austrica* s.l., *G. bulgarica*, *G. ciliata*, *G. albanica* and *G. crispata*) and on Mt Rila five species (*G. engadinensis*, *G. ciliata*, *G. austrica* s.l., *G. bulgarica* and *G. germanica*). In the mountains of Central Bosnia, in Western and Central Serbia and in Bulgaria (Mts Vitoša, Pirin, Central parts of Mt Stara Planina, Rhodopes) three species were registered. The largest number of species occupy the greatest and highest mountain massifs, such as Mts Prokletije, Šarplanina with Korab and Rila with Pirin, which all show the greatest geological diversity.

In the Balkan Peninsula grow four endemic species of the genus *Gentianella*, all in the Sect. *Gentianella* (*G. liburnica, G. albanica, G. laevicalyx* and *G. pevalekii*). In addition, in the Balkan Peninsula, another two species are present, *G. crispata* and *G. bulgarica,* which are distributed mainly in the Peninsula with small disjunctions beyond the borders of the Peninsula. These species in the broader sense may be regarded as endemics (subendemics). The disjunct area of the range of *G. crispata* is situated in Calabria and is an example of Amphi-Adriatic disjunction, whereas a part of the range of *G. bulgarica* is situated in the Southern Carpathians.

G. liburnica and *G. albanica* belong to the group *Anisodonta*, and their ranges overlap mainly in the Western part of the Peninsula. In contrast, *G. albanica,* with its range in the South East Dinaric Alps (Mt Prokletije) as well as in Mts Paštrik and Korab, is clearly separated chorologically and distant from other species of the group *Anisodonta* in the Peninsula.

G. laevicalyx is the only representative of the group *Campestris* in the Peninsula and its range is more than 500 km distant from the range of its closest relative, *G. campestris* in the Alps. *G. pevalekii* is an endemic taxon of the South East Dinaric Alps and its range overlaps with that of the closest relative *G. crispata*.

The occurrence of endemic species of the genus *Gentianella* in the Balkans indicates that the Dinaric Alps and also other high mountains of the Central part of the Peninsula were second, only to the Alps, as the most important center of speciation and evolution of the genus *Gentianella* in Europe. Also, the genus *Gentianella* may serve as a good example that there is a close floristic-genetic relation between Alpine and Balkan floras, which developed during several successive glaciations. It should be noted, however, that some of the previously mentioned endemic species are morphologically and chorologically very well differentiated from their relatives in the Alps. It seems likely that species such as *G. bulgarica* and *G. crispata* had evolved long before the glacial period.

5.4 Concluding Remarks

The genus *Gentiana* in the Balkan Peninsula is represented by 17 taxa ranked as species and subspecies, accounting for 45 % of the total number of its taxa occurring in Europe. Representatives of this genus in the Balkan Peninsula are classified into the 7 sections *Gentiana, Pneumonanthe, Cruciata, Frigida, Chondrophylla, Ciminalis* and *Calathianae*. The largest number of species inhabits high mountain areas of the Dinaric Alps (Mts Vranica, Durmitor, Prokletije), Rhodopes, and the Balkan mountains in Bulgaria (Mts Vitoša, Rila, Pirin, Western and Central parts of Mt Stara Planina), as well as Northern parts of the Scardo-Pindic mountains (Mts šarplanina, Korab). In addition, the lower (below 2,000 m) calcareous mountains of the Western part of the Peninsula (Mts Velebit, Risnjak, Plješevica) are characterized by a considerable number of species of the genus *Gentiana* owing to the floristic influence from the neighboring mountain range of the Alps.

In the Balkan Peninsula, endemism within the genus *Gentiana* is not prominent, in contrast to a considerable number of endemic species of the same genus in the Alps. A true endemic species of the Balkan Peninsula is *G. pneumonanthe* subsp. *nopscae*, which grows in humid habitats of Kosovo Province and Northern Albania. However, *G. tergestina* may be also regarded as an endemic species of the Balkan Peninsula, taking into account that its range extends from Trieste Karst via the Dinaric Alps to the mountains of South West Macedonia, Albania, and Bulgaria. *G. dinarica*, should be also mentioned, since it is a subendemic species of the Balkan Peninsula, given that it is mainly distributed in the Dinaric Alps, from Mt Velebit to the Prokletije massif, whereas a disjunct part of its range is situated in the Central Apennines. Moreover, the distribution of this species points to the ancient amphi-adriatic orophytic relation and highlights the significance of the Dinaric Alps in the speciation of certain groups of plants from the section *Ciminalis* and *Calathianae* within the genus *Gentiana*. In any case, the scarcity of endemic species of this genus in the Balkan Peninsula should confirm the fact that the Peninsula had not been an important speciation center of the genus *Gentiana*, but instead, it is a region into which species migrated from the Alps or the Carpathians. Those migrations must have been particularly operative during the ice ages. The descent of the snow line "pushed" the orophytic flora Southwards, and the subsequent migration of this flora toward the Balkan mountains was facilitated considerably.

In the Balkan Peninsula, the genus *Gentianella* is represented by 13 species, which is almost a half the total number of its taxa recorded in Europe. Taxa of this genus in the Peninsula are classified into two sections, the *Crossopatalae* and *Gentianella*. The presence of the representatives of the section *Comastoma* of the genus *Gentianella* is controversial. The largest number of species was recorded in the mountain massifs of Prokletije, Šarplanina, and Korab, on Mt Pirin and the Western Rhodopes. Additionally, a considerable number of species grow in the mountains of Central Bosnia, Northern Montenegro, Central and Eastern Serbia (Mts Kopaonik, Stara Planina), and Western Bulgaria (Mts Stara Planina, Rila).

Unlike *Gentiana*, the genus *Gentianella* is distinguished by a remarkable number of endemic species. Thus, almost 50 % of the representatives of the genus *Gentianella* that occupy the Balkan Peninsula are either endemic or subendemic taxa. The endemic species, *G. liburnica*, *G. albanica, G. laevicalyx* and *G. pevalekii*, belong to the Sect. *Gentianella*.

The species *G. liburnica* and *G. albanica* belong to the group *Anisodonta*; their ranges are delimited from the ranges of other taxa from the same group. The range of *G. liburnica* is restricted to some mountains of Slovenia and to the Northern Mt Velebit, on which, however, it overlaps with the range of *G. anisodonta*. The range of *G. albanica* is in the South East Dinaric Alps (Mt Prokletije), as well as Mts Paštrik and Korab, so that in the Balkan Peninsula it is clearly separated spatially and distantly from the range of other species of the group *Anisodonta*, such as *G. anisodonta, G. liburnica*, and *G. engadinensis*.

G. laevicalyx, as the only representative of the group *Campestris* in the Peninsula, is distinguished by a small range, which is restricted to the mountains of Northern Montenegro and it is more than 500 km from the range of its closest relative, *G. campestris,* in the Alps. *G. pevalekii* is an endemic taxon of the South East Dinaric Alps. In those mountains, its range overlaps, in part, with the range of its closest relative, *G. crispata*.

The species *G. crispata* and *G. bulgarica* are distributed mainly in the Balkan Peninsula, each with a small disjunct part of the range beyond the borders of the Peninsula. Thus, those species can be regarded as subendemics. The species *G. crispata* is characterized by an amphi-adriatic disjunction, since the largest part of its range is situated in the Balkan Peninsula, and a small, isolated part is in Calabria, on the Apennine Peninsula. The part of the range of the species *G. bulgarica* beyond the Balkan Peninsula is located in the Southern Carpathians; this disjunction is indicative of floristic relations between the Carpathians and the mountains of the Central and Eastern parts of the Peninsula.

The distribution of endemic species of the genus *Gentianella* calls to mind the fact that the Dinaric Alps, together with other high mountains of the Central part of the Peninsula, were the foremost center of speciation and evolution of the genus *Gentianella* in Europe, this center being second in its significance only to the Alps. The relation between the flora of the Alps and that of the Balkan Peninsula is strongly expressed within the genus *Gentianella*, and was established primarily during successive glaciations. In contrast, some endemic species, such as *G. bulgarica* and *G. crispata,* are very well morphologically and chorologically differentiated from their Alpine relatives. In fact, so much so that this differentiation might easily have occurred long before the glacial periods.

Certain interdependent features of the Balkan Peninsula gave rise to some of the Peninsula's floristic specificities, and at the same time account for the exceptional richness in species of the genera *Gentiana* and *Gentianella*. High mountains and mountain massifs of the Peninsula, all notable for their geological heterogeneity, are situated in close proximity to the Alps on their West, and to the Carpathians in the North East. With the lowering of the snow line during major climatic events and the ensuing plant migration Southwards, the proximity of the Balkans mountains, their height and geological diversity, turned them into most suitable habitats for the migrating alpine plants (orophytes) of the neighboring high mountain ranges. It is noteworthy that larger and higher Balkan mountains, particularly those with more geological diversity, harbor more representatives of both *Gentiana* and *Gentianella*. Certainly, the floristic relations between the Balkan mountains on one side, and the Alps and the Carpathians on the other, are now fairly limited compared to the periods of glaciations when these contacts were more direct with fewer obstacles, and enabled the expansion of many species of both genera from these mountain massifs into the Balkans. In favor of these suggestions is the distribution of both genera which sharply declines South of the mountain line, Mts Šarplanina-Pirin-Rhodopes (between latitudes 41° and 42° N) and becomes uneven and sporadic further South, toward the South of Greece. This mountain line, is at the same time, the Southernmost point which the expanding mountain glaciations reached in the Balkan Peninsula during the glacial periods, and which the migrating plants did not cross. Such distribution of both *Gentiana* and *Gentianella* is in agreement with the fact that the majority of mid-European orophytes, Arcto-Alpine as well as boreal species, do not cross the glaciations' maximum extension line on the high mountains of the Scardo-Pindic and the Rhodopes mountain massifs (Horvat 1953; Stevanović et al. 2009).

In the Balkan Peninsula, the present knowledge on both *Gentiana* and *Gentianella* carries the burden of unresolved complex taxonomic and chorologic problems. First in line is the delimitation between the species *Gentiana clusii* and *G. verna* in the Western part of the Peninsula. Also, within the Sect. *Calathianae*, in the group *Gentiana verna*, the relations among the subspecies *verna*, *tergestina* and *balcanica* are neither taxonomically nor chorologically resolved. A taxonomical concept was proposed by Hammerli (2007), according to which *G. verna* does not inhabit the Balkans, but instead there lives *G. tergestina* with the type subspecies *tergestina* and the subspecies *balcanica*. This concept demands a serious check of rather abundant plant material, since broad zones of overlap exist among the distribution areas of the respective subspecies.

A specific problem exists for the representatives of the group *Gentianella germanica* in the Balkan Peninsula. Taxonomic controversies regarding this group are due to hybridization, introgression and also to the expression of seasonal dimorphism, occurring among its species, all of which results in the exceptional morphologic variability within this group. Studies in molecular phylogeny could help to resolve these difficulties, provided they are performed on large samples of plant material. Also, detailed morphometric analyses should be carried out, but again, a representative number of specimens would have to come from as many populations as possible, and covering the entire ranges of different taxa of this complex in the Balkan Peninsula.

Acknowledgments This study was supported by the funding from the Ministry of Education, Science and Technological Development, Republic of Serbia (Grant no. 173030, project: "Plant biodiversity of Serbia and the Balkans—assessment, sustainable use and protection").

References

Aeshiman D, Lauber K, Moser DM, Theurillat JP (2004) Flora Alpina, vol 2. Hauptverlag, Bern

Bjelčić Ž, Mayer E (1973) Zur taxonomie des *Gentianella crispata*-komplexes. Österr Bot Zeit 122:353–358

Degen A (1937) Flora Velebitica, vol 2. Verlag der Ungar, Akademie der Wissenschaften, Budapest

Dimitrov D (2011) *Gentiana frigida* L. In: Red data book of the Republic of Bulgaria, vol 1. Bulgarian Academy of Sciences, Ministry of Environment and Water, Sofia. http://e-ecodb.bas.bg/rdb/en/vol1/Genfrigi.html#map. Cited 24 June 2012

Doluhanov AG (1967) *Gentiana* L. In: Grossheim AA (ed) Flora of caucasus, umbelliferae-scrophulariaceae, vol 7. Nauka, Leningrad, pp 206–215

Euro+Med (2011) Euro+Med Plantbase—the information resource for Euro-Mediterranean plant diversity. http://www.emplantbase.org/home.html. Cited 24 June 2012

Evstatieva L (2011) *Gentiana punctata* L. In: Red data book of the Republic of Bulgaria, vol 1. Bulgarian Academy of Sciences, Ministry of Environment and Water, Sofia. http://e-ecodb.bas.bg/rdb/bg/vol1/Genpunct.html#map. Cited 21 June 2012

Gaži-Baškova V (1994) *Gentiana acaulis* L. In: Šugar I (ed) Crvena knjiga biljnih vsta Republike Hrvatske. Ministarstvo graditeljstva i zaštite okoliša, Zavod za zaštitu prirode, Zagreb, pp 218–219

Greimler J (2010) Revision of *Gentianella austriaca* s.l. and *G. rhaetica* in Slovenia. Hladnikia 25:3–12

Greimler J, Hermanowski B, Jang CG (2004) A re-evaluation of morphological characters in European *Gentianella* section *Gentianella* (Gentianaceae). Plant Syst Evol 248:143–169

Greimler J, Park JM, Schneeweiss H (2011) *Gentianella* (Gentianaceae): a model taxon for evolution in the Alps. Taxon 60:427–435

Greuter W, Burdet HM, Long G (1986) Med-Checklist 3, Dicotyledones (Convolvulaceae-Labiatae). Conservatoire et Jardin Botaniques de la ville de Geneve, Optima, Geneva

Grossheim AA (1952) *Gentiana* L. In: Shishkin BK, Bobrov EG (eds) Flora USSR, vol 18. Academy of Science USSR, Moskow, Leningrad, pp 538–620

Hammerli M (2007) Molecular aspects in systematics of *Gentiana* Sect. *Calathianae* Froel. http://doc.rero.ch/lm.php?url=1000,40,4,20071218101507-RK/these_HaemmerliM.pdf. Cited 12 July 2012

Hartvig P (1991) Gentianaceae. In: Strid A, Tan K (eds) Mountain flora of Greece, vol 2. Edinburgh University Press, Edinburgh, pp 1–9

Hayek A (1927) Prodromus florae peninsulae Balcanicae. Fed Rep Beih 30:1–336

Hayek A (1930) Prodromus florae peninsulae Balcanicae. Fed Rep Beih 30:337–576

Hess HE, Landolt E, Hirzel R (1972) Flora der Schweiz und angrenzender Gebiete, vol 2. Birkhäuser Verlag, Basel, Stuttgart

Hirc D (1909) Revizija hrvatske flore (Revisio florae croaticae). Rad Jugosl Akad Znan 179:1–62

Ho TN, Lui SW (1990) The infrageneric classification of *Gentiana* (Gentianaceae). Bull Br Mus (Nat Hist) Bot Ser 20:169–192

Ho TN, Pringle JS (1995) Gentianaceae B. Juss. In: Wu ZY, Raven PH (eds) Flora of China, vol 16. Science Press, Beijing and Missouri Botanical Garden, St. Louis, pp 1–139

Horvat I (1953) Prilog poznavanju raširenja nekih planinskih biljaka u jugoistočnoj Evropi. Godišnj Biol Inst Sarajevo 5:199–218

Hultén E (1961) Alpigenous and Arctogenous plants. Ranges of circumpolar Arctic-Montane plants. Recent Advances in Botany 1, Toronto

Hungerer KB, Kadereit JW (1998) The phylogeny and biogeography of *Gentiana* L. sect. *Ciminalis* (Adans.) Dumort.: a historical interpretation of distribution ranges in the European high mountains. Perspect Plant Ecol 1:121–135

IOPI (1996-2003) International Organization for Plant Information. http://plantnet.rbgsyd.nsw.gov.au/iopi/iopigpc1.htm/. Cited 23 June 2012

Ivanova D (2011) *Gentianella amarella* (L.) Börner. In: Red data book of the Republic of Bulgaria, vol 1. Bulgarian Academy of Sciences, Ministry of Environment and Water, Sofia. http://e-ecodb.bas.bg/rdb/bg/vol1/Genamare.html. Cited 2 July 2012

Javorka S (1921) Uj adatok Albániá flórájához (Novitates florae Albaniae). Botanikai közlemenijek 19:17–30

Javorka S (1926) Adatok Albania flórájahoz-Additamenta ad floram Albaniae. A Mag Tudomán Akad Balkán-kutatás tudomán eredmén 3:74–89

Jovanović-Dunjić R (1973) Gentianaceae B. Juss. In: Josifović M (ed) Flora of the Republic of Serbia, vol 5. Belgrade, Serbian Academy of Science and Art, pp 403–433

Kožuharov S, Petrova A (1982) Gentianaceae Juss. In: Jordanov D (ed) Flora PR Bulgaria, vol 8. Bulg Acad Sci Press, Sofia, pp 396–408

Kurtto A, Weber HE, Lampinen R, Sennikov AN (eds) (2010) Atlas Florae Europaeae. Distribution of vascular plants in Europe. Rosaceae (*Rubus*), vol 15. The Committee for Mapping the Flora of Europe and Societas Biologica Fennica Vanamo, Helsinki

Kušan F (1969) Biljni pokrov Biokova. Prirod Istraž, Ser Acta Biol 5:5–224

Lakušić R, Grgić P (1969) Rasprostranjenost i ekologija vrste *Valeriana pancicii* Hal. et Bald., *Valeriana bertiscea* Panč., *Asperula dorfleri* Wettst. i *Gentiana laevicalyx* Rohlena. Ekologija 4:237–245

Marhold K (2011) Gentianaceae. In: Euro+Med Plantbase-the information resource for Euro-Mediterranean plant diversity. http://ww2.bgbm.org/EuroPlusMed/PTaxonDetail.asp?NameCache=Gentianaceae&PTRefFk=7200000. Cited 25 June 2012

Markgraf F (1931) Pflazen aus Albanien, 1928. Denkschriften der Akademie der Wissenschaften in Wien, Mathematisch-naturwissenschaftliche Klasse 102. Holder-Pichler-Tempsky, AG, Wien und Leipzig

Markišić H (2002) Priroda Rožaja- Floristične i vegetacijske specifičnosti. In: Martinović MŽ, Markišić H (eds) Priroda Rožaja. Rožaje, pp 247–366

Matevski V, Wraber T (2010a) *Gentiana* L. In: Matevski V (ed) The flora of the Republic of Macedonia, vol 2(1). Mac Acad Sci Art Skopje, pp 31–40

Matevski V, Wraber T (2010b) *Gentianella* Moench. In: Matevski V (ed) The flora of the Republic of Macedonia, vol 2(1). Mac Acad Sci Art Skopje, pp 41–46

Mayer E (1969a) Zur Kenntnis der Gattung *Gentianella* Moench in Jugoslawien I.Der G. anisodonta-Komplex. Österr Bot Zeit 116:393–399

Mayer E (1969b) Zur Kenntnis der Gattung *Gentianella* Moench in Jugoslawien. II. Der *G. aspera- G. germanica-* und *G. austriaca*-Komplex. Biološki Vestnik 16:23–28

Mayer E, Bjelčič Ž (1974a) *Gentiana* L. In: Čović B (ed) Flora Bosnae et Hercegovinae. Zemaljski muzej Bosne i Hercegovine 4:50–53

Mayer E, Bjelčič Ž (1974b) *Gentianella* Moench. In: Čović B (ed) Flora Bosnae et Hercegovinae. Zemaljski muzej Bosne i Hercegovine 4:54–57

Meusel H, Jäger E, Raushert S, Weinert E (1978) Vergleichende Chorologie der Zentraleuropäischen Flora, Band II, Karten. Gustav Fisher Verlag, Jena

Mišić L (1965) Biljnogeografsko rasprostranjenje vrste *Gentiana dinarica* Beck. God Inst Biol Univ Sarajevo 18:199–208

Nikolić T (1997) Flora Croatica. Index Florae Croaticae. Pars 2. Nat Croat 6(1):1–232

Nikolić T (ed) (2012) Flora croatica red book. Department of Botany, Faculty of Science, University of Zagreb, Croatia. http://hirc.botanic.hr/fcd/CrvenaKnjiga/. Cited 16 June 2012

Pavletić Z (1994) *Gentiana clusii* Perr. et Song. In: Šugar I (ed) Crvena knjiga biljnih vsta Republike Hrvatske. Ministarstvo graditeljstva i zaštite okoliša, Zavod za zaštitu prirode, Zagreb, pp 222–223

Pignatti S (1982) Flora d'Italia, vol 2. Edizioni Agricole, Bologna

Pritchard NM (1977) *Gentianella* and *Gentiana*. In: Davis PH (ed) Materials for a flora of Turkey XXIV. Notes Roy Bot Gard Edinburgh 35:312–313

Pritchard NM (1978a) *Gentiana* L. In: Davis PH (ed) Flora of Turkey, vol 6. Edinburgh University Press, Edinburgh, pp 183–190

Pritchard NM (1978b) *Gentianella* Moench. In: Davis PH (ed) Flora of Turkey, vol 6. Edinburgh University Press, Edinburgh, pp 190–193

Pritchard NM, Tutin TG (1972) *Gentianella* Moech. In: Tutin TG, Heywood VH, Burges NA, Moore DM, Valentine DH, Walters SM, Webb DA (eds) Flora Europaea, vol 3. Cambridge University Press, Cambridge, pp 63–67

Rohlena J (1912) Fünfter Beitrag zur flora von Montenegro. Sitzungsber Königl Böhm Ges Wiss Prag, Math-Naturwiss Cl 5:1–143

Rohlena J (1937–1938) Neue Pflanzen aus Montenegro. Fedde Repert 42:150–152, 319–320

Rohlena J (1942) Conspectus Florae Montenegrinae. Preslia 20–21:1–506

Rossi M (2011) Taxonomy, phylogeny and reproductive ecology of *Gentiana lutea* L. Ph.D. in biodiversity and evolution, faculty of mathematical, physical and natural sciences, Department of Experimental Evolutionary Biology, University of Bologna

Schiman-Czeika H (1967) *Gentiana* L. In: Rechinger KH (ed) Flora Iranica, vol 41. Akademische Druck u Verlagsanstalt, Graz, pp 10–22

Stevanović V, Vukojičić S, Šinžar-Sekulić J, Lazarević M, Tomović G, Tan K (2009) Distribution and diversity of Arctic-Alpine species in the Balkans. Plant Syst Evol 283:219–235

Thiers B (2013) Index Herbariorum: A global directory of public herbaria and associated staff. New York Botanical Garden's Virtual Herbarium. http://sweetgum.nybg.org/ih/. Cited 20 June 2012

The Plant List (2010) Version 1. http://www.theplantlist.org/. Cited 23 June 2012

Tomović G (1999) *Gentiana pneumonanthe* L. subsp. *nopscae* (Jáv) T.Wraber. In: Stevanović V (ed) The red data book of Flora of Serbia 1. Extinct and critically endangered taxa. Ministry of environment of the Republic of Serbia, Faculty of Biology, University of Belgrade, Institute for Protection of Nature of the Republic of Serbia, Belgrade, pp 151–152

Topa E (1961) Gentianaceae. In: Savulescu T (ed) Flora Republici Populare Romine, vol 8. Acad Repub Popul Romine, Bucaresti, pp 430–479

Turrill WB (1929) The plant life of the Balkan Peninsula. A phytogeographic study. Clarendon Press, Oxford

Tutin TG (1972) *Gentiana* L. In: Tutin TG, Heywood VH, Burges NA, Moore DM, Valentine DH, Walters SM, Webb DA (eds) Flora Europaea, vol 3. Cambridge University Press, Cambridge, pp 59–63

Vangeli J, Ruci B, Mullaj A (1994) Red book. Threatened and rare plant species of Albania. Acad Sci, Inst Biol Res, Tirana

von Hagen KB, Kadereit JW (2001) The phylogeny of *Gentianella* (Gentianaceae) and its colonization of the southern hemisphere as revealed by nuclear and chloroplast DNA sequence variation. Org Divers Evol 1:61–79

Wettstein R (1896) Die Europäischen arten der gattung *Gentiana* aus der section *Endotricha* Froel. und ihr entwicklungsgeschichtlicher Zusammenhang. C Gerold, Wien

Wraber T (1990) De *Gentiana nopscae* Jávorka notula. Biol Vestn 38:39–46

Yuan YM, Küpfer P (1995) Molecular phylogenetics of the subtribe *Gentianinae* (Gentianaceae) inferred from the sequences of internal transcribed spacers (ITS) of nuclear ribosomal DNA. Plant Syst Evol 196:207–226

Yuan YM, Küpfer P, Doyle JJ (1996) Infrageneric phylogeny of the genus *Gentiana* (Gentianaceae) inferred from nucleotide sequences of the internal transcribed spacers (ITS) of nuclear ribosomal DNA. Amer J Bot 83:641–652

Chapter 6
Biochemical and Morphological Determination of Flower Color in Gentianaceae

Ewa Młodzińska

Abstract Since Charles Darwin wrote his excellent book about flower-color polymorphism within the same species (Darwin 1877), studies on plant dyes have become one of the most popular research subjects in botany. After nearly 200 years of investigation into the biochemistry and physiological factors involved, the pathway leading to the accumulation of pigments in petals is well understood. Flavonoids are the major pigments that confer flower color, and their biosynthesis is now characterized at the genetic and enzymatic level in many decorative plants, including Eustoma, Carnation, Petunia, and Snapdragon. This chapter summarizes the current status of all the key biosynthesis steps and secondary modification of the pigments and co-pigments, which are responsible for the diversity of color patterns observed in Gentian species. A brief overview is also given about the cellular and subcellular localization of plant pigments in order to facilitate understanding of the molecular basis of color patterning in flowers. More specifically, anthocyanin vacuolar inclusion structures (AVIs) have been shown to affect the final color by accumulating high concentrations of anthocyanins. The shape and morphology of petal epidermal cells may play an important role in determining flower color in Gentianaceae.

6.1 Introduction

Representatives of the family Gentianaceae are found in subtropical and temperate climates and the color of their flowers corresponds to the natural environment in which they reside in reference to the classification described by Beale et al. (1941). Accordingly, plants inhabiting subtropical zones tend to have red flowers due to

---

E. Młodzińska (✉)
Department of Plant Molecular Physiology, Institute of Experimental Biology,
ul. Kanonia 6/8 50-328 Wrocław, Poland
e-mail: ewa.mlodzinska@uni.wroc.pl

J. J. Rybczyński et al. (eds.), *The Gentianaceae - Volume 1: Characterization and Ecology*, DOI: 10.1007/978-3-642-54010-3_6,

the presence of pelargonidin and cyanidin, whereas blue flowers with delphinidin derivatives usually predominate in plants of mountainous areas. These pigments (pelargonidin, cyanidin, and delphinidin) belong to the anthocyanins which are primarily responsible for most of the red, purple, pink and blue colors of flowers. Other color pigments are carotenoids, chlorophylls, and betalains (Griesbach 2005; Grotewold 2006).

Wild plants of the genus *Gentiana* produce very impressively bright-colored flowers from white, and creamy-yellow (*Gentiana lutea* L.) to vivid blue (*Gentiana triflora* L.). Furthermore, *Lisianthus nigrescens* has almost black colored petals and is known as the blackest flower in the world (Nicholson 1999). Thanks to unusual colors and the decorative shapes of flowers, gentians are very popular ornamental plants, especially in Japan. Cultivars of gentians are used in rock gardens (*Gentiana dinarica, Gentiana punctata* L.) and also as potted plants and cut flowers (*Eustoma* sp.). A wide variety of flower color in Gentianaceae is the result of the accumulation of secondary metabolites, such as carotenoids and anthocyanins, as well as their subsequent modification by glucosylation and acylation. In addition, many factors such as co-pigmentation, metal ion complexation, the pH of the vacuolar lumen and temperature, determine the intensity and shade of color (Davies 2004; Młodzińska 2009). Pigmentation of the petals may also be affected by the shape of epidermal cells. Moreover, color of the flowers may also change in response to pollination when the pH in petal cells changes, or pigments are synthesized and degraded (Griesbach 2005; Miller et al. 2011).

6.2 Anthocyanin Color and Biosynthesis in Gentian Plants

Anthocyanins are synthesized via the phenylpropanoid pathway in the cytosol and the biosynthesis of these secondary metabolites is conserved among higher plants. These compounds have been studied extensively in petunia, snapdragon, maize, and Arabidopsis (Tornielli et al. 2009; Holton and Cornish 1995; Mol et al. 1998). Over the past few years, considerable progress has been made in our understanding of the molecular and enzymatic stages in anthocyanin biosynthesis. All encoding enzymes involved in the main pathway of flavonoid biosynthesis, as well as a few regulatory genes for transcriptional regulation, have been cloned and classified from several cut flowers. Genetics and biochemistry research are focused mostly on two species of gentian, namely *G. triflora* and *Eustoma grandiflorum* (Lisianthus) in order to elucidate the flavonoid pathway which is responsible for providing many color variants including, pink, mauve, purple, and blue in gentian flowers (Tanaka et al. 1996; Shamir-Oren et al. 1999). It is worth noting that floricultural gentians do not have yellow or red colored petals (Nakatsuka et al. 2011). Among six common anthocyanidins (pelargonidin, cyanidin, peonin, delphinidin, petunidin, and malvidin), only three, namely cyanidin, pelargonidin, and delphinidin, are accumulated in gentians. The pink flower color in *E. grandiflorum* is derived primarily from pelargonidin glycosides and the mauve flowers arise

Fig. 1 Gentian flowers (*Gentiana ciliata*) containing blue gentiodelphin—diacylated delphinidin based anthocyanins

from cyanidin and pelargonidin glycosides (Markham and Ofman 1993; Shamir-Oren et al. 1999). Despite the diversification of colors, most gentian flowers are blue and their pigmentation is determined by delphinidin derivatives (Fukuchi-Mizutani et al. 2003). The intensively studied example of delphinidin derivatives is gentiodelphin (Fig. 1) containing aromatic acyl groups (delphinidin 3-O-glucosyl-5-O-(6-O-caffeolyl-glucosyl)-3′-O-6-O-caffeoyl-glucoside) (Goto et al. 1982; Hokosawa et al. 1997; Fukuchi-Mizutani et al. 2003). In general, blue color is associated with a greater pH inside vacuoles. However, gentiodelphin possess a unique blue color at a wide range of pH as a result of intramolecular interaction between anthocyanidin and two caffeic acids which, in turn, stabilize the structure of this dye (Yoshida et al. 2000).

The biosynthetic steps leading to anthocyanin accumulation in gentians are well defined, with reviews on the molecular biology being published during the last few years (Tanaka et al. 1996; Nakatsuka et al. 2005; Nishihara et al. 2008). There are also several publications available on some of the specific enzymes from gentian, including glucosyltransferases (Nakatsuka et al. 2008a), acyltransferases (Fujiwara et al. 1998), and flavonoid hydroxylases (Tanaka et al. 1996). Nakatsuka et al. (2005) described in detail the putative flavonoid biosynthesis pathway in gentians that leads to the production of gentiodelphin and gentiocyanin. Pink-colored gentians accumulate predominantly gentiocyanin (Fig. 2), whereas blue-colored flowers accumulate gentiodelphin (Hosokawa et al. 1995; Hokosawa et al. 1997).

Fig. 2 *Pink-colored* flowers of Gentiana (*Centaurium erythraea*) contain cyanidin derivatives—gentiocyanin

To date, pelargonidin derivatives have not been detected in wild gentian flowers (Tanaka et al. 1996). However, Shamir-Oren et al. (1999) produced through cross-breeding two pink lines of lisianthus (*E. grandiflorum*) which contain a high concentration of pelargonidin (almost 100 %) in their petals. About 19 different anthocyanins have been identified in flowers from the family Gentianaceae (Andersen and Jordheim 2006). Several steps of the putative anthocyanin biosynthetic pathway in gentian plants are illustrated in Fig. 3.

The first enzyme in the biosynthesis of flavonoids is chalcone synthase (CHS) which catalyzes the condensation of one molecule of 4-coumaroyl CoA and three molecules of maloyl CoA to tetrahydroxychalcone (naringrninchalcone). Only one gene, *GtCHS1*, was cloned from *G. triflora* encoding a putative protein which exhibited 88 % homology to chalcone synthase (CHSA) from *Petunia hybrida* (Kobayashi et al. 1998). Several sequences for binding protein regulators MYB and MYC were found in the promoter region of *GtCHS1*. These transcriptional factors are involved in the regulation of anthocyanin biosynthesis in Arabidopsis, Petunia, and Antirrhinum (Ramsay et al. 2003; Quattrocchio et al. 1993, 2006; Schwinn et al. 2006). It seems likely that *MYB* and *MYC* genes might contribute to control of pigment biosynthesis in *Gentiana* (Kobayashi et al. 1998). It has been reported that suppression of the CHS gene, using genetic engineering, induced either entirely white or white-patterned flowers in petunia (Tsuda et al. 2004;

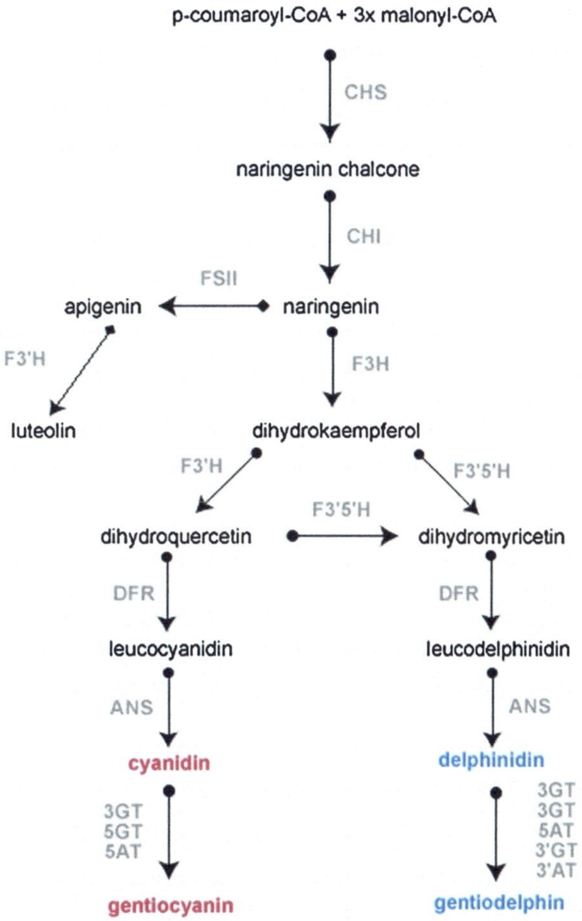

Fig. 3 Scheme of putative anthocyanin biosynthetic pathway in gentian flowers (Nakatsuka et al. 2005, modified). The final color depends on various factors described in the text. The enzymes are written in *gray* letters. Abbreviations: *CHS* chalcone synthase; *CHI* chalcone isomerase; *FSII* flavone synthase II; *F3H* flavanone 3-hydroxylase; *F3'H* flavonoid 3′-hydroxylase; *F3′5′H* flavonoid 3′5′-hydroxylase; *DFR* dihydroflavonol 4-reductase; *ANS* anthocyanidin synthase; *3GT* UDP-glucose: anthocyanin 3-O-glucosyltransferase; *3GT* UDP-glucose:anthocyanin 3-O-glucosyltransferase 5AT-anthocyanin 5-aromatic acyltransferase; *5′GT* UDP-glucose: anthocyanin 5′-O-glucosyltransferase; *3′AT* anthocyanin 3′aromatic acyltransferase

Tornielli et al. 2009). Similarly, suppression of the CHS gene with antisense technology in gentians resulted in white flowers (Nishihara et al. 2006). An interruption of CHS using RNA interference (RNAi) as a method was more efficient than the antisense approach and resulted in pure white flowers in *G. triflora* and *G. scabra* and, in both cultivars, the concentration of anthocyanin was reduced considerably in terms of gene silencing compared with wild type (Nakatsuka et al. 2008b).

Naringeninchalcone is converted to naringenin by naringeninchalcon. It has been shown that stimulation of chalcone isomerase (CHI) activity in petals was correlated with increasing color intensity from pink to purple in *E. grandiflorum* (Davies et al. 1993). Naringenin is hydroxylated by flavanone 3-hydroxylase (F3H) leading to the formation of dihyrokempferol.

Flavonoid 3′hydroxylase (F3′H) and flavonoid 3′5′hydroxylase (F3′5′H) catalyze the hydroxylation to dihydroquercetin and dihydromyricetin, respectively.

These two enzymes are membrane-bound monooxygenases belonging to the monooxygenase family, required for determination of the structure of anthocyanins and, thus, their colors (Tanaka et al. 2008). It is well known that blue color is enhanced with increasing hydroxylation in the B ring of anthocyanidins which is catalyzed by the two enzymes F3′H and F3′5′H. Many ornamental plants such as roses, carnations, and chrysanthemums do not accumulate delphinidins because of the absence of F3′5′H, which is a key enzyme involved in delphinidin synthesis (Katsumoto et al. 2007; Tanaka et al. 2008; Tanaka et al. 2009). *Gentiana scabra* with insertion of transposable elements in the F3′5′H gene, exhibited pink color instead of blue, due to the accumulation of cyaniding-based anthocyanins (Nakatsuka et al. 2006). Transgenic gentian plants with RNAi suppression of the F3′5′H gene also showed decreased delphinidin and increased cyanidin derivatives (Nakatsuka et al. 2008b).

The next step of flavonoid biosynthesis is the reduction of dihydroflavonols to the leucoanthocyanidins by dihydroflavonol 4-reductase (DFR). The latter has a strong preference for dihyromyricetin and promotes accumulation of delphinidin-derived anthocyanins (Davies 2009). Gentian DFR may not have reducing ability and, therefore, pelargonidin has not been isolated from gentian flowers (Tanaka et al. 1996). Dihydroflavonols may also be converted to flavones and flavonols, respectively, by flavone synthase (FSI or FSII) and flavonol synthase (FLS). Flavones and flavonols are generally colorless or have pale colors, but they affect the color of anthocyanins through co-pigmentation (Griesbach 2005). In gentians, apigenin and luteolin were detected as main flavones (Nakatsuka et al. 2005). In two white-flowered cultivars of *G. triflora* (cv. Homoi) and hybrid *G. triflora* x *G. scabra* (cv. Polano White), flavone analyses showed that white apigenin derivates accumulated at greater concentration than luteolin. Moreover, pigment analysis in both Homoi and Polano White revealed undetectable amounts of anthocyanins. It has been suggested that gentian plants contain only small concentrations of flavonols or lack flavonols in petals (Tanaka 1996; Nakatsuka et al. 2005). Davies et al. (1993) described two lines of *Lisianthus*, W48 and W42, with cream-white and pure white flowers, respectively. The line W48 contained more flavonols in relative to line W42. The transcript level of DFR in line W48 was greatly reduced, which could be the reason why anthocyanidin products were not found in this line (Davies et al. 1993).

The final enzyme involved in converting colorless anthocyanidins into colored anthocyanins is anthocyanidin synthase (ANS), which belongs to the 2-oxoglutarate-dependent dioxygenase enzyme family (Tanaka et al. 2008). A gene encoding ANS has been isolated from *G. triflora* (Nakatsuka et al. 2005), but a transcript of ANS was not obtained after the RT-PCR reaction from white flowers of cv. Homoi. These results indicated that deficiency of gentian ANS might be responsible for a white phenotype (Nakatsuka et al. 2005). It has also been demonstrated that suppression of the *ANS* gene by RNAi in *G. triflora* resulted in pale blue flower pigmentation. It could be implied that expression of the *ANS* gene was not completely inhibited by RNAi, thus resulting in slight colored flowers (Nakatsuka et al. 2008b). Some of the examples of flower color modification by regulating flavonoid biosynthesis in gentian, as well as in other popular ornamental

Fig. 4 *Gentiana asclepiadea* with pure white flowers (*Photo* by courtesy of Michał Śliwiński from The Department of Biodiversity and Plant Cover Protection, Wrocław University)

species such as a rosa and petunia, have been summarized by Nishihara and Nakatsuka (2011). Most modifications of structural genes of anthocyanin biosynthesis in gentian change colors from blue to white, pale blue or magenta. *Gentiana asclepiadea*, which grows wild in the Karkonosze Mountains of Poland, usually has blue colored flowers. However, a few white flowers have been also observed (Fig. 4). It has not been explained so far, whether the white color of the flowers is a result of spontaneous mutations in genes encoding key enzymes in the synthesis of delphinidin derivatives, or whether it is connected with changes of mineral nutrients in the soil.

As described earlier, the pigmentation of gentian flowers is regulated temporally and spatially by the level of transcription of each corresponding anthocyanin biosynthetic gene, including F3H, F3′5′H, DFR, ANS, 3GT, 5GT, and 5AT. Moreover, studies of model plants (Arabidopsis, petunia) have revealed that transcriptional regulation of those structural genes is controlled by the complex of MYB, basic helix-loop-helix (bHLH) and WD40 proteins (Hichri et al. 2011). In 2008, Nakatsuka et al. identified two new transcriptional factors, termed GtMYB3 and GtbHLH1, which are involved in the regulation of gentiodelphin biosynthesis in gentian petals (Nakatsuka et al. 2008c). They demonstrated that expression profiles of *GtMYB3* and *GtbHLH1* were similar to the anthocyanin biosynthetic genes, including *GtCHS, GtF3′5′H, Gt5AT*. In addition, it was found that white-flowered gentians contain several mutations (insertion of single nucleotide) in *GtMYB3*. Loss of function of this transcriptional factor caused deficiency in anthocyanin accumulation in white flowers, providing evidence that MYB3 may regulate biosynthesis of gentiodelphin in gentians (Nakatsuka et al. 2008c; Nishihara and Nakatsuka 2011). It will be useful to carry out further work to identify other transcriptional factors in gentian flowers, because this information might assist in developing new biotechnological tools in the modification and generation of plants with optimized anthocyanin content.

6.2.1 Modification of Anthocyanidins

Anthocyanidins of most flowers may be modified in glycosylation, acylation, and methylation reactions. These modification patterns are unique for different plant species and play an important role in the diversity of flower coloration (Tanaka et al. 2008, 2009). The knowledge about enzymes involved in secondary modifications of anthocyanins in gentian is still increasing, especially in those steps which lead to production of the blue color. Most publications available are focused on glycosylation and acylation steps in gentian.

6.2.1.1 Glycosylation

Glycosylation is catalyzed by two enzymes, namely anthocyanidin UDP-glucosyl 3-O-glucosyltransferase (3GT) and anthocyanidin 5-O-glucosyltransferase (5GT), respectively. In general, GT enzymes catalyze the transfer of a glycosyl moiety from UDP-sugar to the hydroxyl group of the A or B rings in the anthocyanins (Gachon et al. 2005). Glycosylation has been studied biochemically and genes have been isolated from maize, snapdragon, barley, grape, and petunia (Fukuchi-Mizutani et al. 2003; Nakatsuka et al. 2008a). The 3GT enzyme has also been characterized in *G. triflora*, and heterologous expression in *Escherichia coli* confirmed its strong substrate specificity for anthocyanidins (pelargonidin, cyanidin, and delphinidin). It has also been reported that 3GT has wide specificity and can glucosylate dihydroflavonols and flavonols. However, in gentian, 3GT converts anthocyanins more efficiently than flavonols (Tanaka et al. 1996). Usually, 3-glucosylation occurs prior to 5-glucosylation, and the latter is catalyzed by 5-O-glucosyltransferase (5GT). Nakatsuka et al. (2008a) reported that *G. triflora* has a 5GT and these workers isolated the corresponding cDNA clone Gt5GT7, exhibiting a high level of the identity with 5GT from other plant species. Interestingly, the best substrate among tested anthocyanidin 3-glycosides was delphinidin, which is the precursor of gentiodelphin. In addition, a unique anthocyanin 3′-glucosyltransferase (3′GT) has been purified from petals of *G. triflora* and a cDNA encoding 3′GT has been cloned and expressed in petunia.

Enzymatic studies revealed that 3′GT exhibits strict substrate specificity towards the 3′hydroxy moiety of delphinidin, but shows no activity toward the 3′hydroxy group of cyanidin or pelargonidin without a 3′hydroxyl group (Fukuchi-Mizutani et al. 2003). The phylogenetic analysis based on the amino acid sequence shows that gentian 3′GT belongs to a different GT clade than 3GT and 5GT. It has been proposed that glucosylation at the 3′hydroxy group of delphinidin via 3′GT is probably the final step which contributes to accumulation of gentiodelphin (Fukuchi-Mizutani et al. 2003).

6.2.1.2 Acylation

Anthocyanin aromatic 5-acyltransferase (5AT) catalyzes the transfer of aromatic acyl groups from CoA esters to the glycosyl moiety of anthocyanins. Acyl glucose-dependent AT activity has been documented in gentian, and the gene encoding the anthocyanin 5-aromatic acyltransferase was first isolated from petals of *G. triflora* (Fujiwara et al. 1998). Anthocyanins containing plural aromatic acyl residues (p-coumaroyl, caffeoyl, feruoloyl) are called polyacylated anthocyanins, and exhibit blue a color as a result of intramolecular stacking. The first complete structure of the polyacyled anthocyanin-gentiodelphin from blue *G. makinoi* was published in 1982 (Goto et al. 1982). It has been confirmed experimentally that aromatic acylation of anthocyanins changes their color to blue and makes them more sustainable (Yoshida et al. 2009). Most blue flowers contain in their petals, aromatically acylated delphinidin derivatives. The best example is a gentian anthocyanin, namely gentiodelphin, in which the caffeolyl moiety in the B-ring plays a valid role in color development and stability. The bright blue color seen in *Clitoria ternatea* (butterfly pea) is also associated with accumulation of polyacylated anthocyanins, such as ternatin A1, and loss of acyl groups leads to a change from blue to a mauve color (Honda and Saito 2002).

6.2.1.3 Co-Pigmentation and Metal Complexation

Intramolecular co-pigmentation is formed by stacking of flavonols or flavones with anthocyanins. The structure of the co-pigment and its concentration also affect the pigment color and intensity. In such complex structures, anthocyanins take on more intense and stable colors. For example the substitution of the flavonol co-pigments with flavone glycosides in lisianthus changed the color of delphinidin from purple to blue (Davies 2004). It has been known for some time that anthocyanin color depends on the presence of the metal ions aluminum (Al), iron (Fe) and magnesium (Mg). In a blue-flowered species, such as *Commelina communis* (dayflower), co-pigment complexes interact with two Mg ions (Griesbach 2005). However, until now there have been no reports about such metal complexes in blue flavonoid pigmentation in Gentianaceae.

6.3 Factors Influencing Flower Color in Gentiana Species

6.3.1 Shape of Epidermal Cells

The stability and color of anthocyanins are determined by several factors including pH, temperature, light, the presence of co-pigments, metals, and sugars. Furthermore, color in petals depends on the morphology of the epidermal cells which have a different shape, such as being conical or flat on their surface. Noda et al. (1994) described how the shape of epidermal cells affect flower color. It has been postulated that conical cells enhance color intensity and brightness by increasing light absorption by pigments, in contrast to flat epidermal petal cells. Petals with flat epidermal cells appear more pale (Noda et al. 1994). The gene named *MIXTA* was identified and its role confirmed in the transcriptional control of cell shape. A mutant *mixta* of snapdragon (*Antirrhinum majus*) shows flat cell form in the inner epidermal cells and, therefore, the flowers are pink instead of magenta, formed by stacking of flavonols. Butterfly pea encoding MYB-related transcription factors have been reported to regulate the development of conical shape in petal epidermal cells in *Petunia hybrida* (*PhMYB1*) and *Arabidopsis thaliana* (*AtMYB16*). The flowers of the *phmyb1* mutant are lighter than those of the wild type plant and petals are curved backwards and downwards (Baumann et al. 2007). However, no transcriptional factors involved in the regulation of cell shape in gentian flowers have been identified to date. Thus, determination of the endogenous factors which are responsible for the morphological features of gentian flowers can be an interesting route to follow in the future. In most flowers, the induction of anthocyanin biosynthesis occurs during the initial period of cell elongation in petals. The unique exception is lisianthus (*E. grandiflorum*), because anthocyanin synthesis occurs when the petals have reached their final size (Nakatsuka et al. 2005).

6.3.2 pH

In general, there is a strong correlation between flower color and pH in the vacuole. Differences in vacuolar pH have a dramatic impact on anthocyanin coloration. As the pH becomes more acidic, so there are more red flowers, and the more alkaline the pH the more flowers become blue. There is a protein believed to be involved in the H^+ pump (PH5 belonging to P3A- ATPases) which is localized in the vacuolar membrane, which can adjust the pH in the vacuolar lumen. Mutation of *PH5* in petunia reduces vacuolar acidification resulting in blue flower color (Verweij et al. 2008). In petunia, seven loci have been defined (*PH1–PH7*), mutations of which caused a shift toward the blue color of flowers and increased pH in petals. The transcription factor encoded by *PH4* activates vacuolar acidification and acts in conjunction with the recent cloning transcription regulators of anthocyanin biosynthesis AN1 and AN2 (Mol et al. 1998; Quattrocchio et al.

2006). The only known investigation concerning the pH of lisianthus was described by Asen et al. (1986), who determined the pH of epidermal petal cells as approximately 6.5. A few years later Markham and Ofam (Markham and Ofman 1993) reported that pH in purple petals of lisianthus was in the range 4.75–5.2. Determining the mechanism of pH control in gentian petals will require the cloning and functional characterization the structural genes which regulate vacuolar pH in blue flowers. If the controlled manipulation of co-pigmentation and regulation of pH are elucidated in gentian species, then the dream about black and blue roses may come true.

6.3.3 Temperature, Light, and Sugar

Besides pH, the most common environmental factors influencing pigment color are temperature and light. There are a few studies which have shown that light induces anthocyanin accumulation in green tissues through the activation of anthocyanin biosynthetic genes. The role of light has been demonstrated in *CHS* gene expression in petunia flowers. When petunia flowers were grown in the dark *CHS* expression was strongly inhibited and anthocyanin production was limited (Weiss and Halevy 1991). In contrast, direct illumination is not necessary for *CHS* activation and anthocyanins accumulation in *E. grandiflorum* flowers (Kawabata et al. 1995). The effect of sugar on anthocyanin biosynthesis genes has also been investigated in detached flowers of lisianthus. The transcript levels of CHS, chalcone isomerase (CHI) and dihydroflavonol 4-reductase (DFR) were enhanced in lisianthus petals in sucrose concentration. Moreover, 0.5 M sucrose also increases petal growth in lisianthus. Similar results were obtained for petunia flowers and it has been shown that an exogenous sucrose supply promoted the expression of the *CHS* gene (Kawabata et al. 1999). However, it is still not clear whether sugar does act to promote anthocyanin gene expression as a specific signaling molecule, or as a metabolic precursor.

Little is known about the role of temperature in the regulation of anthocyanin synthesis in gentian flowers. Mu et al. (2010) described an alpine herbaceous species, *G. leucomelanea* which has two flower colors, namely blue and white. They suggested that temperature affects flower color in this species and that it is responsible for flower color polymorphism. In the experiments at low temperature (12 °C), the frequency of the white flowers was significantly greater than that of the blue flowers. Consistently, at the high temperature (20 °C), the percentage of white flowers decreased and more blue flowers formed. These results implied that flower color diversity is dependent on temperature in this species. It is interesting, that some of commercial cultivars of *E. grandiflorum* are blue flowering and heat tolerant plants, e.g., "Maurine Blue". Another example of heat-tolerant lisianthus is "Maurine Dawn" with pink/white-bicolored flowers, which was characterized in detail by Harbaugh and Scott in 2005.

6.3.4 Anthocyanic Vacuolar Inclusions (AVIs)

Flavonoids are synthesized in the cytosol but are transported mainly to the vacuole for storage. Anthocyanins are also deposited in the vacuole among flavonoids. If the biosynthetic pathway of anthocyanins is now well understood, the transport mechanism has not been yet clarified. Two possible models for the transport of anthocyanins to the vacuole have been proposed. The first mechanism involves the activity of glutathione S-transferase and ABC transporters for sequestration anthocyanins into vacuole and this strategy was observed in petunia, maize, and Arabidopsis. The second mechanism occurs via pro-vacuolar vesicles which transfer anthocyanins directly from the endoplasmic reticulum to the acidic environment of the vacuole (Pourcel et al. 2011). This vesicle-mediated transport has been observed in lisianthus (Zhang et al. 2006). After transport to the vacuole, the highly pigmented bodies with regular and irregular shape have been formed in the epidermal cells of petals and they have been called **A**nthocyanic **V**acuolar **I**nclusions (AVIs; Markham et al. 2000). It has been reported that AVIs accumulate high concentrations of anthocyanins, especially cyanidin 3-glucosides and delphinidin acylated 3, 5-diglucosides, and they might assist in the accumulation of a high concentration of anthocyanins in flowers and thereby they create darker colors. (Markham et al. 2000; Pourcel et al. 2011). As mentioned above, the AVIs were found in flowers of violet *E. grandiflorum* and further investigation revealed that this member of the family Gentianaceae contains 1,4 % anthocyanins in its petals (Markham et al. 2004). In comparison, *L. nigrescen* (Gentianaceae) contains an uncommonly high concentration of anthocyanins (24 % of flower dry mass) and it is established as the blackest flower in the world (Nicholson 1999).

6.4 Carotenoids in Gentians

Carotenoids are one of the largest groups of pigments found in plants, and their essential function is harvesting of light and protecting the photosynthetic apparatus against photo-oxidation. They are also responsible for furnishing flowers, fruit, and some roots with yellow, orange, and red colors (Tanaka et al. 2008; Farré et al. 2010). Carotenoiods are classified into carotenes and oxygenated xanthophylls (Davies 2004). The most widespread xanthophylls, lutein and violaxanthin, are present in yellow and orange fruit and vegetables, as well as in yellow flowers of *G. lutea*. Simultaneously, three major carotenoids, β-carotene, lutein, and zeaxanthin, have been separated and identified in yellow-flowered *Gentiana lutea* (Zhu et al. 2003). The importance of carotenoids as a component of human nutrition and health (precursors of vitamin A, natural antioxidants) and their role as substrates in the synthesis of phytohormones has generated interest in the molecular and biochemical basis of carotenoid synthesis and its endogenous regulation. The carotenogenic synthesis pathway in flowering plants has been reviewed recently (Hirschberg

2001; Farre et al. 2010, Zhu et al. 2010). In brief, carotenoids are tetraterpenoides and the biosynthesis starts from the isomeric C_5 precursor's isopenthyl diphosphate (IPP) and the condensation of four IPPs leads to geranylgeranylpyrophosphate (GGPP). The next condensation of two GGPP is catalyzed by phytoene synthase (PSY), yielding the first carotenoid phytone C_{40}- the backbone of all plant carotenoids (Hirschberg 2001; Tanaka et al. 2008). Enhanced expression of the *PSY* was observed during flower development of *G. lutea*. This upregulation of phytoene synthase was correlated with accumulation of carotenoids and induction of genes encoding enzymes involved in the early steps of carotenoid biosynthesis (Zhu et al. 2002). Phytoene then undergoes a series of desaturation reactions catalyzed by the two desaturases, phytoene desaturase (PDS) and carotene desaturase (ZDS), respectively. Finally, desaturation reactions convert phytoene to tetra-*cis*-lycopene which is isomerized by carotenoid isomerase (CRTISO) to the first pigmented carotenoid, bright red, translycopene. Lycopene is a substrate for two competing enzymes, lycopene β-cyclase (LCYB) and lycopene ε-cyclase (LCYE), which produce orange-pigmented β-carotene and α-carotene, respectively. β-carotene, is in turn, converted to the yellow dye termed zeaxanthin, and α-carotene is hydroxylated to yellow lutein. Zeaxanthin enters the xanthophyll cycle and can be converted to yellow antheraxanthin and violaxanthin, which is modified further to neoxanthin by neoxanthin synthase (NXS) (Hirschberg 2001; Tanaka et al. 2008). The carotenoid biosynthesis pathway in gentian plants is summarized in Fig. 5.

Gentiana lutea is a model among the Gentianaceae for carotenogenesis due to its bright yellow flowers, which are rich in β-carotene and xanthophylls (Zhu et al. 2002). Genes from *G. lutea* encoding the major enzymes required for carotenoid biosynthesis have been cloned and enzyme activities have been examined (Zhu et al. 2002, 2003, 2010). It has been shown that *PSY* and *ZDS* were strongly upregulated during flower development. It is widely known that carotenoid accumulation is controlled by transcriptional or posttranscriptional regulation of genes which are responsible for carotenoid accumulation and/or degradation. One potential way to enhance carotenoids in plants is by increasing the concentration of their precursors. Consequently, phytoene synthase (PSY) is a useful target for upregulation. It has been established that a knock-out mutant of the *PSY1* gene in tomato resulted in fruit without carotenoids and the second isoform of *PSY* (*PSY2*) was not to able to compensate for disrupted *PSY1* (Farré et al. 2010).

The flower petals of most plants accumulate yellow xanthophylls, but the carotenoid composition in flowers can vary from species to species. Pale yellow petals of *Eustoma* sp. and yellow petals of *G. lutea* have similar carotenoid composition to those from leaves, as they accumulate lutein, violaxanthin, neoxanthin, β-carotene, antheraxanthin and zeaxanthin (Ohmiya 2011). The carotenoid composition also changes also from early to later stages of flower development, i.e., lutein content in flower buds (stage S1) was approximately 32 % and at stage S5 (fully mature and pigmented flower) was 42 % of total carotenoids (Zhu et al. 2002).

Despite of our knowledge about carotenoids and their role in health and nutrition, many carotenoids remain unknown and still unidentified. There are also

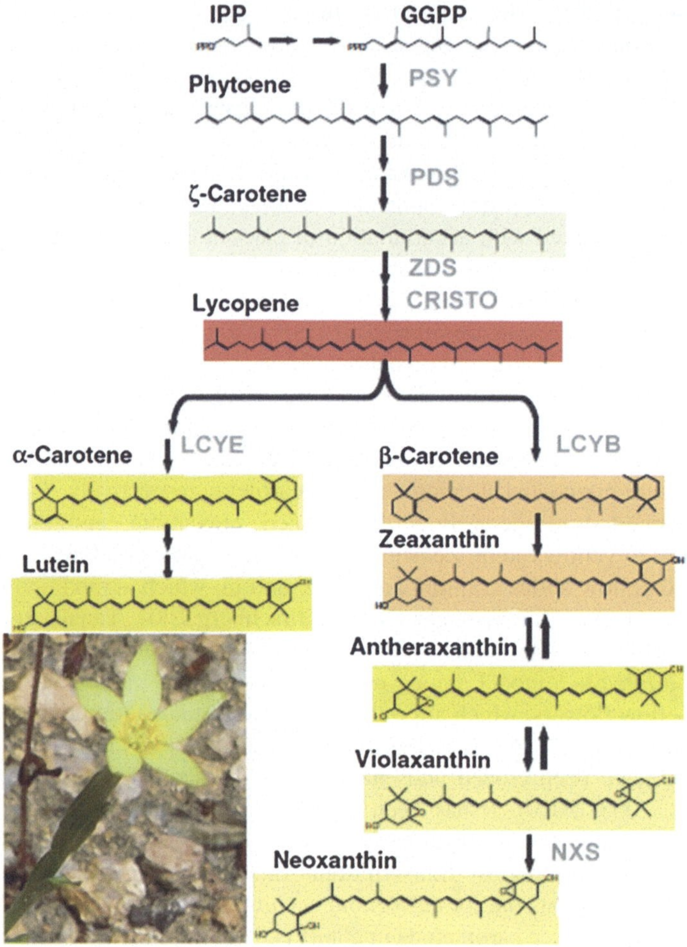

Fig. 5 Formation of major carotenoids and key enzymes which are involved in regulation of carotenoids composition in gentian plants (Tanaka et al. 2008, modified). *PSY* phytoene synthase; *PDS* phytoene desaturase; *ZDS* carotene desaturase; *CRISTO* carotenoid isomerase; *LCYB* lycopene *β*-cyclase; *LCYE* lycopene ε-cyclase; *NXS* neoxanthin synthase

many carotenoids whose biosynthesis has not been characterized. However, there has been significant progress in the development of transgenic plants producing more carotenoids and some interesting work has been also carried out with some plants with modified carotenoid biosynthetic pathways that produce specific carotenoids, for example astaxanthin, and that express unique colors such as blood-red. Future research using genetic engineering methods may be undertaken to create red-flowered gentians.

6.5 Conclusions

Natural anthocyanins and carotenoids have powerful color properties. The diversity of floral morphology and pigmentation of Gentianaceae make members of this family attractive for studies of the basic chemistry and genetics of flower color. Anthocyanins, in particular, have been the target of biotechnological efforts to generate new, or to alter, existing plant colors. Creating ornamental indoor and outdoor plants with novel colors, such as perfect sky-blue flowers, could be achieved in the future by introducing gentiodelphin transgenes and/or transgenes for co-pigmentation.

Acknowledgements Many thanks to Ewa Szczęśniak and Michał Śliwiński (The Department of Biodiversity and Plant Cover Protection, Wrocław University, Wrocław, Poland) for providing the photographs of Gentian flowers.

References

Andersen ØM, Jordheim M (2006) The anthocyanins. In: Andersen ØM, Markham KR (eds) Flavonoids chemistry, biochemistry and applications. CRC Press, Boca Raton

Asen S, Griesbach RJ, Norris KH, Leonhard BA (1986) Flavonoids from *Eustoma grandiflorum* flower petals. Phytochemistry 25:2509–2513

Baumann K, Perez-Rodriguez M, Bradley D, Venail J, Bailey P, Jin H, Koes R, Roberts K, Martin C (2007) Control of cell and petal morphogenesis by R2R3 MYB transcription factors. Development 134:1691–1701

Beale GH, Price JR, Sturgess VC (1941) A survey of anthocyanins. VII. The natural selection of flower colour. Proc. R. Soc. Lond. Ser. B, Biol Sci 130:113–126

Darwin CR (1877) The different forms of flowers on plants of the same species. John Murray, London

Davies KM (2004) Plant pigments and their manipulation. Annual plant reviews, vol 14. Blackwell, Oxford

Davies KM (2009) Modifying anthocyanin production in flowers. In: Gould K, Davies KM, Winefield (eds) Anthocyanins biosynthesis, functions and applications. Springer, Science + Business Media LLC, Berlin

Davies KM, Bradley JM, Schwinn KE, Markham KR, Podivinsky E (1993) Flavonoid biosynthesis in flower petals of five lines of lisanthus (*Eustoma grandiflorum* Grise). Plant Sci 95:67–77

Farré G, Sanahuja G, Naqvi S, Bai Ch, Capell T, Zhu Ch, Christou P (2010) Travel advice on the road to carotenoids in plants. Plant Sci 179:28–48

Fujiwara H, Tanaka Y, Yonekura-Sakakibara K, Fukuchi-Mizutani M, Nakao M, Fukui Y et al (1998) cDNA cloning, gene expression and subcellular localization of anthocyanin 5-aromatic acyltransferase from *Gentiana triflora*. Plant J 16:421–431

Fukuchi-Mizutani M, Okuhara H, Fukui Y, Nakao M, Katsumoto Y, Yonekura-Sakakibara K, Kusumi T, Hase T, Tanaka Y (2003) Biochemical and molecular characterization of a novel UDP-glucose:anthocyanin 3-O-glucosyltransferase, a key enzyme for blue anthocyanin biosynthesis, from Gentian. Plant Physiol 132:1652–1663

Gachon CM, Langlois-Meurinne M, Saindrenan P (2005) Plant secondary metabolism glycosyltransferases: the emerging functional analysis. Trends in Plant Sci 10:542–549

Goto T, Kondo T, Tamura H, Imagawa H, Iino A, Takeda K (1982) Structure of gentiodelphin, an acylated anthocyanin isolated from *Gentiana makinori*, that is stable in dilute aqueous solution. Tetrahedron Lett 23:3695–3698

Griesbach R (2005) Biochemistry and genetics of flower colour. Plant Breed Rev 25:89–114

Grotewold E (2006) The genetics and biochemistry of floral pigments. Annu Rev Plant Biol 57:761–780

Harbaugh BK, Scott JW (2005) "*Maurine dawn*"-A heat tolerant lisianthus with pink/white bicolored flowers. Hort Sci 40:858–860

Hichri I, Barrieu F, Bogs J, Kappel CH, Delrot S, Lauvergeat V (2011) Recent advanced in the transcriptional regulation of flavonoid biosynthetic pathway. J Expl Bot 28:1–19

Hirschberg J (2001) Carotenoid biosynthesis in flowering plants. Curr Opin Plant Biol 4:210–218

Hokosawa K, Fukushi E, Kawabata J, Fujii CH, Ito T, Yamamura S (1995) Three acylated cyaniding glucosides in pink flowers of Gentiana. Phytochem 40:941–944

Hokosawa K, Fukushi E, Kawabata J, Fujii CH, Ito T, Yamamura S (1997) Seven acylated anthocyanins in blue flowers of Gentiana. Phytochem 45:167–171

Holton TA, Cornish EC (1995) Genetics and biochemistry of anthocyanin biosynthesis. Plant Cell 7:1071–1083

Honda T, Saito N (2002) Recent progress in the chemistry of polyacylated anthocyanins as flower color pigment. Heterocycles 56:633–692

Katsumoto Y, Fukuchi-Mizutani M, Yuko Fukui Y, Brugliera F, Holton TA, Karan M, Nakamura N, Yonekura-Sakakibara K, Togami J, Pigeaire A, Tao G, Nehra NS, Lu Ch, Dyson B, Tsuda S, Ashikari T, Kusumi T, Mason JG, Tanaka Y (2007) Engineering of the rose flavonoid biosynthetic pathway successfully generated blue-hued flowers accumulating delphinidin. Plant Cell Physiol 48:1589–1600

Kawabata S, Ohta SM, Kusuhara Y, Sakiyama R (1995) Influences of low light intensities on the pigmentation of *Eustoma grandiflorum* flowers. Acta Hortic 405:173–178

Kawabata S, Kusuhara Y, Li Y, Sakiyama R (1999) The regulation of anthocyanin biosynthesis in *Eustoma grandiflorum* under low light conditions. http://www.paper.edu.cn

Kobayashi H, Oikawa Y, Koiwa H, Yamamura S (1998) Flower specific expression directed by the promoter of a chalcone synthase gene from *Gentiana triflora* in *Petunia hybrida*. Plant Sci 131:173–180

Markham KR, Bloor SJ, Nicholson R, Riviera R, Shelmuck M, Kevan PG, Michener C (2004) Black flower coloration in wild *Lisianthius nigrescens*: Its chemistry and ecological consequences. Z Naturforsch 59:625–630

Markham KR, Gould KS, Winefield CS, Mitchell KA, Blorr SJ, Boase MR (2000) Anthocyanic vacuolar inclusions–their nature and significance in flower coloration. Phytochem 55:327–336

Markham KR, Ofman JD (1993) Lisianthus floavonoid pigments and factors influencing their expression in flower color. Phytochem 34:679–685

Miller R, Owens SJ, Rorslett B (2011) Plants and colour: flowers and pollination. Opt Laser Technol 43:282–294

Młodzińska E (2009) Survey of plant pigments: Molecular and environmental determinants of plant colors. Acta Biol Crac s. Bot 51:7–16

Mol J, Grotewold E, Koes R (1998) How genes paint flowers and seeds. Trends Plant Sci 3:212–217

Mu J, Li G, Sun S (2010) Petal color, flower temperature, and behavior in an alpine annual herb *Gentiana leucomelaena* (Gentianaceae). Artic, Antarct Alpine Res 42:219–226

Nakatsuka T, Nishihara M, Mishiba K, Yamamura S (2005) Temporal expression of flavonoid biosynthesis related genes regulates flower pigmentation in gentian plants. Plant Sci 168:1309–1318

Nakatsuka T, Nishihara M, Mishiba K, Hirano H, Yamamura S (2006) Two different transposable elements inserted in flavonoid 3', 5-hydroxylase gene contribute to pink flower coloration in *Gentiana scabra*. Mol Genet Genom 275:231–241

Nakatsuka T, Sato K, Takahashi H, Yamamura S, Nishihara M (2008a) Cloning and characterization of the UDP-glucose:anthocyanin 5-O-glucosyltransferase gene from blue-flowered gentian. J Exp Bot 59:1241–1252

Nakatsuka T, Mishiba KI, Abe Y, Kubota A, Kakizaki Y, Yamamura S, Nishihara M (2008b) Flower color modification of gentian plants by RNAi-mediated gene silencing. Plant Biotech 25:61–68

Nakatsuka T, Haruta KS, Pitaksutheepong C, Abe Y, Kakizaki Y, Yamamoto K, Shimada N, Yamamura S, Nishihara M (2008c) Identification and characterization of R2R3-MYB and bHLH transcription factors regulating anthocyanin biosynthesis in gentian flowers. Plant Cell Physiol 49:1818–1829

Nakatsuka T, Saito M, Yamada E, Nishihara M (2011) Production of picotee-type flowers in Japanese gentian by CRES-T. Plant Biotech 28:173–180

Nicholson R (1999) The blackest flower in the world. Nat Hist 108:60–63

Nishihara M, Nakatsuka T, Hosokawa K, Yokoi T, Abe Y, Mishiba K, Yamamura S (2006) Dominant inheritance of white-flowered and herbicide-resistant traits in transgenic gentian plants. Plant Biotech 23:25–31

Nishihara M, Nakatsuka T, Mizutani Fukuchi M, Tanaka Y, Yamamura S. (2008) Gentians: from gene cloning to molecular breeding. In: Floricultural and Ornamental Biotechnology V. Edited by da Silva T Global Science Books, UK pp. 57-67

Nishihara M, Nakatsuka T (2011) Genetic engineering of flavonoid pigments to modify flower color in floricultural plants. Biotech Lett 33:433–441

Noda KI, Glover BJ, Linstead P, Martin C (1994) Flower color intensity depends on specialized shape cell controlled by a Myb related transcription factor. Nature 369:661–664

Ohmiya A. (2011). Diversity of carotenoid composition in flower petals. JARQ 45: 163–171. http://www.jircas.affrc.go.jp/english/publication/jarq/45-2/45-02-05.pdf

Pourcel L, Irani NG, Lu Y, Riedl K, Schwartz S, Grotewold E (2011) The formation of anthocyanic vacuolar inclusions in *Arabidopsis thaliana* and implications for the sequestration of anthocyanin pigments. Mol Plant 3:78–90

Quattrocchio F, Wing J, Leppen HTC, Mol JNM, Koes RE (1993) Regulatory genes controlling anthocyanin pigmentation are functionally conserved among plant species and have distinct sets of target genes. Plant Cell 5:1497–1512

Quattrocchio F, Verweij W, Kroon A, Spelt C, Mol J, Koes R (2006) PH4 of petunia is an R2R3 MYB protein that activates vacuolar acidification through interactions with basic-helix-loop helix transcription factors of the anthocyanin pathway. Plant Cell 18:1274–1291

Ramsay NA, Walker AR, Mooney M, Gray JC (2003) Two basic–helix-loop-helix genes (MYC 146 and GL3) from Arabidopsis can activate anthocyanin biosynthesis in a white flowered *Matthiola incana* mutant. Plant Mol Biol 52:679–688

Shamir-Oren M, Sachray-Shaked L, Levi-Nissim A, Ecker R (1999) Anthocyanin pigmentation of lisianthus flower petals. Plant Sci 140:81–86

Schwinn K, Venail J, Shang Y, Mackay S, Alm V, Butelli E, Oyama R, Bailey P, Davies K, Martin C (2006) A small family of MYB-regulatory genes controls floral pigmentation intensity and pattering in the genus Antirrhinum. Plant Cell 18:831–851

Tanaka Y, Yonekura Y, Fukuchi-Mizutani M, Fukui Y, Fujiwara H, Ashikari T, Kusumi T (1996) Molecular and biochemical characterization of three anthocyanin synthetic enzymes from *Gentiana triflora*. Plant Cell Physiol 37:711–716

Tanaka Y, Sasaki N, Ohmiya A (2008) Biosynthesis of plant pigments: anthocyanins, betalains and carotenoids. Plant J 54:733–749

Tanaka Y, Brugliera F, Chandler S (2009) Recent progress of flower colour modification by biotechnology. Intl J Mol Sci 10:5350–5369

Tornielli G, Koes R, Quattrocchio F (2009) The genetics of flower color. In: Gerats T, Strommer J (eds) Petunia, evolutionary, development and physiological genetics. Springer Science + Business Media LLC, Berlin, pp 269–298

Tsuda S, Fukui Y, Nakamura N, Katsumoto Y, Yonekura-Sakakibara K, Fukuchi-Mizutani M, Ohira K, Ueyama Y, Ohkawa H, Holton TH, Kusumi T, Tanaka Y (2004) Flower color modification of *Petunia hybrida* commercial varieties by metabolic engineering. Plant Biotech 21:377–386

Verweij W, Spelt C, Di Sansebastiano GP, Vermeer J, Reale L, Ferranti F, Koes R, Quattrocchio F (2008) An H+ P-ATPase on the tonoplast determines vacuolar pH and flower colour. Nature Cell Biol 10:1456–1462

Weiss D, Halevy AH (1991) The role of light reaction in the regulation of anthocyanin synthesis in petunia corollas. Physiol Plant 81:127–133

Yoshida K, Toyama Y, Kameda K, Kondo T (2000) Contribution of each cafferoyl residue of the pigment molecules of gentiodelphin to blue color development. Phytochem 54:85–92

Yoshida K, Mori M, Kondo T (2009) Blue flower color development by anthocyanins: from chemical structure to cell physiology. Nat Prod Rep 26:884–915

Zhu C, Yamamura S, Koiwa H, Nishihara M, Sandmann G (2002) cDNA cloning and expression of carotenogenic genes during flower development in *Gentiana lutea*. Plant Mol Biol 48:277–285

Zhu C, Yamamura S, Nishihara M, Koiwa H, Sandmann G (2003) cDNAs for the synthesis of cyclic carotenoids in petals of *Gentiana lutea* and their regulation during flower development. Biochim Biophys Acta 1625:305–308

Zhu C, Bai C, Sanahuja G, Yuan D, Farré G, Naqvi S, Shi L, Capell T, Christou P (2010) The regulation of carotenoid pigmentation in flowers. Arch Bioch Biophys 504:132–141

Zhang H, Wang L, Deroles S, Bennett R, Davies K (2006) New insight into the structures and formation of anthocyanic vacuolar inclusions in flower petals. BMC Plant Biol 6:1–14

Chapter 7
Karyology of European Species of Genus *Gentiana* L.

Vitaliy M. Mel'nyk, Nadia M. Drobyk, Maryana O. Twardovska and Viktor A. Kunakh

Abstract This chapter summarizes the data on chromosome numbers of European *Gentiana* species. Karyological details are considered for some sections and individual genus members in the context of their propagation. Based on a comparison of the results of our own studies with those in the literature, chromosome numbers were confirmed in *G. lutea*, *G. punctata*, *G. acaulis*, *G. verna*, *G. cruciata* and *G. pneumonanthe*, as well as those of one race of *G. asclepiadea*. The necessity for further karyological investigations is discussed in order to elucidate the evolution of *Gentiana* species and the classification of individual taxons.

7.1 Introduction

Gentiana L., the largest genus of the family *Gentianaceae* Juss., includes about 400 species. *Gentiana* species are typical mountain plants, most of which grow at over 1,000 m above sea level. Some adapt themselves to growth on the plains. Most gentians are propagated in Asia (312 species), fewer in South and Central America (35), and Europe (27–29 species), with more limited numbers in South America (3), Africa (2; only in Morocco), and Eastern Australia (1; Tutin 1972; Ho and Liu 1990; Yuan et al. 1996). However, the number of species raised is increasing globally, and in Europe, in particular, new taxons are being found (Hsieh et al. 2007; Prosser and Bertolli 2008).

The scope of the genus and its classification are considered differently by monographers. This may be explained by the presence of aggregate species within

V. M. Mel'nyk (✉) · M. O. Twardovska · V. A. Kunakh
Institute of Molecular Biology and Genetics, National Academy of Sciences of Ukraine, Akad. Zabolotnogo Street 150, Kyiv 03680, Ukraine
e-mail: v.m.melnyk@imbg.org.ua

N. M. Drobyk
Volodymyr Hnatiuk Ternopil National Pedagogical University, M. Kryvonis Street 2, Ternopil 46027, Ukraine
e-mail: drobyk.n@gmail.com

J. J. Rybczyński et al. (eds.), *The Gentianaceae - Volume 1: Characterization and Ecology*, DOI: 10.1007/978-3-642-54010-3_7,

the genus and difficulties in identifying genus boundaries resulting from peculiarities of gentian evolution that have led to appearance of numerous parallel series of convergent forms. Currently, there are several genus classifications based on morphological, anatomical, environmental-geographic, and ontogenetic features (Kusnezow 1894; Tutin 1972; Tzvelev 1978; Zuyev 1990; Ho and Liu 1990).

We adhere to the genus *Gentiana* being divided into sections according to Ho and Liu (1990), since it is based on the most complete studies on species composition. Within the genus, 15 sections are distinguished; 14 of them are widespread in Asia, 8 in Europe, and 6 in America. The section *Ciminalis* (Adans.) Dumort. is present exclusively in Europe, *Gentiana*, and *Cruciata* Gaudin in Europe and Asia, *Pneumonanthe* (Gled.) Gaudin, *Frigida* Kusn., *Dolichocarpa* T.N. Ho in Europe, Asia and America, *Calathianae* Froelich. in Europe, Asia, America, and Africa and *Chondrophyllae* Bunge in Europe, Asia, America, Australia, and Africa.

According to the classification of Ho and Liu (1990), European *Gentiana* amount to 27 species, while 29 are described in *Flora Europea* (Tutin 1972), of which 17 are endemics, while the rest are widespread. In Europe, most of the species are encountered in mountain massifs such as the Alps, Pyrenees, and Carpathians (Ho and Liu 1990). Given the numerous unresolved problems concerning genus systematics and evolution, it is important to study the karyology of *Gentiana* species. Such studies were initiated in 1921 (Stolt 1921) and are ongoing.

Currently, bulk data from cytogenetic studies of *Gentiana* from Europe, Asia, America, and Africa is collected. Based on chromosome number, it has been suggested that there are two cytological centers of variability, one in the European Alps, the other in the Himalayas. Another such center relating to extensive species variability (many endemic ones, in particular) may be in mountainous regions of China (Yuan 1993).

Cytological investigations into European members of the genus *Gentiana*, appear to be fragmentary, ambiguous, and sometime contradictory (Bolkhovskikh et al. 1969; Takhtajan 1990). Karyological studies on species from many habitats, particularly unexplored habitats, would provide answers relative to questions concerning karyotype evolutionary pathways. This chapter summarizes data available on the karyology of European species of *Gentiana*.

7.2 Summary of Cytological Data

7.2.1 Section Cruciata Gaudin (1828)

Aptera Kusnezow (1893); *Erythaliae* Bunge (1829).

Two species are propagated in Europe, namely *G. decumbens* L. and *G. cruciata* L. Two different chromosome sets were provided for *G. decumbens* from Russia, these being $2n = 26$ (Krogulevich 1978) and $2n = 52$ (Löve and Löve 1986). $2n = 52$ was identified for *G. cruciata* from various parts of the area, as well as for plants cultivated in Canada (Table 7.1).

For other species from the section *Cruciata*, except for *G. macrophylla*, two values are listed for diploid chromosome numbers ($2n = 52, 26$), with a basic chromosome number ($x = 13$) suggesting section cytogenetic stability (Yuan 1993).

7.2.2 Section Frigida Kusn. (1893)

Two species are distributed in Europe, these being *G. frigida* Haenke and *G. froelichii* Jan ex Reichchenb. The references provide the chromosome numbers $2n = 24$ for *G. frigida* and $2n = 42$ for *G. froelichii* (Table 7.1). For other non-European species, one additional chromosome number of $2n = 26$ has been reported (Takhtajan 1990).

7.2.3 Section Gentiana

Coelanthae Froelich (1796).

Five species prevail in Europe, namely *G. burseri* Lapeyr., *G. lutea* L., *G. pannonica* Scop., *G. punctata* L., *G. purpurea* L. Each of these species from various habitats of Europe has a diploid chromosome number of $2n = 40$ (Table 7.1).

In *G. lutea*, the literature reports $2n = 40$ (Table 7.1), although other values for chromosome numbers of this species have been recorded, including $n = 17$–18 (Wóycicki 1935) and 21, and $2n = 42$ (Stolt 1921). Discrepancies in reporting *G. lutea* chromosome numbers may result from difficulties in counting.

Most research dealing with karyological studies of *G. punctata*, give a chromosome number of $2n = 40$. Only Mattick, studying *G. punctata* plants from Austrian habitats (Tischler 1950) described numbers of $n = 17$–20. As in the case of *G. lutea*, various values for *G. punctata* may result from difficulties in chromosome counting. It is recognized that most gentians appear to be natural polyploids (Rork 1949). For species in the section *Gentiana*, reports give the basic chromosome number of $x = 5$ with reference to natural octaploids ($2n = 8x = 40$; Favarger 1949; Skalińska 1951).

7.2.4 Section Pneumonanthe (Gled.) Gaudin (1828)

Europe abounds with the two species *G. pneumonanthe* L. and *G. asclepiadea* L. Citations for *G. pneumonanthe* give one chromosome number of $2n = 26$, for the first time reported by Scheerer (1939) for plants from Germany. The same diploid

Table 7.1 Chromosome numbers for some European *Gentiana* L. species

Taxon	Locality	2*n*	Author
Section *Cruciata*			
G. decumbens	Russia	26	Krogulevich (1978)
	Russia	52	Löve and Löve (1986)
G. cruciata	Switzerland	52	Favarger (1949)
	Poland	52	Skalińska (1951)
	Spain	52	Löve and Löve (1986)
	Slovak	52	Májovský et al. (1974a)
	Byelorussian	52	Semerenko (1985)
	Georgia	52	Gagnidze et al. (2006)
	Cultivated in Canada	52	Rork (1949)
	Ukraine	52	Strashniuk et al. (2008)
Section *Frigida*			
G. frigida	Czechoslovakia	24	Murin (1974)
	Tatra Mts., Poland	24	Skalińska (1951)
G. froelichii	Slovenia	42	Favarger (1965)
	Slovenia	42	Lovka et al. (1971, 1972)
	Slovenia	42	Löve and Löve (1986)
Section *Gentiana*			
G. burseri	Spain	40	Löve and Löve (1975a, 1986)
	Pyrenees	40	Küpfer and Favarger (1967)
	Pyrenees	40	Favarger and Küpfer (1968)
G. pannonica	Czechoslovak	40	Holub et al. (1970)
	Austria	40	Löve and Löve (1986)
G. purpurea	Norway	40	Engelskjon (1979)
	Norway	40	Löve and Löve (1986)
	Switzerland	40	Favarger (1949)
	Scandinavia	40	Knaben and Engelskjon (1967)
G. lutea	Alps	40	Küpfer (1974)
	Alps, Switzerland	40	Favarger (1949, 1952)
	Tatra Mts., Poland	40	Skalińska (1951)
	Pyrenees	40	Löve and Löve (1975a, 1986)
	Greece	40	Strid and Andersson (1985)
	Slovenia	40	Löve and Löve (1986)
	Bulgaria	40	Andreev (1982)
	–	42	Stolt (1921)
	–	$n = 17–18$	Wóycicki (1935)
	Carpathians, Ukraine	40	Strashniuk et al. (2008)
G. punctata	Alps, Switzerland	$n = 20$	Favarger (1952)
	Tatra Mts., Poland	40	Skalińska (1951)
	Slovak	40	Murin (1974)
	Austria	40	Löve and Löve (1986)
	–	40	Delay (1970)
	Austria	$n = 17–20$	Mattick in book Tischler (1950)
	Carpathians, Ukraine	40	Strashniuk et al. (2008)

(continued)

Table 7.1 (continued)

Taxon	Locality	2*n*	Author
Section *Pneumonanthe*			
G. pneumonanthe	Germany	26	Scheerer (1939)
	Spain	26	Löve and Löve (1975a)
	Spain	26	Fernández et al. (1985)
	Spain	26	Küpfer (1968)
	Poland	26	Pogan et al. (1987)
	France	26	van Loon and de Jong (1978)
	Ukraine	26	Strashniuk et al. (2008)
G. asclepiadea	Slovak	36	Uhríková in work Májovský et al. (1974a)
	Lechtenstein	36	Löve and Löve (1986)
	Alps, Switzerland	44	Favarger (1949)
	Tatra Mts. and Carpathians, Poland	44	Skalińska (1951)
	Slovak	44	Murin (1974)
	Bulgaria	44	Nikolov (1991)
	Caucasus	44	Gagnidze et al. (1992)
	Cultivated in USA	44	Rork (1949)
	Tatra Mts., Poland	$n = 16$	Wóycicki (1932)
	Carpathians, Ukraine	36	Strashniuk et al. (2008)
Section *Calathianae*			
G. bavarica	Slovenia	28	Löve and Löve (1986)
	Germany	28	Mattick in book Tischler (1950)
	Switzerland	30	Favarger (1965)
	Switzerland	30	Müller (1974, 1982)
G. brachyphylla	Germany	26	Mattick in book Tischler (1950)
	Switzerland	28	Favarger (1965)
	Switzerland	28	Müller (1974, 1982)
	Switzerland	28	Scholte (1977)
	Pyrenees	28	Löve and Löve (1986)
G. sierrae (synonym *G. brachyphylla*)	Morocco, Spain	30	Müller (1982)
G. favratii (subspecies *G. brachyphylla*)	Switzerland	32	Müller (1974)
	Switzerland	28–32	Favarger (1969b)
G. nivalis	Switzerland	14	Müller (1974, 1982)
	Switzerland	14	Favarger (1949, 1952)
	Switzerland	14	Favarger and Küpfer (1968)
	Norway	14	Knaben (1950)
	Norway	14	Laane (1969)
	Iceland	14	Löve (1953)
	Iceland	14	Löve and Löve (1986)
G. pumila	Austria	20	Müller (1982)
	Alps	20	Favarger (1965)
	Slovenia	21	Löve and Löve (1986)

(continued)

Table 7.1 (continued)

Taxon	Locality	2*n*	Author
G. rostanii	Italy	28	Löve and Löve (1986)
	Italy	30	Müller (1974)
	France	30	Müller (1982)
	Alps	30	Favarger (1969a, b)
G. terglouensis	Slovenia	42	Löve and Löve (1986)
	Austria	38, 40	Müller (1982)
G. utriculosa	Italy	$n = 11$	Favarger (1952)
	Slovenia	28	Löve and Löve (1986)
	Switzerland, Italy	22	Müller (1982)
G. verna	Austria	28	Mattick in book Tischler (1950)
	Spain	28	Löve and Löve (1975b, 1986)
	Switzerland	28	Müller (1974, 1982)
	Poland	28	Skalińska (1951)
	France	28	Müller (1982)
	Carpathians, Ukraine	28	Strashniuk et al. (2008)
G. verna (ssp. *balcanica*)	Greece	30	Müller (1982)
G. verna (ssp. *pontica*)	Iran	30	Müller (1982)
Section *Ciminalis*			
G. acaulis	Spain	36	Löve and Löve (1975b)
	Slovenia	36	Löve and Löve (1986)
	Yugoslavia	36	Mededović et al. (1984)
	Cultivated in USA	36	Rork (1949)
	Carpathians, Ukraine	36	Strashniuk et al. (2008)
G. alpina	Switzerland	36	Favarger (1949)
	Pyrenees	36	Favarger and Küpfer (1968)
	Alps	36	Küpfer (1974)
	Spain	36	Löve and Löve (1975b, 1986)
G.angustifolia	France	36	Favarger (1965)
	Pyrenees	36	Favarger and Küpfer (1968)
	Italy	36	Löve and Löve (1986)
G. clusii	Switzerland	36	Favarger (1949)
	Slovenia	36	Lovka et al. (1972)
	Slovenia	36	Löve and Löve (1986)
	Slovak	36	Májovský et al. (1974a)
	Slovak	36	Murin (1974)
	Tatra Mts., Poland	36	Skalińska (1951)
G. dinarica	Italy	36	Löve and Löve (1986)
	Yugoslavia	36	Mededović et al. (1984)
G. ligustica	Alps Maritimes	36	Löve and Löve (1986)
G. occidentalis	Spain	36	Löve and Löve (1975b, 1986)

(continued)

Table 7.1 (continued)

Taxon	Locality	$2n$	Author
Section *Dolichocarpa*			
G. prostrata	Austria	36	Löve and Löve (1986)
	Switzerland	36	Favarger (1952)
	Chukotka	36	Zhukova (1982)
	Colorado (USA)	36	Löve and Löve (1975a)
	Alaska	32–36	Johnson and Packer (1968)
Section *Chondrophyllae*			
G. boryi	Spain	20	Küpfer (1968)
	Pyrenees	20	Löve and Löve (1986)
	Spain	26	Löve and Löve (1975b)
G. pyrenaica	Pyrenees	26	Küpfer and Favarger (1967)
	Pyrenees	26	Löve and Löve (1986)
	Spain	26	Löve and Löve (1975b)
	Bulgaria	26	Nikolov (1991)
	Caucasus	26	Gagnidze et al. (1992)

chromosome number was given by other authors, when exploring plants of this species from various parts of the area (Table 7.1). For this species, as in the section *Pneumonanthe* as a whole, the chromosome number $x = 13$ is reported, presuming *G. pneumonanthe* to be natural diploid ($2n = 2x = 26$; Löve and Löve 1975a; Yuan 1993). An exception in this section may be *G. asclepiadea* (Rork 1949; Májovský et al. 1974a, b; Yuan et al. 1996), for which two values for chromosome numbers are known, these being $2n = 36$ and 44. The haploid chromosome set ($n = 16$) for *G. asclepiadea* from Polish Tatra (Wóycicki 1932) differs from chromosome numbers reported by other authors and even for specimens from the same mountain area. It is possible that Wóycicki (1932) wrongly identified chromosome numbers.

The presence within *G. asclepiadea* of plants with various chromosome numbers ($2n = 36$ and 44) may indicate the existence of different cytological races of this species (Rork 1949).

For *G. asclepiadea*, the basic chromosome numbers of $x = 9$ (Löve and Löve 1986) and $x = 11$ (Favarger 1949) suggest this species is a natural tetraploid ($2n = 4x = 36$ and $2n = 4x = 44$).

7.2.5 Section Calathianae Froelich (1796)

Cyclostigma Grisebach (1838).

In Europe, eight species are propagated, these being *G. bavarica* L., *G. brachyphylla* Vill., *G. nivalis* L., *G. pumila* Jacq., *G. rostanii* Reuter ex Verlot, *G. terglouensis* Hacq., *G. utriculosa* L., and *G. verna* L. Various values have been reported for chromosome numbers, ranging from $2n = 14$, 20, 21, 22, 24, 28, 30–33, 38, 40, to 42.

For *G. bavarica* from various habitats, chromosome sets of $2n = 28$ and 30 have been documented; for *G. brachyphylla* $2n = 26$ and 28. Other chromosome numbers include $2n = 30$ for *G. sierrae* Briq. (synonym *G. brachyphylla*) from Morocco and Spain, and $2n = 28$–32 for *G. favratii* (Rittener) Favrat (subspecies *G. brachyphylla*). Only a single chromosome number of $2n = 14$ was reported for *G. nivalis* from various habitats. Several values have been reported for *G. pumila* ($2n = 20$ and 21), *G. rostanii* ($2n = 28$ and 30), and *G. terglouensis* ($2n = 38$, 40, 42). A haploid number of $n = 11$ and diploid ones ($2n = 22$ and 28) were cited for *G. utriculosa*. For *G. verna* from various habitats, $2n = 28$; only for members of this species from Greece (ssp. *balcanica*) and Iran [ssp. *pontica* (Soltok.) Hayek.] diploid chromosome numbers of $2n = 30$ were reported.

G. verna appears to be a natural teteraploid with a basic chromosome number of $x = 7$ ($2n = 4x = 28$; Favarger 1952; Müller 1982).

Overall, for the section *Calathianae*, at least seven basic chromosome numbers ($x = 5$, 6, 7, 8, 11, 15, 19; Müller 1982) have been recorded, suggesting their karyotypic variability. Species from this section with the same basic chromosome number are not always comparable morphologically (Yuan 1993).

7.2.6 Section Ciminalis (Adans.) Dumort (1827)

Megalanthe Gaudin (1828); *Thylacites* Grisebach (1845).

Seven species are propagated in Europe, these being *G. acaulis* L., *G. alpina* Vill., *G. angustifolia* Vill., *G. clusii* E.P. Perr. and Song., *G. dinarica* G. Beck., *G. ligustica* R. de Vilmorin and Chopinet, *G. occidentalis* Jakowatz. All have an identical chromosome number of $2n = 36$.

The single value of the diploid chromosome number for every species from various habitats may suggest cytogenetic conservation in this section, for which there is the basic chromosome number of $x = 9$. Species from this section are recognized as being natural tetraploids ($2n = 4x = 36$; Rork 1949; Löve and Löve 1975b).

7.2.7 Section Dolichocarpa T. N. Ho (1985)

The species in this section are distributed mostly in Asia (Ho 1985), although *G. prostrata* Haenke occurs exclusively in Europe with a range of chromosome numbers of $2n = 32$–36 (Table 7.1).

The literature concerning other species in the section *Dolichocarpa*, indicate various diploid chromosome numbers of $2n = 12$, 14, 16, 18, 20, 22, 24, 30, 32–36 (Yuan 1993), thus suggesting genetic variability in this section.

7.2.8 Section Chondrophyllae Bunge (1829)

Europe abounds with the three species *G. boryi* Boiss., *G. pyrenaica* L., and *G. laciniata* Kit. ex Kanitz. Chromosome numbers of $2n = 20$ and 26 are listed for *G. boryi*, and $2n = 26$ for *G. pyrenaica*. *Flora Europaea* (Tutin 1972) suggests that *G. laciniata* is a synonym of *G. pyrenaica*, although *G. laciniata* was reported as a species as early as 1862 by Kit. ex Kanitz. and is provided as a species for Flora of the Ukrainian Carpathians (Zerov 1957; Chopyk 1976; Didukh 2009). Values of chromosome numbers for *G. laciniata* are absent from the literature.

Hence, for European species of the section *Chondrophyllae*, two values for chromosomes are known, these being $2n = 20$ and 26. Non-European species of this section show a broad range of diploid chromosome numbers of $2n = 12$, 14, 18, 20, 24, 26, 36, 38, 40, 44, 48, 60, 96–98 (Yuan 1993), indicating karyotypic variability in this section.

7.3 Discussion and Conclusions

The results have been summarized of karyological studies for European *Gentiana* species. Examination of the bibliographical sources revealed karyotype conservation in the section *Ciminalis*, which is endemic to Europe (seven species are propagated exclusively in Europe), and in almost endemic *Gentiana* (one of five species also occurs in Asia). Most variation is represented in the European *Calathianae* (two from eight species are propagated not only in Europe, but also in America and Marocco); 13 different diploid chromosome sets are listed for European species.

Comparison of variation in chromosome numbers of sections with a genus including one to three species (*Dolichocarpa*—one species, *Cruciata*, *Pneumonanthe*, *Frigida*—two species each, *Chondrophyllae*—3 species) permitted identification of karyotypically conservative and variable sections. The section *Pneumonanthe*, whose species (except for *G. asclepiadea*) have only a single chromosome number of $2n = 26$, may be considered as conservative. Sections *Cruciata* and *Frigida* are relatively conserved with three different chromosome numbers. Among the variable sections are *Dolichocarpa* and *Chondrophyllae*, for which 11 and 14 different diploid chromosome numbers have been found.

Chromosome numbers for *Dolichocarpa*, *Chondrophyllae*, and *Calathianae* are represented by the typical dysploid chromosome series, which could possibly derive from structural chromosome rearrangements and gene mutations (Yuan 1993). Consideration of the literature revealed that most European *Gentiana* species appear to be natural polyploids. Dysploidization, in combination with polyploidization, could play an important role in karyological evolution of *Gentiana* (Yuan et al. 1996), which could possibly explain evidence for cytological races in each European species of the section *Calathianae* (except for *G. nivalis*), as well as in *G. decumbens* (section *Cruciata*), *G. boryi* (section *Chondrophyllae*), and *G. asclepiadea* (section *Pneumonanthe*).

The results of our cytogenetic analysis of seven *Gentiana* species from the Ukraine support the data of other workers and suggest the conservation of chromosome numbers in these species. The chromosome number for *G. asclepiadea* ($2n = 36$) identified by us corresponds to that proposed for one of the chromosome races of this species. Further karyological studies on *Gentiana* species, including those from other localities should facilitate discussion of the details of genus evolution, resolving classification problems of its constituent taxons.

Acknowledgments This work was supported by the Queen Jadwiga Fund of the Jagiellonian University in Cracow, Poland. The authors thank Prof. Andrzej J. Joachimiak (Department of Plant Cytology and Embryology, Jagiellonian University, Poland) for his support and valuable suggestions, and Volodymyr Adonin (Department of Cell Population Genetics, Institute of Molecular Biology and Genetics National Academy of Sciences of Ukraine) for assistance in translating the text into English.

References

Andreev N (1982) IOPB chromosome number reports LXXVI. Taxon 31:575–576
Bolkhovskikh Z, Grif V, Matvejeva T, Zakharyeva O (1969) Chromosome numbers of flowering plants. Nauka, Leningrad, pp 1–926
Bunge ADE (1829) Conspectus generis *Gentianae*, imprimis specierum rossicarum. Nouveaux Mémoires de la Société Impériale des Naturalistes de Moscou 1:199–256
Chopyk VI (1976) The High Mountain Flora of the Ukrainian Carpathians. Naukova Dumka Press, Kyiv, pp 1–267
Delay J (1970) Orophytes. Inform Annuelles Caryosyst et Cytogenet 4:1–16
Didukh YaP (ed) (2009) Red Data Book of Ukraine. Vegetable Kingdom. Globalconsulting, Kyiv
Dumortier B-C (1827) Florula Belgica, operis majoris prodromus. Staminacia. J. Casterman, Tornaci Nerviorum, p 51
Engelskjon T (1979) Chromosome numbers in vascular plants from Norway, including Svalbard. Opera Botanica 52:1–38
Favarger C (1949) Contribution à l'étude caryologique et biologique des Gentianacées. Bull Societe Botanique Suisse 59:62–86
Favarger C (1952) Contribution à l'étude caryologique et biologique des Gentianacées. II. Bull Societe Botanique Suisse 62:244–257
Favarger C (1965) Notes de caryologie Alpine. IV. Bull Societe Neuchâteloise des Sciences Naturelles 88:5–60
Favarger C (1969a) Reports. In: Löve A (ed) IOPB chromosome number reports XXII. Taxon 18:433–442
Favarger C (1969b) Notes de caryologie Alpine. V. Bull Societe Neuchâteloise des Sciences Naturelles 92:13–30
Favarger C, Küpfer P (1968) Contribution à la cytotaxonomie de la flore alpine des Pyrénées. Collectanea Botanica (Barcelona) 7:325–352
Fernández I, Diéz MJ, Pastor J (1985) Números cromosómicos para la flora española. Lagascalia 13:299–302
Froelich JA (1796) De *Gentiana* libellus sistens specierum cognitarum descriptiones cum observationibus. Erlangen, pp. 15–57
Gagnidze R, Küpfer P, Yuan Y-M (1992) Chromosome numbers of some *Gentianaceae* from the Caucasus. Bull Societe Neuchâteloise des Sciences Naturelles 115:47–52

Gagnidze RI, Griniashvili TN, Dzhindzholia LD (2006) Chromosome numbers of some species of the *Georgian flora*. Botanicheskij Zhurnal 91:1928–1929
Gaudin J (1828) Flora helvetica, vol 2. Orelli, Fuesslini et Sociorum, Zürich, pp 269–270
Grisebach AHR (1838) Genera et Species gentianearum adjectis observationibus quibusdam phytogeographicis. Tübingen, Cotta JG, Stuttgart, p 259
Grisebach AHR (1845) Gentianaceae. In: de Candolle A (ed) Prodromus Systematis Naturalis Regni Vegetabilis. vol 9. Fortin, Masson, et Sociorum, Paris, pp. 39–141
Ho T-N (1985) A study on the *Gentiana* of China. IV. Bulletin of Botanical Research. Harbin 5:1–22
Ho T-N, Liu S-W (1990) The infrageneric classification of *Gentiana* (*Gentianaceae*). Bull British museum (Natural History). Botany 20:169–192
Holub J, Mesicek J, Javurkova V (1970) Annotated chromosome counts of Czechoslavak plants (1–15). Folia Geobotanica et Phytotaxonomica 5:339–368
Hsieh TY, Hsu TC, Kono Y, Ku SM, Peng CI (2007) *Gentiana bambuseti* (*Gentianaceae*), a new species from Taiwan. Bot Stud 48:349–355
Johnson AW, Packer JG (1968) Chromosome numbers in the flora of Ogotruk Creek, N.W. Alaska. Botaniska Notiser 121:403–456
Knaben G (1950) Chromosome numbers of Scandinavian arctic-alpine plant species. I. Blyttia 8:129–155
Knaben G, Engelskjon T (1967) Chromosome numbers of Scandinavian arctic-alpine plant species. II Acta Borealia. A. Scientia 21:1–57
Krogulevich RE (1978) Karyological analysis of the species of the flora of eastern Sayana. In: Malyshev LI, Peshkova GA (eds) Flora of the Prebaikal. Nauka, Sibirskoe Otdelenie, Novosibirsk, pp 19–48
Küpfer P (1968) Nouvelles prospections caryologiques dans la flore orophile des Pyrénées et de la Sierra Nevada. Bulletin de la Societe Neuchâteloise des Sciences Naturelles 91:87–104
Küpfer P (1974) Recherches sur les liens de parenté entre la flore orophile des Alpes et celle des Pyrénées. Boissiera 23:1–322
Küpfer P, Favarger C (1967) Premières prospections caryologiques dans la flora orophile des Pyrénées et de la Sierra Nevada. C R Hebd Seances Acad Sci 264:2463–2465
Kusnezow NI (1893) Nueu asiatische und amerikanische Gentianen. Trudy Imperatorskago S.-Peterburgskago Botaniceskago Sada 13:55–64
Kusnezow NI (1894) Subgenus *Eugentiana*, genus *Gentiana* tournefort. Systematical, morphological and geographical working. Trudy Imperatorskago S.-Peterburgskago Obshchestva estestvoispytatelei. Otdelenie botaniki 4:1–507
Laane MM (1969) Further cytological studies in Norwegian vascular plants. Blyttia 27:5–17
Löve A, Löve D (1975a) Reports. In: Löve A (ed). IOPB chromosome number reports L. Taxon 24:671–678
Löve A, Löve D (1975b) The Spainish gentians. Anales del instituto botanico A.J. Cavanilles 32:221–232
Löve A, Löve D (1986) IOPB chromosome number reports XCIII. Taxon 35:897–903
Löve D (1953) Cytotaxonomical remarks on the *Gentianaceae*. Hereditas 39:225–235
Lovka M, Susnik F, Löve A, Löve D (1971) Reports. In: Löve A (ed). IOPB chromosome number reports XXXIV. Taxon 20:785–797
Lovka M, Susnik F, Löve A, Löve D (1972) Reports. In: Löve A (ed). IOPB chromosome number reports XXXVI. Taxon 21:333–346
Májovský J et al (1974a) Index of chromosome numbers of Slovakian flora (part 3). Acta Facultatis Rerum Naturalium Universitatis Comenianae, Botanica 22:1–20
Májovský J et al (1974b) Index of chromosome numbers of Slovakian flora (part 4). Acta Facultatis Rerum Naturalium Universitatis Comenianae, Botanica 23:1–23
Mededović S, Šiljak-Yakovlev S, Mišić L (1984) Comparative biosystematic study on the *Gentiana aculis* L. and *Gentiana dinarica* Beck. species. Godisnjak Bioloskog Instituta (Universiteta) Sarajevo 37:79–90
Müller G (1974) IOPB chromosome number reports XLIV. Taxon 23:373–380

Müller G (1982) Contribution à la cytotaxonomie de la section *Cyclostigma* Griseb. du genre *Gentiana*. Feddes Repertorium 93:625–722

Murin A (1974) Index of chromosome numbers of Slovakian flora (part 4). Acta Facultatis Rerum Naturalium Universitatis Comenianae, Botanica 23:1–23

Nikolov NA (1991) Chromosome numbers of Bulgarian angiosperms from north pirin mountain: reserve "bajuvi dupki-dzindzirica". Fitologiya (Sofia) 41:70–75

Pogan E, Jankun A, Turała-Szybowska K (1987) Further studies in chromosome numbers of Polish angiosperms. Part XX. Acta Biol Crac. Series: Botanica 29:1–17

Prosser F, Bertolli A (2008) A new species of *Gentiana* sect. *Calathianae* (*Gentianaceae*) from the Brenta Group, European Alps, Italy. Willdenowia 38:423–431

Rork CL (1949) Cytological studies in the *Gentianaceae*. Amer J Bot 36:687–701

Scheerer H (1939) Chromosomenzahlen aus der schleswig-holstenischen flora I. Planta 29:636–642

Scholte G (1977) IOPB chromosome number reports LVI. Taxon 26:257–274

Semerenko LV (1985) Chromosome numbers in some species of flowering plants of Byelorussian flora. Botanicheskij Zhurnal 70:992–994

Skalińska M (1951) Cytological studies in *Gentiana*-species from the Tatra and Pieniny Mts. Bulletin de la Academie Polonaise des Sciences, Serie des Sciences Biologiques 1:119–137

Stolt KAH (1921) Zur embriology der gentianaceen und menyanthaceen. Svenska Vetenskaps Academiens Handlingar 61:3–55

Strashniuk NM, Twardovska MO, Mel'nyk VM (2008) Karyology of some European species of *Gentiana* L. (*Gentianaceae*). Ukrain Bot J 65:836–848

Strid A, Andersson IA (1985) Chromosome numbers of Greek mountain plants. An annotated list of 115 species. Botanische Jahrbücher für Systematik 107:203–228

Takhtajan AL (ed) (1990) Chromosome numbers of flowering plants of flora USSR. Aceraceae-Menyanthaceae. Nauka, Leningrad, pp 426–429

Tischler G (1950) Die Chromosomenzahlen der Gefässpflanzen Mitteleuropas. Gravenhage, Junk, pp 1–263

Tzvelev NN (1978) Gentianaceae. In: AnA Fedorov (ed) Flora of the European Part of the SSSR. Nauka, Leningrad, pp 57–86

Tutin TG (1972) Gentianaceae. In: Tutin TG, Haywood VH, Burges NA, Moore DM, Valentine DH, Walters SM, Webb DA (eds) Flora Europea, vol 3. Cambridge Univ Press, Cambridge, pp 59–63

van Loon JC, de Jong H (1978) Reports. In: Löve A (ed) IOPB chromosome number reports LIX. Taxon 27:53–61

Wóycicki Z (1932) Nieco szczegółów z zakresu rozwoju pylników i pyłku u niektórych przedstawicieli rodzaju *Gentiana*. I. *Gentiana asclepiadea* L. (Quelques détails du dévelopement des anthéres et du pollen chez certain réprésentants du genre *Gentiana*. I. *Gentiana asclepiadea* L.). Acta Soc Bot Pol 9:7–30

Wóycicki Z (1935) Nieco szczegółów z zakresu rozwoju pylników i pyłku u niektórych przedstawicieli rodzaju *Gentiana*. III. *Gentiana lutea* L. (Zur Entwicklungsgeschichte der Antheren and des Pollens bei einigen Repräsentanten der Gattung *Gentiana*. III. *Gentiana lutea* L.). Acta Soc Bot Pol 12:207–226

Yuan Y-M (1993) Karyological studies on *Gentiana* sect. *Cruciata* (*Gentianaceae*) from China. Caryologia 46:99–114

Yuan Y-M, Küpfer Ph, Doyle JJ (1996) Infrageneric phylogeny of the genus *Gentiana* (*Gentianaceae*) inferred from nucleotide sequences of the internal transcribed spacers (ITS) of nuclear ribosomal DNA. Am J Bot 83:641–652

Zerov DK (ed) (1957) Flora URSR, vol 8. Publish AN URSR, Kyiv, pp 236–256

Zhukova PG (1982) Chromosome numbers of some north-eastern Asia plant species. Botanicheskij Zhurnal 67:360–365

Zuyev VV (1990) On the systematic of the *Gentianaceae* family in Siberia. Botanicheskij Zhurnal 75:1296–1305

Chapter 8
Twenty-first Century Centauries: An Updated Review on *Centaurium* Hill and Allies (Gentianaceae)

Guilhem Mansion

Abstract Centauries have a long tradition in human medicine and botanical accounts. Early indications of centauries refer to the famous works of Theophraste (fourth century BC), Pline, and Dioscoridis (first century), who dedicated the herb to the Centaur Chiron. First preserved, albeit idealistic, representations date back to the sixth century. Since typification of *Gentiana centaurium* and *Chironia* by Linnaeus (1753), the classifications of Centauries were debated by renowned botanists until the end of the twentieth century. This chapter presents an overview of the current and accepted delimitations of centauries, as well as new insights into their taxonomic and evolutionary history.

8.1 Introduction: Centauries Through the Centuries

In Greek mythology, Centaurs were half-horse half-human, generally infamous, wild and lusty creatures. Chiron, their most famous representative, was nonetheless highly intelligent, cultured, and wise, and proved an ideal tutor and teacher for many Greek heroes and divinities, including Asclepius the God of Medicine (Grant and Hazel 1973). The first accounts of "*Kentaurion*" or "*Centaurium*" as a healing herb dedicated to Chiron's memory were attributed to Theophraste (Fourth Century BC), and further mentioned by Dioscoridis and Pline (First Century), whose works remained absolute references in plant and medical knowledge for more than fifteen centuries (Dioscorides and Ruel 1552; Pliny et al. 1554; Theophrastus et al. 1552). The first preserved representation of *Kentaurion* is to be found in the famous "*Juliana Anicia Codex*" or "*Vienna Dioscorides*", an early illuminated version of Dioscorides' "*De Materia Medica*." The somehow stylized plant unambiguously

G. Mansion (✉)
Botanischer Garten und Botanisches Museum Berlin-Dahlem, Freie Universität Berlin, Königin-Luise-Straße 6–8 14195 Berlin, Germany
e-mail: g.mansion@bgbm.org

J. J. Rybczyński et al. (eds.), *The Gentianaceae - Volume 1: Characterization and Ecology*, DOI: 10.1007/978-3-642-54010-3_8,

represents a gentian currently known as *Centaurium erythraea* Raf., or small Centaury. In reference to "*Kentaurion*" and "*Centaurium*," the general name "Centauries" is used here to describe a natural group of gentians morphologically similar to *Centaurium* Hill (Mansion 2004), and taxonomically included in the subtribe Chironiinae of the Gentianaceae-Chironieae (Struwe et al. 2002).

The taxonomic history of Centauries is complex and can be viewed as a succession of use and misuse of names such as *Centaurium*, *Erythraea*, *Chironia* or *Cicendia*. The taxonomic entity *Centaurium* appeared for the first time in the late sixteenth century (Dodoens and Plantini 1583), and was later cited by Bauhin (1623), Tournefort (1700) and Ray (1724). Nevertheless, in the first edition of his "*Species Plantarum*," Linnaeus (1753) placed centauries into both *Gentiana* or *Chironia*, the former encompassing the current genera *Blackstonia* Huds., *Centaurium* Hill, *Cicendia* Adans., and *Schenkia* Griseb., the latter including *Chironia* L., *Orphium* Mey. and *Sabatia* Adans. Hill (1756), in his British Herbal, challenged Linnaeus and proposed to use the name *Centaurium* to include *Gentiana centaurium* L., but also the "little yellow marsh centaury" (now *Cicendia filiformis*) and the "perfoliate yellow centaury" (now *Blackstonia perfoliata*). Hill was not followed by his contemporaries. Indeed, de Necker (1790) proposed "*Erithrea*" (an invalidly published name under the current rules of nomenclature), and "*Erythraea*" as correctly proposed by Borckhausen (1796) was then used in most important systematic work on Gentianaceae (Bentham and Hooker 1876; Brown 1810; Don 1838; Endlicher 1838; Gilg 1895; Grisebach 1839, 1845), before it finally became a synonym of *Centaurium*, following the rule of priority (Robyns 1954).

Some authors never used the names *Centaurium* or *Erythraea*, but instead expanded Linnaeus' or Lamarck's concepts of *Chironia* (Lamarck and Poiret 1783; Linnaeus 1753). Willdenow (1797), for instance, transferred most Centauries placed in the Linnaean *Gentiana* (except *Gentiana exaltata*, now *Zeltnera* G. Mans.) into *Chironia*, making the latter a polyphyletic assemblage of 16 species of African (*Chironia*, *Orphium*), American (*Sabatia*), and Eurasian (*Centaurium*, *Schenkia*) origins. Brown (1810) was finally the first to understand the artificial circumscription of *Chironia* L. and to delimit the genus on a sound footing (i.e., keeping only the African species in it).

Grisebach's contributions (1839, 1845, 1853) to the delimitation and understanding of Centauries was inestimable. Not only did he provide the first monographic treatise of Gentianaceae, he also proposed new generic combinations (e.g., *Gyrandra*, *Schenkia*, or *Zygostigma*) to accommodate the diversity of Centauries. Although pertinent, the two first genera were not used until their reinstatement in 2004 (Mansion 2004; Mansion and Struwe 2004). While perceiving the artificial nature of *Centaurium* Hill (treated by him as *Erythraea*) by segregating *Gyrandra* and *Schenkia*, Grisebach failed to unravel the independent nature of the American species of *Centaurium* (now in *Zeltnera* G. Mans.). He indeed included them in different sections (*Euerythraea* or *Trichostylus*), or genera such as *Cicendia* Adans.

Overall, Grisebach's concept of *Cicendia* remained inadequate. Of the eight, generally annual, divaricate, and slender species, treated as *Cicendia* in his Gentianaceae monograph (Grisebach 1839), only two (*C. filiformis* and *C. quadrangularis*) actually corresponded to the genus definition given by Adanson (1763). In 1845, Grisebach revised his first opinion and reduced *Cicendia* to six species by transferring two of them into *Microcala*, a genus proposed earlier by Hoffmannsegg and Link (1820). Unfortunately, none of the new combinations proved to be correct. *Cicendia*, thereby circumscribed, contained species belonging to the current *Exaculum*, *Sabatia*, *Zeltnera*, or *Exacum* (now in tribe Exaceae), while *Microcala* encompassed the two authentic *Cicendia* species. Bentham and Hooker (1876) reduced Grisebach's *Cicendia* to one species (the current *Exaculum pusillum*), but kept *Microcala* as defined earlier. Gilg (1895) followed his predecessors, perpetuating the long-standing confusion between *Cicendia* (under "*Microcala*") and *Exaculum* (under "*Cicendia*"). As a result, both genera were often treated as synonyms or considered as sister taxa (Struwe et al. 2002).

Besides taxonomic confusion, the systematic delimitation of "natural groups" in Centauries was, and still remains, a hot topic, mainly due to the difficulty in inferring morphological synapomorphies for genetic clades (Mansion and Struwe 2004). The main progress in the classification of Centauries was established through general systematic treatments of the Gentianaceae by Endlicher (1838), Grisebach (1839, 1845), Bentham and Hooker (1876), Knoblauch (1894), and Gilg (1895). Gilg's contribution, based mainly on pollen characters, was used as a reference for at least a Century.

During the twentieth century, no significant improvement was made regarding the systematics of Centauries. Instead, most authors concentrated on the realization of monographs of Gilg's groups, using either morphology (*Chironia*; Marais and Verdoorn 1963; *Geniostemon*; Turner 1994), cytogeography (*Centaurium* pp; Broome 1973; *Sabatia*; Perry 1971; *Blackstonia* and *Centaurium* pp; Zeltner 1970) or phytotaxonomy (*Blackstonia* and *Centaurium* pp; Van der Sluis 1985). Most of these studies, however, suffered biased geographic taxon sampling and lack of independent evidence that could help to avoid subjective interpretations in systematic treatments. Therefore, it is not before the premise of the DNA revolution, and the sudden availability of thousands of nonmorphological characters, that substantial improvement could be made. At the dawn of the twenty-first century, pioneer studies on *Gentiana* (Gielly and Tabarlet 1996; Yuan and Küpfer 1995) were soon followed by more detailed work of other groups, including the Centauries (Mansion 2001). At the same time, a new global account on the Gentianaceae (Struwe et al. 2002) placed Centauries in the subtribe Chironiinae, but did not include important results on *Centaurium* s.l., published soon after the book was released (Guggisberg et al. 2006; Mansion 2004; Mansion and Struwe 2004; Mansion and Zeltner 2004; Mansion et al. 2005; Zeltner and Mansion 2003).

In this volume on the Gentianaceae, work on Centauries has been reviewed since Mansion (2001). As a robust framework for discussion, a phylogenetic tree has been used encompassing more than 200 sequences of nuclear ribosomal DNA (nrDNA), including unpublished ones. Overall, this chapter presents an overview of the current and accepted delimitations of Centauries, as well as insights into their taxonomic and evolutionary history. It also provides a new light on poorly known taxa and suggests guidelines for future work.

8.2 Analytical Procedures Employed and Results

Based on previous phylogenetic studies (Mansion and Struwe 2004; Yuan et al. 2003), the genera *Canscora* and *Microrphium* were used as outgroups. Total DNA extraction, PCR amplification, and sequencing of the Internal Transcribed Spacer (ITS) region of nrDNA followed protocols described in Mansion and Struwe (2004). Samples from the rare and endangered species *Schenkia sebaeoides* came from the Hawaiian Plant DNA Library (Morden et al. 1996; Randell and Morden 1999). A total of 222 ITS sequences were aligned using Muscle (Edgar 2004), with additional manual corrections in PhyDe (Müller et al. 2005). Newly generated sequences were deposited in GenBank.

The aligned matrix (222 taxa, 481 characters) was analyzed using the respective maximum parsimony (MP) and Bayesian inference (BI) approaches. The MP analysis, using a Fitch criterion, was performed with the version 4.0b10 of PAUP (Swofford 2002). Heuristic searches were conducted with a ratchet batch-file, including 200 iterations, each of them with 25 % of the positions randomly weighted (weight = 2), and 100 random additions, generated with PRAP (Müller 2004). Branch support was calculated with the bootstrap (BS) method, using 10,000 replicates, TBR branch swapping, 10 random-additions, multrees option OFF, and resampling all characters.

The BI analyses were conducted with MRBAYES (Huelsenbeck and Ronquist 2001), using six simultaneous runs of Metropolis-coupled Markov Chain Monte Carlo (MC3), under a GTR + G model of sequence substitution selected using the Akaike Information Criterion in MRMODELTEST (Nylander 2004). Each chain was run in parallel for 5 million generations, saving one tree each 1,000 generation, keeping a default temperature parameter value of 0.2. The MC3 runs were repeated twice, and the first 10 % of the saved trees were discarded as burn-in. The remaining trees were used to produce a majority-rule consensus tree and to calculate the posterior probability (pp) values.

Ratchet parsimony analyses inferred 5,202 most parsimonious (MP) trees, with the following metrics: Length (*L*) = 1,194, Consistency Index (CI) = 0.475, and Retention Index (RI) = 0.915 (Fig. 8.1). Independent Bayesian analyses performed under the GTR + G model of nucleotide substitution, yielded congruent topologies and similar posterior probability (pp) values for each separate 50 % majority-rule consensus tree. As both analyses inferred similar topologies, the next section discusses the general structure depicted by the MP strict consensus tree, and mention only minor differences when appropriate.

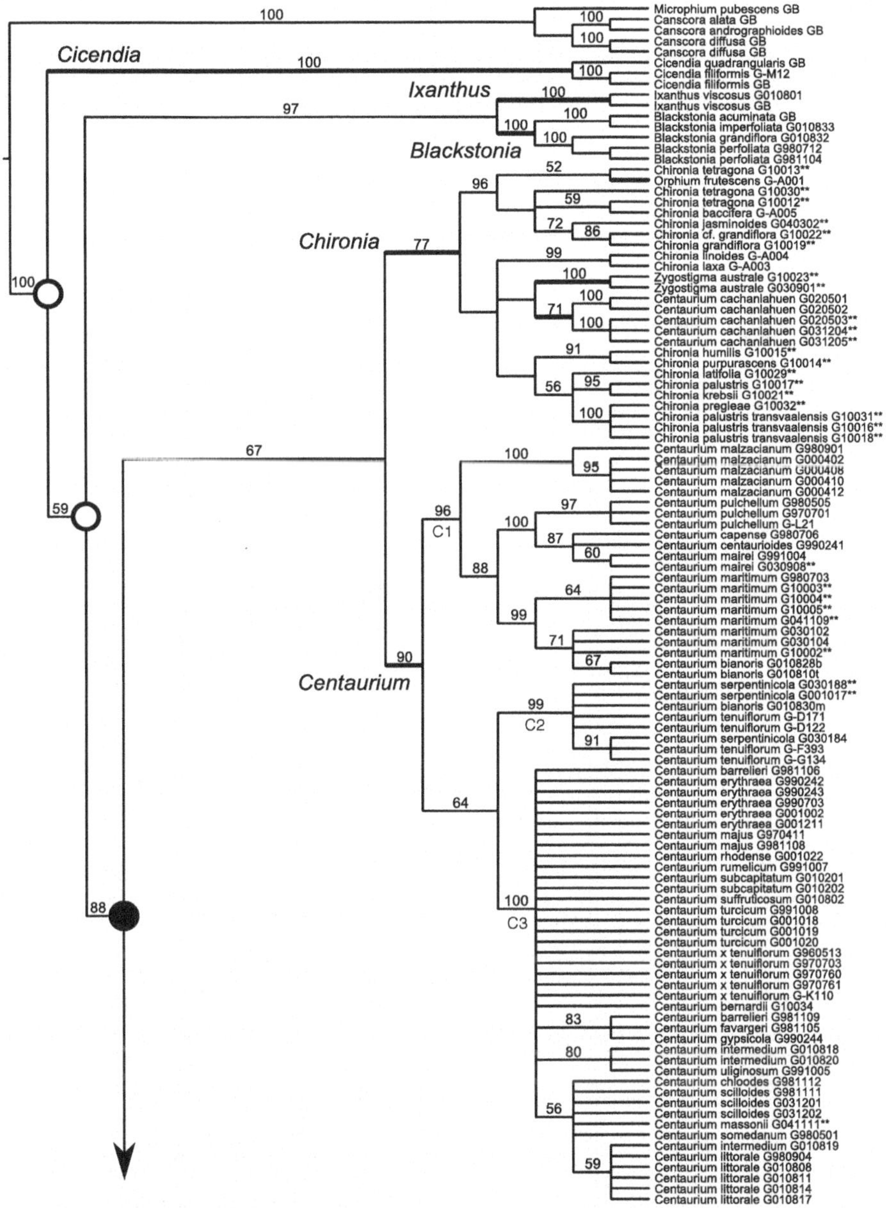

Fig. 8.1 Strict consensus of 5,202 trees inferred from MP analysis of 222 ITS sequences. Numbers above branches are bootstrap values. Number after the taxon name refers to the DNA accession number (*G* Published sequences produced by G. Mansion, *GB* sequences directly retrieved from GenBank, ** unpublished sequences). *White* and *black* dots refer to "yellow centauries" and "pink centauries," respectively

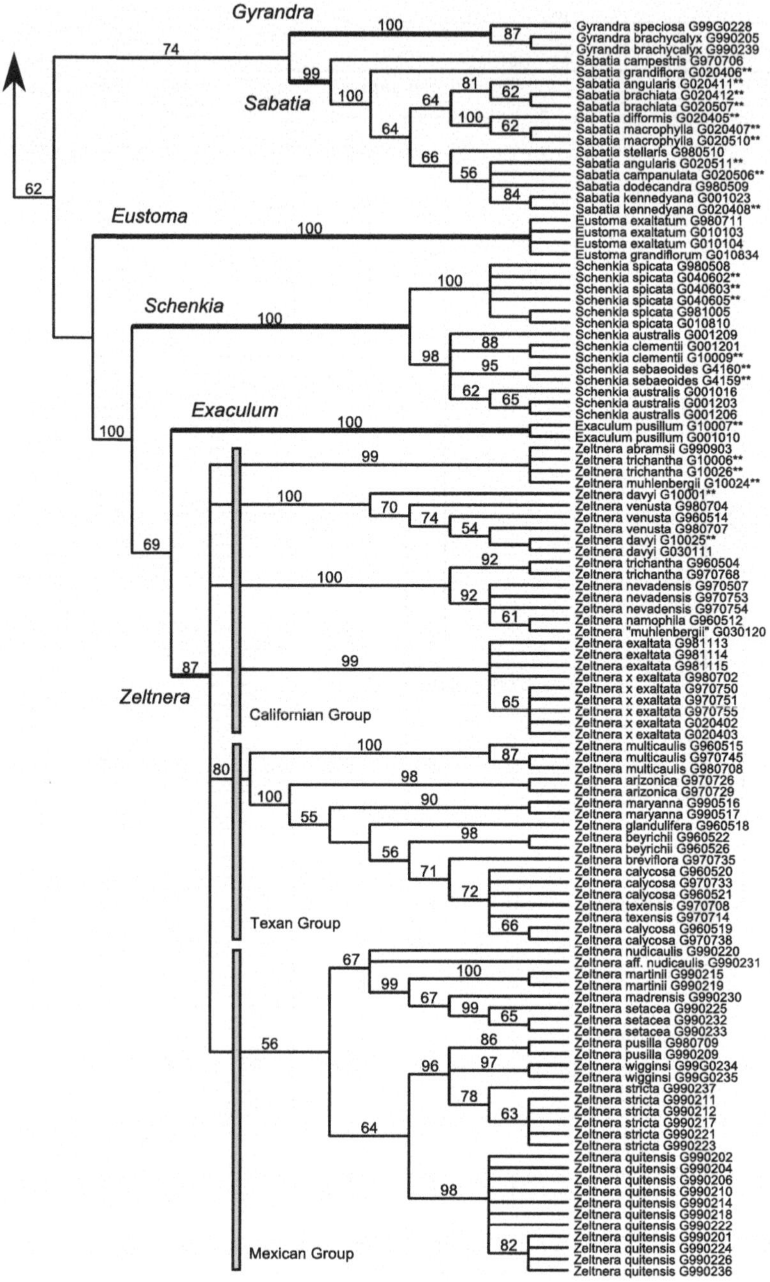

Fig. 8.1 continued

8.3 Discussion

8.3.1 The ITS Region and Centauries: A Long History

Sequences of the ITS region of nrDNA have been widely used in cladistic inference of plant groups since the pioneer work of Baldwin (1992). The advantages and limitations of such a molecular region have been discussed (Alvarez and Wendel 2003; Feliner and Rossello 2007), and important reasons for its success and widespread utilization include the satisfactory level of sequence variation, generally allowing phylogenetic reconstruction and evolutionary studies at the species level, the incomplete concerted evolution that in some cases permits the identification of hybrids or introgressive forms, and the lack of alternative variable regions within the respective chloroplast and mitochondrial plant genomes. Finally, the large production of ITS sequences by taxonomically trained scientists favor potential exchange of material and further collaboration between groups using the same analytical tools.

On the whole, the ITS region has proven successful in the Gentianaceae, starting at the infrageneric level within *Gentiana* (Yuan and Küpfer 1995, 1997; Yuan et al. 1996), and quickly extending toward the generic and tribal levels, with the Chironiae (Mansion 2001; Mansion and Struwe 2004; Thiv et al. 1999), Exaceae (Kissling et al. 2009; Yuan et al. 2003), or Gentianeae (Chassot et al. 2001; von Hagen and Kadereit 2001). The use of ITS alone reveals only one aspect of the history of Centauries and the present results are discussed in the light of other available evidence, including phylogenetic results from maternally inherited DNA regions (cpDNA), karyology, or morphology.

8.3.2 Molecular Evidence on the Centauries

The ITS strict consensus tree, rooted with *Canscora* and *Microrphium* (Canscoriinae), inferred a strongly supported Chironiineae (BS100), with two early divergent clades comprising plants with yellow corollas, namely *Cicendia* (BS100) and (*Blackstonia–Ixanthus*) (BS 100), respectively (Fig. 8.1a: "yellow Centauries"). A third, more derived clade (BS 88) comprises all the remaining species, with pink to lilac or reddish corollas (Fig. 8.1a: "pink Centauries"). Within the pink Centauries, *Centaurium* forms a strongly supported lineage (BS90) sister to a *Chironia* clade (BS77), the latter being paraphyletic with respect to *Orphium*, *Centaurium cachanlahuen*, and *Zygostigma*. A second moderately supported clade (BS62) encompasses *Gyrandra–Sabatia* (BS74), *Eustoma* (BS100), and a derived *Schenkia–Exaculum–Zeltnera* lineage (BS100).

8.3.2.1 The Cicendia Adans. Clade

The taxonomy, phylogenetic position, and affinities of the genus have long been questioned. The confusing use of the respective names *Cicendia* Adans., *Exaculum* Caruel, and *Microcala* Hoffmans. & Link (the latter a nomenclatural synonym of *Cicendia* Adans.) by different authors, resulted in the intimate association of both *C. filiformis* and *E. pusillum* by most European botanists. The fact that those genera are morphologically similar (slender herbs, sparsely branched, with minute flowers), with comparable ecological affinities (Favarger 1960), further reinforced the feeling that both entities shared a common ancestor.

Molecular inferences based on both ITS (Fig. 8.1a) and cpDNA data fully reject this assumption. *Cicendia* is actually an early divergent lineage within the subtribe, while *Exaculum* belongs to a more derived assemblage encompassing the Mediterranean *Schenkia* and the North American *Zeltnera* (Fig. 8.1b). Other differences between the respective *Cicendia* and *Exaculum* have been discussed elsewhere (Mansion and Struwe 2004).

8.3.2.2 The (Blackstonia Huds.–Ixanthus Griseb.) Clade

Close phylogenetic relationships between the Mediterranean *Blackstonia* and the Canarian endemic *Ixanthus* were suggested early on by Grisebach (1839) and later confirmed by molecular data (Mansion 2001; Mansion and Struwe 2004; Thiv et al. 1999).

Blackstonia comprises four annual, diploid or tetraploid species of mainly Mediterranean distribution (Zeltner 1970). The genus was long known as *Chlora*, a pre-Linnean name first used by Renealmus (1611) and published by Adanson (1763). Nonetheless, Hudson proposed the name *Blackstonia* one year before in his "*Flora Anglica*" (Hudson 1762), and his combination has thus priority. The most comprehensive study on the genus was performed by Zeltner (1970) who recognized four species, four subspecies, and two varieties, based on cytogeographic evidence. ITS-based inference supports two sister lineages, *B. acuminata* – *B. imperfoliata* and *B. grandiflora* –*B. perfoliata*, respectively, in accordance with Zeltner's hypotheses.

The monotypic ***Ixanthus***, described by Grisebach (1839), contains perennial, basally woody, and sparsely branched plants, which can reach 2 m in height. A detailed morphological study of the genus is provided by Thiv et al. (1999). The striking difference in gross morphology between the insular (*Ixanthus*) and continental (*Blackstonia*) lineages has been exemplified in several other genera, including, e.g., *Echium*, *Lotus*, or *Sonchus* (Allan et al. 2004; Bramwell 1972; Kim et al. 1996; Mansion et al. 2009).

8.3.2.3 The *Centaurium* Hill Clade

Centaurium was described by Hill (1756), but was only typified two centuries later by Gillett (1963), based on *Centaurium littorale*. This important time frame reflects the taxonomic complexity of a group in which morphological plasticity and important reticulate patterns hinder clear delimitations (Mansion et al. 2005; Melderis 1972; Zeltner 1970).

The inferred gene tree depicts two rather unbalanced clades with a first lineage of six species (C1, BS96; Fig. 8.1a), sister to a clade encompassing all the remaining species (BS64). The latter can be further divided into a "*C. tenuiflorum*" clade (C2, BS99; Fig. 8.1a) and a "*C. erythraea*" clade (C3, BS100; Fig. 8.1a).

Within C1, the diploid *C. maritimum* ($n = 10$) is unique among *Centaurium* (and other pink Centauries) due to the yellow color of its corolla, a character which was used to define the monotypic Section *Xanthea* (Reichenbach 1834). Despite an apparent morphological homogeneity throughout its distribution area, molecular data support the division of *C. maritimum* into two well-separated clades (BS99). One of the subclade contains populations sampled on the Levant Island, where the subspecies *shuttleworthianum* was described (Rouy 1891; Rouy et al. 1893). The individuals, nonetheless, did not match perfectly Rouy's detailed description of the taxa. The same subclade also contains *C. bianoris*, an allopolyploid studied in detail by Guggisberg et al. (2006). The hybrid species was shown to have arisen from *C. maritimum* as seed parent and diploid *C. tenuiflorum* subsp. *acutiflorum* as pollen parent. Heterogeneous ITS sequences were further distributed within the respective parental clades, as shown here. Overall, the genetic subdivision of *C. maritimum* seems to contain a geographic signal.

Clade C1 also exhibits a well-supported "*C. pulchellum*" subclade (BS100), consisting of four species with a basic chromosome number of $x = 9$. The tetraploid *C. pulchellum* is the most widespread species within the genus, naturally occurring from Finland to China, but also frequently naturalized around the world. The morphologically similar *C. capense*, endemic to Baja California, forms nonetheless a clade with two hexaploid species, (a) *C. mairei*, which replaces *C. pulchellum* in the S part of the Mediterranean basin, from the Canary Islands to India and, (b) *C. centaurioides* ($2n = 6x = 56$), a putative allopolyploid, occurring in India.

Finally, the strong genetic separation between the two aforementioned clades (*C. maritimum* and *C. pulchellum*) and the sister *C. malzacianum* clade (BS100) contradicts the subspecific ranking of *C. malzacianum* (*C. centaurioides* subsp. *malzacianum*), as proposed previously (Zeltner 1985). Evidence (Mansion et al. 2005) suggests an allopolyploid origin for *C. malzacianum* ($n = 28$) between *C. maritimum* ($n = 10$, maternal parent) and *C. pulchellum* ($n = 18$, paternal parent). However, the current distribution of the species involved would not support a recent formation for this potential allopolyploid.

The strongly supported clade C2 (BS99; Fig. 8.1a) encompasses diploid and tetraploid accessions of *C. tenuiflorum*, *C. serpentinicola*, and *C. bianoris*. In this case, the ITS signal contradicts both cpDNA-based gene trees (Mansion et al. 2005), and gross morphology, which supports the inclusion of *C. tenuiflorum*

($x = 10$) and *C. pulchellum* ($x = 9$; clade C1) in subsection *Parviflora* (Zeltner 1970). *Centaurium serpentinicola* ($n = 20$) is characterized by a robust and erect habit, large flowers, and shows morphological affinities with *C. erythraea* and *C. tenuiflorum* (Carlström 1986), so that a hybrid origin has been postulated for this species (Zeltner 1991). The polyphyletic distribution of the respective accessions in the ITS cladogram (BS91) gives some support to this hypothesis.

Lastly, the largely unresolved clade C3 (BS100; Fig. 8.1a) comprises mainly species that taxonomically fit Sections *Centaurium* and *Caespitosa* (Zeltner 1970). The current gene tree depicts weak sequence divergence among species related to *C. erythreae* Raf., a morphologically very variable group including three chromosome valences. Diploid taxa, including *C. bernardii* (included here for the first time) and *C. suffruticosum* are infrequent and morphologically well differentiated, whereas diploid *C. rumelicum* resembles the tetraploid *C. erythraea*. The latter has been subdivided in several varieties whose status remains questionable (Gilmour 1937). Finally, the hexaploid *C. turcicum* exhibits a scabrid indumentum and elliptic to linear leaves, a feature also encountered in some species of subsection *Vulgaria* Melderis.

Subsection *Vulgaria*, which consists of five biennial species with a rosette of linear, sometimes scabrid leaves, does not form a clade. Nonetheless, most of its diploid species (*C. barrelieri*, *C. favargeri*, *C. gypsicola*) form a well-supported assemblage (BS83; Fig. 8.1a), with *C. barrelieri* and *C. gypsicola* exhibiting showy corollas, and sometimes confounded with *C. majus* and *C. rhodense*. The taxonomy of most large-flowered *Centaurium* remains confusing (Gonzalez 1979). Tetraploid species of *Vulgaria* (*C. littorale* and *C. chloodes*) are grouped with some Atlantic species, including *C. scilloides* and *C. somedanum* (BS 56; Fig. 8.1a). *Centaurium scilloides* is a perennial diploid species ($n = 10$) of disjunct Atlantic distribution, recently introduced on the Pacific coast of Chile (Zeltner and Mansion 2003). Hexaploid populations have been discovered on the Azores Islands (Louis Zeltner, personal communication), along with potential hybrid forms that are currently being investigated (G. Mansion, unpublished data). The rare and threatened tetraploid ($n = 20$) is a perennial that resembles morphologically *C. scilloides*, and could represent a link between the diploid and hexaploid cytotypes.

The extensive gene flow between *C. littorale* (including *C. uliginosum*) and *C. erythraea*, producing allotetraploid and hexaploid taxa, along with unstabilized derivatives of the F1 hybrids (with chromosome numbers ranging from $2n = 41$ to $2n = 59$), (here under the name "*C. intermedium*") have been described previously in detail (Mansion et al. 2005; Ubsdell 1976a, b). Further implication of *C. erythraea* in the formation of potential hybrids has also been highlighted with a taxon provisionally named *C. xtenuiflorum* (Mansion 2001; Mansion et al. 2005). This potential allopolyploid, now naturalized in most countries with Mediterranean climates, including Australia, Chile, and Western North America (G. Mansion and L. Zeltner, personal observations) tends to occupy ecological niches of indigenous gentians including *Schenkia australis* (Australia), *C. cachanlahuen* (Chile), and *Zeltnera muhlenbergii* (California). The latter species has been confounded with *C. xtenuiflorum* in many North American floras, resulting in taxonomic chaos (Pringle 2010).

8.3.2.4 The Paraphyly of Chironia L.

Chironia was proposed by Linnaeus (1753) to include eight species, of which only four actually corresponded to the modern definition of the genus. Major works to the comprehension of the group include Brown's contribution (1810) who clearly defined the generic delimitations, Schoch's first monograph of the whole genus (Schoch 1904), describing 33 species, the critical review of Prain (1908), and detailed treatments of south African species by Marais and Verdoorn (1963). Currently, *Chironia* encompasses some 31 species and subspecies distributed throughout tropical and subtropical Africa, with an important diversity in the Fynbos vegetation of the Cape region. Most recent studies, however, still erroneously include Madagascar in the genus distribution (Leistner 2000; Struwe et al. 2002). The so-called "Madagascan" *Chironia* have been transferred by Klackenberg (1986) into the tribe Exaceae, under the name *Ornichia* (an anagram of *Chironia*).

Early phylogenetic studies (Mansion 2001; Mansion and Struwe 2004) indicated the potential paraphyly of *Chironia* with the South-African *Orphium* and the South-American endemic *Centaurium cachanlahuen*. The current study, encompassing 18 accessions of *Chironia* (13 species), confirms this pattern and further depicts the high paraphyly of *Chironia* (BS77) with respect to three different taxa including *Orphium*, *C. cachanlahuen*, and *Zygostigma*, whose phylogenetic position is inferred for the first time (Fig. 8.1a). Overall, ITS reconstruction depicted a number of well-supported subclades, but failed to clearly infer relationships between them. Incomplete taxon-sampling and potential reticulate patterns, which are frequent among centauries, are likely explanations to both the general lack of resolution and the presence of polyphyletic species (e.g., *C. tetragona*, albeit with low BS).

The relationships between ***Orphium*** and *Chironia* have long been discussed (Prain 1908). *Orphium*, first described by Linnaeus (1753) under *Chironia frutescens*, was proposed by Meyer (1837) for suffrutescent and pubescent plants, with a smooth calyx, and a nectar disc (missing in *Chironia*). *Orphium* was accepted in most taxonomic treatments of the family (Bentham and Hooker 1876; Gilg 1895; Grisebach 1845). However, Prain (1908) noticed that the sum of characters separating *Orphium* from *Chironia* were equal to the ones separating *C. baccifera* from other *Chironia*, demonstrating that most generic characters described for *Orphium* are also encountered in *Chironia*.

The unique phylogenetic position of the South American ***Centaurium cachanlahuen***, and its affinities with the African *Chironia*, was inferred first by Mansion and Struwe (2004), who further noticed the similarity between the style and stigma structures of the respective taxa. The current ITS gene tree weakly supports the sister relationships between *C. cachanlahuen* and *Zygostigma*, another genus endemic to South America (BS < 50; pp $= 0.68$).

Grisebach (1839) described ***Zygostigma*** (2 spp.) as a genus with intermediate morphological characters between *Centaurium* (under *Erythraea*) and *Sabatia*. Recent treatments consider *Zygostigma* as monospecific (Filippa and Barboza 2006), but none mention potential relationships with either *Centaurium* or *Chironia*, despite similar floral structure, including corolla color and curved or spirally twisted anthers at postanthesis (anthers in *Zygostigma* have long been considered to be immutable, which is incorrect). Karyological data, however, supports affinities between *C. cachanlahuen* and *Zygostigma australe*, both species sharing the chromosome number of $n = 36$ (Trunz et al. 2012; Zeltner and Mansion 2003). This number is also close to the unique one reported for *Chironia baccifera* ($n = 34$; Weaver and Rüdenberg 1975).

Overall, the current state of knowledge supports affinities between *C. cachanlahuen*, *Chironia*, *Orphium*, and *Zygostigma*. In the absence of a complete phylogeny of *Chironia* including species from Tropical Africa, it remains difficult to resolve unambiguously the taxonomy of the group, except for the exclusion of *C. cachanlahuen* from *Centaurium* Hill. From an evolutionary point of view, it seems that a single dispersal from a *Chironia*-like ancestor better explains the proposed African–American disjunction. A more detailed biogeographic study is in preparation (G. Mansion, unpublished data).

8.3.2.5 The (Gyrandra Griseb.–Sabatia Adans. Clade) and Eustoma Griseb

The current phylogenetic reconstruction, presenting for the first time an extended sampling of *Sabatia*, strongly supports a (*Gyrandra–Sabatia*) clade, but overall remains inconclusive regarding the position of *Eustoma*. The MP strict consensus tree reveals affinities between *Eustoma* and the (*Schenkia–Exaculum–Zeltnera*) clade (albeit with BS < 50; Fig. 8.1b), while the BI reconstruction moderately supports an [*Eustoma*–(*Gyrandra–Sabatia*)] clade (pp $= 0.73$; not shown).

Nonmolecular evidence better argues for a common origin of the three lineages. Indeed, as noted by Mansion and Struwe (2004), most taxa share a chromosome number of $2n = 72$, with a high dysploid variation in *Sabatia*, large corollas (except *Sabatia arenicola* and *Gyrandra brachycalyx*) when compared with other centauries (except some *Chironia*), and a stamen insertion near the sinus of the corolla lobes. Moreover, their respective geographic distribution in the Caribbean, Mexico, and eastern North America, may support vicariant patterns from a common ancestral area.

Eustoma occurs in the Southern United States, Mexico, and Greater Antilles. *Eustoma* differs mainly from *Gyrandra* and *Sabatia* by the degree of anther coiling, the stigma shape, and pollen structure that convinced Gilg (1895) to place the genus in a separate tribe Tachiinae, close to *Lisianthius*. Shinners (1957) described three species based on quantitative traits, and further mentioned the occurrence of potential introgression between the described taxa. More recent treatments (Pringle 2012), however, consider the genus as monotypic, with two subspecies *E. russellianum* subsp. *russelianum* and subsp. *exaltatum*, respectively. The ITS signal depicts one strongly supported (BS100) but unresolved clade, overall supporting both low genetic distance between the respective taxa and/or potential occurrence of introgressive patterns. A detailed phylogeographic overview of the group is required.

Grisebach's genus ***Gyrandra*** (Grisebach 1839) was never fully accepted. Gray (1876) nonetheless used the name to describe a section of *Centaurium* and his opinion was followed in major treatments of the gentian family (Gilg 1895). In 2004, based on Broome's suggestion to include five Mexican species of *Centaurium* in Gray's section *Gyrandra* (Broome 1973), it was proposed to resurrect Grisebach's generic name for those specific centauries (Mansion 2004). Within *Gyrandra*, all species are relatively rare, some being known from only a few populations (*G. pterocaulis*), and a complete phylogeny is still lacking. So far, molecular analyses performed on three species of *Gyrandra* (*G. brachycalyx*, *G. pauciflora*, and *G. speciosa*) support the monophyly of the genus (Fig. 8.1b; G. Mansion, unpublished data).

Sabatia was described by Adanson (1763), based on *S. dodecandra* (Wilbur 1989). As currently circumscribed, the genus contains ca. 19 annual to perennial, herbaceous species, mainly distributed in wetland habitats of Eastern North America (Perry 1971; Wilbur 1955). Intensive studies based on morphological and karyological data, crossing experiments, and breeding systems (Grisebach 1845; Perry 1971; Wilbur 1955) resulted in the systematic division of the genus into three sections namely *Campestres*, *Pseudochironia* (sometimes considered a different genus, *Lapithea*), and *Sabatia*, the latter further split into four subsections.

The inferred phylogenetic tree, encompassing Sections *Campestres* and *Sabatia* (including all subsections), but not *Pseudochironia*, temporary supports the monophyly of the genus, and denotes discrepancies between traditional and molecular approaches. This incongruence is reinforced further by cpDNA-based results, overall suggesting the occurrence of hybridization and/or introgression episodes during the genus history (Mansion and Mathews 2006). The group is currently being investigated more extensively (G. Mansion and C. Mathews, unpublished data).

8.3.2.6 The (Schenkia Griseb.–Exaculum Caruel–Zelnera Mansion) Clade

The ITS-based phylogeny supports a common ancestor for the three genera (BS100), and further depicts a strongly supported *Schenkia* clade (BS100) sister to a rather unbalanced (*Exaculum–Zeltnera*) lineage (BS69; Fig. 8.1b). Published cpDNA cladograms, however, weakly inferred sister relationships between *Schenkia* and *Exaculum* (Mansion and Struwe 2004).

Schenkia currently encompasses five species distributed in the Mediterranean basin and Middle East, Russia, Japan, Australia, and the Hawaiian Islands (Mansion 2004). Despite a sum of characters supporting a generic status, including morphology (spike-like cyme inflorescence, subsessile flowers and subcapitate stigma) and karyology ($n = 11$), Grisebach's genus *Schenkia* (Grisebach 1853) was never used in taxonomic treatments of Centauries. Instead, most authors placed the corresponding plants into *Centaurium* Section *Spicata*, proposed earlier by Grisebach himself (Grisebach 1839).

One reason for the misuse of *Schenkia* could be that Grisebach described his new genus based on a rare Hawaiian endemic *S. sebaeoides*, whose affinities with existing material could not be checked easily. Furthermore, Grisebach did not mention his section *Spicata* as a potential synonym for the new genus, but only pointed out some affinities between his *Schenkia* and *Erythraea spicata* (Grisebach 1853).

The current phylogeny expands that of Mansion and Struwe (2004), but further includes the genus type species (*S. sebaeoides*), and newly found tetraploid populations of *S. spicata* (Louis Zeltner, personal communication), a primarily Mediterranean, diploid, species. Molecular evidence strongly supports (BS100) previous delimitation of *Schenkia* (Mansion 2004), by further including the type species in a strongly supported Australian clade (BS98). The latter is sister to an Eurasian clade encompassing both diploid and tetraploid accessions of *S. spicata* (BS100; Fig. 8.1b).

While molecular data and other evidence clearly support the separation of ***Exaculum*** and *Cicendia*, the phylogenetic position of this monotypic genus still remains unresolved. Indeed, *Exaculum* is inferred as either sister to *Zeltnera* (ITS; Fig. 8.1b) or to *Schenkia* (cpDNA; Mansion and Struwe 2004). Before more data become available, existing evidence would not support the inclusion of *Exaculum* in either genus. Indeed, morphological (divaricate [*Exaculum*] versus spike-like cyme [*Schenkia*] inflorescence) and karyological ($n = 10$ [*Exaculum*] vs. $n = 11$, 22 [*Schenkia*]) evidence argues against the inclusion of *Exaculum* in *Schenkia*. *Exaculum* also differs from *Zeltnera* by the chromosome number ($n = 10$ [*Exaculum*] vs. $n = 17$, 20, 21, 22 [*Zeltnera*]), the stamen insertion (throat of the corolla [*Exaculum*] vs. upper half of the corolla tube [*Zeltnera*]), and the coiling of the anthers (slightly recurved [*Exaculum*] vs. highly spiralized [*Zeltnera*]).

The new World endemic ***Zeltnera*** has been recently described to encompass 25 species formerly included in *Centaurium* Hill (Mansion 2004). Based on karyological and molecular data, the genus has further been divided into respective "Californian," "Texan," and "Mexican" groups (Mansion and Zeltner 2004). The present study, including rare Californian species missing in previous work, still supports the monophyly of the last two mentioned assemblages, but has uncovered additional polytomies in the Californian clade (Fig. 8.1b).

Nonmolecular evidence supporting a **Californian *Zeltnera* clade** include morphology (plants generally single-stemmed from the base with small [10–15 mm in diameter] to large [more than 20 mm in diameter] corollas, and seeds larger than those of the remaining *Zeltnera* species [length range from 0.35–0.75 mm]) and karyology ($n = 17$ is a synapomorphy). Nevertheless, the ITS-based cladogram inferred four robust clades (BS99 to BS100), but no clear relationships between them. Interestingly, three species were shown as polyphyletic, including *Z. davyi*, *Z. muhlenbergii*, and *Z. trichantha*. This topological incongruence may reflect either reticulate patterns or simply an insufficient phylogenetic signal. It further adds some complication to the already difficult taxonomy of the group (Broome 1973; Mansion 2001, 2004; Pringle 2010).

Zeltnera davyi is a rare Californian endemic that was sometimes considered a variety of *Z. exaltata* (Jepson 1925), from which it can be easily distinguished by its shorter peduncles and keeled calyx (Mansion 2004). The current gene tree inferred a paraphyletic *Z. venusta* lineage with respect to two accessions of *Z. davyi*, while the third accession diverged at the base of the clade (Fig. 8.1b). The presence of additive polymorphic sites between *Z. davyi* and *Z. venusta*, in this early diverging accession, overall supports reticulate patterns in the clade (Guggisberg et al. 2006, 2009; Mansion et al. 2005).

Zeltnera muhlenbergii is quite a problematic species that has been subject to many unsatisfactory taxonomic treatments resulting in many heterotypic synonyms such as *Erythraea curvistaminea* Wittr., *E. floribunda* Benth., or *E. minima* Howell (Broome 1973; Mansion 2004). A recent taxonomic account clearly summarizes the difficulties and errors surrounding the delimitation of *Z. muhlenbergii* and gives a first clear overview on this group (Pringle 2010). The current ITS cladogram suggests sister relationships of *Z. muhlenbergii* with either the (*Z. trichantha–Z. abramsii*) clade, or the (*Z. namophila–Z. nevadensis*) clade, overall highlighting the potential occurrence of reticulate patterns. In the absence of additional markers (cpDNA signal, chromosome count of *Z. muhlenbergii*, etc.), no further conclusions can be drawn at the present time.

Zeltnera trichantha is also distributed into the previously described *Z. muhlenbergii* clades. Morphological and karyological affinities between *Z. trichantha*, *Z. namophila*, and *Z. nevadensis* have been discussed (Mansion and Zeltner 2004). Molecular affinities with *Z. abramsii*, an alpine species occurring in open meadows of the Sierra Nevada Mountains, are inferred here for the first time. If this pattern is confirmed by other sources of evidence, it would exclude the potential affinities early proposed between *Z. abramsii* and *Z. venusta* (Abrams 1951).

Molecular data strongly support a **Texan *Zeltnera* clade** (BS80; Fig. 8.1b). All species share the same chromosome number ($n = 20$), except the early diverging *Z. multicaulis* ($n = 21$). Chromosome loss likely occurred in the morphologically close *Z. arizonica* and in all taxa now present in the Edwards Plateau region of Texas is found in the northern part of Texas to Oklahoma and is easily recognized by several morphological features (Broome 1973; Mansion 2004). *Zeltnera glandulifera*, a rare endemic of the Trans Pecos region, often bears minute glands on the stem, leaves, and calyx (Turner 1993), probably an environmentally induced character (G. Mansion, personal observations). Another infrequent endemic, *Z. maryanna* can be found on gypsophilous soils at the Texas–New Mexico boundary. Finally, the clade comprising *Z. breviflora*, *Z. texensis* and *Z. calycosa* (BS 70) unifies three entities that are not easily distinguishable in the field. Indeed, flower size and leaf shape, which are the main distinctive characters between these species, are generally of low taxonomic value in *Zeltnera* (Mansion 2004). It is likely that the north Mexican *Z. nesomii*, recently described by Turner (2011), can be added to this group.

The poorly supported **Mexican *Zeltnera* clade** (BS56) can be roughly divided into three groups (Fig. 8.1b). A first clade (BS67) mostly contains species with $n = 21$ chromosomes, except an early diverging, still undescribed taxa presenting some affinities with *Z. nudicaulis* (*Z.* aff. *nudicaulis*, $n = 22$). The remaining species form a robust clade (BS99) including *Z. madrensis*, the lone Mexican species with a large corolla, *Z. setacea*, a reduced version of the former, with very thin leaves, and *Z. martini*, immediately recognizable by its leafless appearance (Mansion 2004). A second well-resolved ITS clade (BS 96; Fig. 8.1b) contains three species with $n = 22$ chromosomes, namely *Z. stricta*, *Z. wigginsii* and *Z. pusilla*. The first two entities are morphologically similar, and can be distinguished only by the branching pattern (narrow for *Z. stricta* vs. divaricate for *Z. wigginsii*) and inflorescence, conspicuously leafy in *Z. stricta* and with few bracts in *Z. wigginsii*. Lastly, *Z. pusilla* shares morphological features with *Z. quitensis*. The latter, however, forms a third and well-supported clade (BS 88; Fig. 8.1b) comprising accessions with both $n = 21$ and $n = 22$ chromosomes, suggesting further segregation within this species.

8.4 Conclusions

This chapter, based on a robust molecular framework, is the most exhaustive and updated review on Centauries since Mansion and Struwe (2004). It provides the first satisfactory gene trees for the poorly known genera *Chironia*, *Sabatia*, and *Zygostigma*, and resolves the phylogenetic position of the rare and endangered species *S. sebaeoides*, *Z. davyi*, and *Z. muhlenbergii*. The results also suggest the occurrence of complex reticulate patterns in many groups with polyphyletic or paraphyletic species (e.g., *Chironia tetragona*, *Z. davyi*, or *Z. trichantha*). Finally, the taxonomic delimitation of *Chironia*, which forms a robust ITS clade with *C. cachanlahuen*, *Orphium*, and *Zygostigma*, is questioned. Overall, this paper argues for more detailed studies, based on both nrDNA and cpDNA sequence data, toward the resolution of a first "species tree" aimed at resolving the evolutionary history and species delimitation in this fascinating group. With the advent of the genomic era, the dream will certainly come true in the near future.

Acknowledgments I thank Prof. Jan Rybczyński for organizing this special issue on the Gentianaceae, Dr. Alessia Guggisberg for revising and improving a first version of this manuscript, and Dr. Louis Zeltner for sharing unpublished results on Centauries. The following institutes are acknowledged for providing information and material; Botanischer Garten und Botanisches Museum Berlin-Dahlem (B), The Natural History Museum, London (BM), Conservatoire et Jardin botaniques de la Ville de Genève (G), Royal Botanic Gardens, Kew (K), University of Texas at Austin (LL), Royal Botanic Gardens, Melbourne (MEL), University of Minnesota (MIN), Ludwig-Maximilians-Universität München (MSB), and Université de Neuchâtel (NEU). "Un grand merci" to Prof. Philippe Küpfer (Neuchâtel, Switzerland) for the opportunity to study this amazing group of plant.

References

Abrams LR (1951) An illustrated flora of the Pacific States, Washington, Oregon and California. Stanford University Press, Stanford

Adanson M (1763) Familles de plantes. Vincent, Paris

Allan GJ, Francisco-Ortega J, Santos-Guerra A, Boerner E, Zimmer EA (2004) Molecular phylogenetic evidence for the geographic origin and classification of Canary Island Lotus (Fabaceae: Loteae). Mol Phys Evol 32:123–138
Alvarez I, Wendel JF (2003) Ribosomal ITS sequences and plant phylogenetic inference. Mol Phyl Evol 29:417–434
Baldwin BG (1992) Phylogenetic utility of the internal transcribed spacer of ribosomal DNA in plants: an example from the Compositae. Mol Phyl Evol 1:3–16
Bauhin C (1623) Pinax Theatri Botanici. King, Basel
Bentham G (1876) Gentianeae. In: Bentham G, Hooker J (eds) Genera Plantarum. Lovell & Reeve, London, pp. 799–820
Borckhausen MB (1796) Über Linne's Gattung Gentiana. In: Roemer DJ (ed) Archiv für die Botanik, vol 1. Schäfer, Leipzig, pp 29–30
Bramwell D (1972) Endemism in the flora of the Canary Islands. In: Valentine DH (ed) Taxonomy, phytogeography and evolution. Academic Press, London, pp 141–159
Broome CR (1973) Systematics of *Centaurium* (Gentianaceae) of Mexico and Central America. PhD Thesis. Duke University, Durham, p 416
Brown R (1810) Prodromus florae Novae Hollandiae et insulae Van-Diemen. Schrag L., Nürnberg
Carlström A (1986) New taxa and notes from the sea Egean area and SW Turkey. Willdenowia 16:73–78
Chassot P, Nemomissa S, Yuan YM, Kupfer P (2001) High paraphyly of Swertia L. (Gentianaceae) in the Gentianella-lineage as revealed by nuclear and chloroplast DNA sequence variation. Plant Syst Evol 229:1–21
de Necker NJ (1790) Elementa botanica. Typis Societatis Typographicae Neowedensis, Parisi is
Dioscorides P, Ruel J (1552) Pedanii Dioscoridis Anazarbei, De medicinali materia libri sex. Apud Balthazarem Arnolletum, Lugduni
Dodoens R, Plantini C (1583) Remberti Dodonaei Stirpium historiae pemptades sex, sive libri XXX. Ex officina Christophori Plantini, Antverpiae
Don G (1838) A general history of the dichlamydeous plants. In: JG, Rivingtonn F (eds) London
Edgar RC (2004) MUSCLE: multiple sequence alignment with high accuracy and high throughput. Nucleic Acids Res 32:1792–1797
Endlicher SL (1838) Genera plantarum secundum ordines naturales disposita. Vienna
Favarger C (1960) Etude cytologique du Cicendia filiformis et du Microcala pusilla (Gentianacees). Bulletin de la Société Botanique de France 107:94–98
Feliner GN, Rossello JA (2007) Better the devil you know? Guidelines for insightful utilization of nrDNA ITS in species-level evolutionary studies in plants. Mol Phylogenet Evol 44:911–919
Filippa EM, Barboza GE (2006) Gentianaceae. Flora Fanerogamica Argentina. PROFLORA (CONICET), Córdoba
Gielly L, Tabarlet P (1996) A phylogeny of the European gentians inferred from chloroplast trnL (UAA) intron sequences. Bot J Lin Soc 120:57–75
Gilg E (1895) Gentianaceae. In: Engler A, Prantl K (eds) Die Naturlichen Pflanzenfamilien. Wilhelm Engelmann, Leipzig, pp 50–108
Gillett JM (1963) The gentians of Canada and Greenland. Research Branch/Department of Agriculture, Canada/Ottawa
Gilmour JS (1937) Notes on the genus Centaurium. Kew Bull 10:497–502
Gonzalez GL (1979) Sobre el Centaurium linariifolium (Lam.) G. Beck. Anales del Jardín Botánico de Madrid 36:123–128
Grant M, Hazel J (1973) Who's who in Greek mythology. Routledge, London
Gray A (1876) Erythraea. In: Brewer WH, Watson S (eds) Polypetalae. California geological survey. Botany of California. J. Wilson, Cambridge
Grisebach AHR (1839) Genera et species Gentianearum adjectis observationibus quibusdam phytogeographycis. J. G. Cotta, Stuttgart and Tubingen
Grisebach AHR (1845) Gentianaceae. In: de Candolle A (ed) Prodromus systematis naturalis regni vegetabilis. Fortin Masson Sociorum, Paris, pp 39–141

Grisebach AHR (1853) Schenkia, novum genus Gentianaerum. Bonplandia 1:226

Guggisberg A, Bretagnolle F, Mansion G (2006) Allopolyploid origin of the Mediterranean endemic, *Centaurium bianoris* (Gentianaceae), inferred by molecular markers. Syst Bot 31:368–379

Guggisberg A, Mansion G, Conti E (2009) Disentangling reticulate evolution in an Arctic-Alpine polyploid complex. Syst Biol 58:55–73

Hill J (1756) The British herbal: an history of plants and trees, natives Britain, cultivated for use, or raised for beauty. Osborne T, Shipton J (eds), London

Hoffmannsegg JC, Link HF (1820) Flore Portugaise ou Description de toutes les plantes qui croissent naturellement en Portugal. Charles Fréderic Amelang, Berlin

Hudson W (1762) Flora anglica. In: Nourse J, Moran C (eds) London

Huelsenbeck JP, Ronquist F (2001) MRBAYES: Bayesian inference of phylogenetic trees. Bioinformatics 17:754–755

Jepson WL (1925) A manual of the flowering plants of California. Associated Student Store, University of California, Berkeley

Kim S-C, Crawford DJ, Francisco-Ortega J, Santos-Guerra A (1996) A common origin for woody Sonchus and five related genera in the Macaronesian islands: molecular evidence for extensive radiation. Proc Nat Acad Sci USA 93:7743–7748

Kissling J, Yuan YM, Kupfer P, Mansion G (2009) The polyphyletic genus Sebaea (Gentianaceae): a step forward in understanding the morphological and karyological evolution of the Exaceae. Mol Phyl Evol 53:734–748

Klackenberg J (1986) The new genus Ornichia (Gentianacae) from Madagascar. Adansonia 2: 195–206

Knoblauch E (1894) Beiträge zur Kenntniss der Gentianaceae. Botanisches Centralblatt 60:321–401

Lamarck J-B, Poiret JL (1783) Encyclopédie méthodique: botanique. Panckoucke, Paris

Leistner OA (2000) Seed Plants of southern Africa: families and genera. Strelitzia 10. National Botanical Institute, Pretoria

Linnaeus C (1753) Species Plantarum. L. Salvii, Stockholm

Mansion G (2001) Phylogenetic relationships within the subtribe Chironiinae (Gentianaceae), with a particular focus on the evolutionary trends and biogeographic patterns of the genus *Centaurium* Hill s. l., PhD Thesis. University of Neuchâtel, Neuchâtel, Switzerland

Mansion G (2004) A new classification of the polyphyletic genus *Centaurium* Hill (Chironiinae, Gentianaceae): description of the New World endemic Zeltnera, and reinstatement of Gyrandra Griseb. and Schenkia Griseb. Taxon 53:719–740

Mansion G, Mathews KG (2006) Evolutionary patterns within the North American endemic Sabatia Adans. (Gentianaceae). Evolution 2006. Stony Brook, New York

Mansion G, Selvi F, Guggisberg A, Conti E (2009) Origin of Mediterranean insular endemics in the Boraginales: integrative evidence from molecular dating and ancestral area reconstruction. J Biogeography 36:1282–1296

Mansion G, Struwe L (2004) Generic delimitation and phylogenetic relationships within the subtribe Chironiinae (Chironieae: Gentianaceae), with special reference to *Centaurium*: evidence from nrDNA and cpDNA sequences. Mol Phyl Evol 32:951–977

Mansion G, Zeltner L (2004) Phylogenetic relationships within the new world endemic *Zeltnera* (Gentianaceae-Chironiinae) inferred from molecular and karyological data. Amer J Bot 91:2069–2086

Mansion G, Zeltner L, Bretagnolle F (2005) Phylogenetic patterns and polyploid evolution within the Mediterranean genus Centaurium (Gentianaceae-Chironieae). Taxon 54:931–950

Marais W, Verdoorn IC (1963) Gentianaceae. In: Dyer RA, Codd LE, Rycroft HB (eds) Flora of southern Africa. Department of Agricultural Technical Services, Republic of South Africa, Pretoria, pp 171–243

Melderis A (1972) Taxonomic studies on the European species of the genus Centaurium Hill. Bot J Linn Soc 65:224–250

Meyer E (1837) Commentariorum de Plantis Africae Australioris. Leopoldum Voss, Leipzig

Morden C, Caraway V, Motley T (1996) Development of a DNA Library for Native Hawaiian plants. Pacific Sci 50:324–335
Müller K (2004) PRAP-computation of Bremer support for large data sets. Mol Phyl Evol 31:780–782
Müller K, Quandt D, Müller J, Neinhuis C (2005) Phyde®: Phylogenetic data editor
Nylander J (2004) MrModeltest 2.1. Evolutionary Biology Centre, Uppsala
Perry JD (1971) Biosystematic studies in the North American genus Sabatia (Gentianaceae). Rhodora 73:309–369
Pliny, Camers J, Gelen S, Herold JB, Frobenium H, Episcopium N (1554) C. Plinii Secundi Historiae Mundi Libri XXXVII. H. Frobenium and N. Episcopium, Basileae
Prain D (1908) Notes on Chironia and Orphium. Bulletin of Miscellaneous Information (Royal Gardens, Kew) 8:341–376
Pringle JS (2010) The identity and nomenclature of the Pacific North American species Zeltnera muhlenbergii (Gentianaceae) and its distinction from Centaurium tenuiflorum and other Species with which it has been confused. Madrono 57:184–202
Pringle JS (2012) Gentianaceae. In: Baldwin BG, Goldman DH, Keil DJ, Patterson RW, Rosatti TJ (eds) The Jepson manual: vascular plants of California. University of California Press, Berkeley
Randell RA, Morden CW (1999) Hawaiian plant DNA library II: endemic, indigenous, and introduced species. Pacific Sci 53:401–417
Ray J (1724) Synopsis methodica stirpium britannicarum. Londini
Reichenbach HG (1834) Icones florae Germanicae et Helveticae. F. Hofmeister, Lipsiae
Renealmus P (1611) Specimem historiae plantarum. Hadrianus Beys, Paris
Robyns A (1954) Essai d'étude systématique et écologique des Centaurium de Belgique. Bulletin du Jardin Botanique de Bruxelle 24:349–398
Rouy G (1891) Suites à la Flore de France de Grenier et Godron. Le Naturaliste: Journal des Echanges et des Nouvelles 11:11–12
Rouy G, Boulay JN, Camus E-G, Foucaud J (1893) Flore de France ou description des plantes qui croissent spontanément en France, en Corse et en Alsace-Lorraine. G. Rouy, Asnières
Schoch E (1904) Monographie der Gattung Chironia L. Beihefte Zum Botanischen Centralblatt 38:177–242
Shinners LH (1957) Synopsis of the genus Eustoma (Gentianaceae). Southwestern Nat 2:38–43
Struwe L, Kadereit JW, Klackenberg J, Nilsson S, Thiv M, Vonhagen KB, Albert VA (2002) Systematics, character evolution, and biogeography of Gentianaceae, including a new tribal and subtribal classification. In: Struwe L, Albert VA (eds) Gentianaceae: systematics and natural history. Cambridge University Press, Cambridge
Swofford DL (2002) PAUP* Phylogenetic Analysis Using Parsimony (*and other methods) Version 4. Sinauer Associates, Sunderland
Theophrastus, Gazēs T, Jordanus J, Gazeium G, Bacquenoius N (1552) Theophrasti De historia plantarum. Apud Gulielmum Gazeium, Lugduni
Thiv M, Struwe L, Kadereit JW (1999) The phylogenetic relationships and evolution on the Canarian laurel forest endemic *Ixanthus viscosus* (Aiton) Griseb. (Gentianaceae): evidence from matK and ITS sequences, and floral morphology and anatomy. Plant Syst Evol 218:299–317
Tournefort JP de (1700) Institutiones Rei Herbarieae. Typographia Regia, Paris
Trunz V, Zeltner L, Küpfer P, Grant JR (2012) Cytotaxonomic notes on the Tribe Helieae (Gentianaceae). Harv Pap Bot 17:43–46
Turner BL (1993) The Texas species of Centaurium (Gentianaceae). Phytologia 75:259–275
Turner BL (1994) Taxonomic revision of Geniostemon (Gentianaceae). Phytologia 76:8–13
Turner BL (2011) A new species of Zeltnera (Gentianaceae) from Tamaulipas. Mexico Phytoneuron 61:1–3
Ubsdell RAE (1976a) Studies and evolution in Centaurium erythrea Rafn. and Centaurium littorale (D. Turner) in the British Isles: cytology. Watsonia 11:33–43

Ubsdell RAE (1976b) Studies and evolution in Centaurium erythrea Rafn. and Centaurium littorale (D. Turner) in the British Isles: taxonomic and biometrical studies. Watsonia 11:7–31

Van der Sluis WG (1985) Chemotaxonomical investigations of the genera blackstonia and centaurium (gentianaceae). Plant Syst Evol 149:253–286

von Hagen KB, Kadereit JW (2001) The phylogeny of Gentianella (Gentianaceae) and its colonization of the Southern hemisphere as revealed by nuclear and chloroplast DNA sequence variation. Org Div Evol 1:61–79

Weaver RE, Rüdenberg L (1975) Cytotaxonomic notes on some Gentianaceae. J Arnold Arboretum 56:211–222

Wilbur RL (1955) A revision of the North American genus Sabatia (Gentianaceae). Rhodora 57:1–104

Wilbur RL (1989) The type species of the genus Sabatia Adans (Gentianaceae). Rhodora 91:167–171

Willdenow KL (1797) Caroli a Linné Species Plantarum. GC Nauk, Berolini [Berlin]

Yuan Y-M, Küpfer P (1995) Molecular phylogenetics of the subtribe Gentianinae (Gentianaceae) infered from the sequences of the internal transcribed spacer (ITS) of nuclear ribosomal DNA. Plant Syst Evol 196:206–226

Yuan Y-M, Küpfer P (1997) The monophyly and rapid evolution of Gentiana sect. Chondrophyllae Bunge s.l. (Gentianaceae): evidence from the nucleotide sequences of the internal transcribed spacer of nuclear ribosomal DNA. Bot J Lin Soc 123:25–43

Yuan Y-M, Küpfer P, Doyle JJ (1996) Infrageneric phylogeny of the genus Gentiana (Gentianaceae) inferred from nucleotide sequences of the internal transcribed spacer (ITS) of nuclear ribosomal DNA. Am J Bot 83:641–652

Yuan Y-M, Wohlhauser S, Moller M, Chassot P, Mansion G, Grant J, Kupfer P, Klackenberg J (2003) Monophyly and relationships of the tribe Exaceae (Gentianaceae) inferred from nuclear ribosomal and chloroplast DNA sequences. Mol Phyl Evol 28:500–517

Zeltner L (1970) Recherches de biosystématique sur les genres Blackstonia Huds. et Centaurium Hill (Gentianaceae). Bulletin de la Société Neuchâteloise des Sciences Naturelles 93:1–164

Zeltner L (1985) Etude cytotaxonomique et cytogéographique du Centaurium pulchellum (Swartz) Druce, sensu lato. Bot Helv 95:47–57

Zeltner L (1991) Contribution à l'étude cytogéographique des genres Blackstonia Huds. et Centaurium Hill. (Gentianaceae) en Turquie, à Rhodes et à Chypre. Bulletin de la Société Neuchâteloise des Sciences Naturelles 114:77–103

Zeltner L, Mansion G (2003) Contribution a l'étude cytogéographique de la sous-tribu Chironiinae (Gentianaceae) au Chili. J de Botanique de la Société Botanique de France 24:55–65

Chapter 9
Genes Expressed in the Overwinter Buds of Gentian (*Gentiana* spp.): Application to Taxonomic, Phylogenetic, and Phylogeographical Analyses

Ken-ichi Tsutsumi and Takashi Hikage

Abstract Vegetative reproduction through overwinter buds is important for maintenance of parental strains with favorable phenotypes in the breeding of perennial *Gentiana* spp. This chapter focuses on the genes and proteins that are enriched in buds. *W*14/15 genes are some of the most abundantly expressed genes in these overwinter buds and encode a novel esterase. They are considered to be involved in stress tolerance and in hormone signaling/metabolism and/or defense systems. The highly polymorphic nature of these genes makes them a useful tool to analyze the phylogeny, taxonomy, phylogeography, and pedigree of *Gentiana* spp. In addition, *W*14/15 can be used as a specific phenotypic marker. The functional role of *W*14/15 in regulating some characters of overwinter buds and some practical uses of these genes are discussed in this review.

9.1 Introduction

Plants belonging to the genus *Gentiana* L. are distributed globally over temperate and alpine areas except Africa (Koehlein 1991; Ho and Liu 2001; Struwe et al. 2002). Gentian plants are cultivated commercially for their ornamental flowers (Koehlein 1991; Yoshiike 1992; Ho and Liu 2001), and dried gentian roots have long been used as ingredients of traditional medicines in Europe and Asia (Jensen and Schripsema 2002; Matsukawa et al. 2006; Ogata et al. 2011).

K. Tsutsumi (✉)
Cryobiofrontier Research Center, Iwate University Morioka, Iwate 020-8550, Japan
e-mail: kentsu@iwate-u.ac.jp

T. Hikage
Hachimantai City Floricultural Research and Development Center Hachimantai, Iwate 028-7592, Japan

J. J. Rybczyński et al. (eds.), *The Gentianaceae - Volume 1: Characterization and Ecology*, DOI: 10.1007/978-3-642-54010-3_9,

Many varieties and ecotypes are known for the genus *Gentiana* and conventional hybridization has developed various cultivars (Koehlein 1991; Yoshiike 1992; Ho and Liu 2001). Successful breeding depends largely on stable maintenance of those parental strains with favorable phenotypes. However, gentians exhibit inbreeding depression after self-pollination, which makes breeding difficult. One approach to facilitate breeding is the vegetative reproduction of parental strains through overwinter buds. Hence, from an agricultural perspective, the identification of genes and proteins that regulate the formation and function of the winter buds is of considerable importance.

Overwinter buds develop and grow from the basal tuber, or uppermost part of the roots, in early summer. They enter a dormant state in Autumn under short day and cold environments, and remain in this state over the winter until reinitiation of growth commences under warm, long-day conditions in Spring. Since so many genes are involved in dormancy, cell growth/division, cold- and even freezing tolerance, and flowering systems must cooperate temporally in the buds during overwintering. However, studies have been limited on the physiological and biochemical functions of the proteins/genes that act in the overwinter buds.

This chapter describes the approach to identify and characterize overwinter bud-abundant proteins/genes that might be involved in the physiological functions of the buds. Focus is on one of the most abundantly expressed proteins for characterization of the encoding gene structure i.e., an esterase termed W14/15, which relates to tolerance of the overwintering buds against abiotic stress, including cold and freezing. The structure of the *W*14/15 gene is highly polymorphic; thus, it can be utilized as a marker of specific phenotypes and pedigree analysis. In addition, *W*14/15 gene polymorphism can be a marker for determination and confirmation of the taxonomic position of gentian plants. Studies on taxonomic criteria at the nuclear DNA level have been limited for *Gentiana*.

9.2 Proteins Enriched in the Overwinter Buds: A Unique Feature of their Expression

In order to understand genes involved in physiological functions of gentian overwintering buds, e.g., cold-tolerance and dormancy, comparative analysis was carried out of proteins in the buds and other organs/tissues of *Gentiana triflora*, using two-dimensional polyacrylamide gel electrophoresis (2D-PAGE), followed by mass spectrometric identification of the protein spots separated on 2D-PAGE (Takahashi et al. 2006). Table 9.1 shows proteins that are enriched in the buds during late August, when there occurs a growth stage of the buds just before the induction of dormancy. The enriched proteins at this stage include those homologous to dehydration-induced protein (spot W8), ethylene-induced esterases (W14 and W15), stress-induced membrane pore protein (W20), enolase (W23), glyoxalase I (W11), thioredoxin peroxidase (W25), and cytosolic ascorbate peroxidases (U11 and U12), which were identified previously and characterized in other plants. These proteins have been reported as stress-inducible or involved in stress tolerance. For example, the bifunctional enzyme enolase acts as transcriptional regulator (LOS2) in the nucleus in response to cold stress (Lee et al. 2002), while cytoplasmic ascorbate peroxidase is involved in hydrogen peroxide metabolism in response to oxidative stress. The latter is one of the major components of chilling stress (Smirnoff 1998; Shigeoka et al. 2002). Glyoxalase I (Glx I) is a key enzyme acting in the glyoxalate pathway; it detoxifies methylglyoxal, which is formed when plants are exposed to several abiotic stresses. Glx I is upregulated in response to cold, salt, and abscisic acid (ABA), suggesting its important role in cold tolerance (Espartero et al. 1995; Zang and Komatsu 2007; Lee et al. 2009; Ganeshan et al.2011). Proteins termed W14 and W15 are orthologs of ethylene-induced esterase from *Citrus sinensis* and *Lycopersicon esculentum*. This enzyme is a member of the alpha/beta-hydrolase fold superfamily (Heikinheimo et al. 1999), and cleaves methyl esters in several methylated plant hormones, including methyl jasmonate (JA), methyl indol-3-acetic acid (IAA) and methyl ABA, suggesting its involvement in signaling, activation, and/or metabolism of plant hormones that links various pathways, including dormancy and abiotic stress (Zong et al. 2001; Stuhlfelder et al. 2004; Yang et al. 2008).

It should be noted that gentians used for analysis were grown in a glasshouse virtually without any stress, so that increased accumulation of such stress-induced proteins was not anticipated for the overwinter buds at a stage before entering dormancy and overwintering. A similar pattern of gene expression has been reported for salt cress, *Thellungiella salsuginea* (*halophila*), a halophytic and cryophytic plant, which is a close relative of *Arabidopsis thaliana* (Taji et al. 2004; Amtmann 2009). In this plant, many genes involved in stress tolerance, which are only stress induced in other plants, including *A. thaliana*, are expressed constitutively under nonstress conditions, resulting in a greater stress tolerance, e.g., to salt, cold, and drought, compared to other plants.

From these observations, it quite likely that the overwinter buds in *G. triflora* are programmed genetically to express constitutively stress-related proteins, including those listed in Table 9.1 under nonstress conditions, for prior acquisition of stress tolerance toward wintering or the dormant state.

9.3 The Gene Coding for Overwinter Bud-Enriched Proteins W14 and W15 and its Multiple Variant Forms

9.3.1 Two Different but Closely Related Proteins, W14 and W15, are Encoded by an Allelic Pair

Among the proteins abundant in the overwinter buds (Table 9.1), two proteins namely W14 and W15, were analyzed for their gene structure, because as mentioned earlier, these proteins are considered to be involved in stress tolerance and/or

Table 9.1 Proteins enriched in overwinter buds of *G. triflora*

Protein (spot no.)	Ortholog	Function and expression
W8	Dehydration-induced protein (*Brassica napus*)	Cold- and drought-inducible
W11	Glyoxalase I (*Citrus paradisi*)	ABA-inducible, salt- and osmotic stress-inducible, antioxidant
W12	Isoflavone reductase (*Arabidopsis thaliana*)	
W14	Ethylene-induced esterase (*Citrus sinensis*)	Ethylene-inducible
W15	Ethylene-induced esterase (*Citrus sinensis*)	Ethylene-inducible
W19	Ribosomal protein S12 (*Hordeum vulgare*)	
W20	Membrane pore protein (*Arabidopsis thaliana*)	Cold- and drought-inducible
W23	Enolase (*Arabidopsis thaliana*)	Transcription factor for cold-inducible genes (*LOS2*)
W25	Thioredoxin peroxidase (*Capsicum annuum*)	Cold-tolerance, antioxidant
W26	Translationally controlled tumor protein (*Arabidopsis thaliana*)	
U11	Ascorbate peroxidase (*Lycopercicon esculentum*)	Cold-tolerance, antioxidant
U12	Ascorbate peroxidase (*Lycopercicon esculentum*)	Cold-tolerance, antioxidant

signaling/metabolism of plant hormones during dormancy and/or overwintering. The amino acid sequences of both W14 and W15 proteins belong to a group of hydrolases/esterases in the alpha/beta-hydrolase fold superfamily (Heikinheimo et al. 1999). Indeed, we confirmed that both of the W14 and W15 proteins synthesized in *Escherichia coli* exhibited esterase activity in vitro (Hikage et al. 2007). They showed high homology to ethylene-induced esterase (Zong et al. 2001), salicylic acid binding protein (methyl salicylic acid cleaving enzyme; Kumar and Klessig 2003), and methyl jasmonate hydrolyzing esterase (Stuhlfelder et al. 2004). *Arabidopsis* orthologs of this enzyme family cleave methyl indole acetic acid (IAA), and to a lesser extent, other methylated plant hormones (Yang et al. 2008). Since those homologous proteins are considered to be involved in phytohormone-mediated signaling and/or defense-related signaling, both of the W14 and W15 proteins may have similar physiological roles in the overwinter buds.

The names W14 and W15 (Table 9.1) were termed originally from two closely positioned, but isolated protein spots on 2D-PAGE plates (Takahashi et al. 2006). Determination of the genes coding for W14 and W15 revealed that the two genes (termed *W14a* and *W15a*, respectively) are very similar except for several insertion/deletion sites and base substitutions (Fig. 9.1). The encoding proteins, W14 and W15, are very closely related; only three amino acid substitutions were found in the entire 259 amino acids. Further, genotype analysis of homozygous doubled haploid lines (DH lines) generated from *G. triflora* var. *japonica* (Doi et al. 2010, 2011), revealed that the *W14a* and *W15a* genes segregated into individual DH lines. Therefore, the two genes locate as an allelic pair on the gentian genome (Hikage et al. 2007). Accordingly, the encoding gene or genomic locus was referred to as *W14/15*. Hereafter, W14 and W15 proteins are designated together as W14/15, or W14/15 for proteins as a whole in variants.

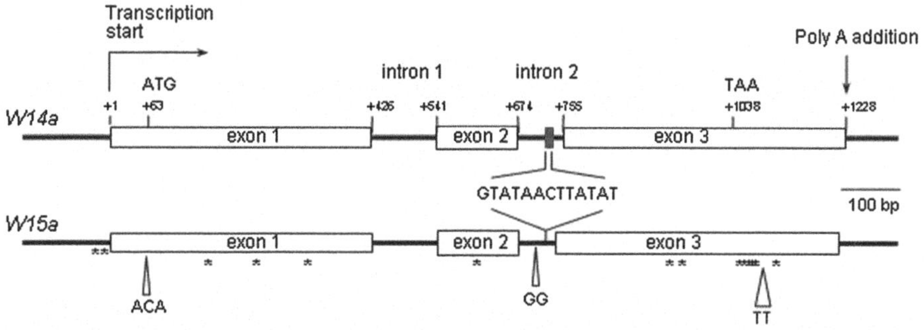

Fig. 9.1 Structure of the *W*14 and *W*15 genes in *G. triflora*. Transcription start site, poly (A) addition site, translation start (ATG), and termination (TAA) sites are shown. Numbers are relative to the transcription start site (+1). *Asterisks* below the *W*15*a* gene indicate positions of nucleotide substituted from the cognate position in the *W*14*a* gene. In intron 2, 13 nucleotides present in the *W*14*a* gene (referred to as insertion 1, see also (Fig. 9.2) were deleted in the *W*15 gene

9.3.2 Variation of the W14/15 Gene Occurs Among Gentiana Species

As mentioned earlier, two very similar but not identical proteins, W14 and W15, were differently encoded by each allele at the same locus. This observation had led the present authors to imagine that the *W*14/15 gene is polymorphic and might have multiple different alleles or haplotypes. Figure 9.2 illustrates the polymorphic feature of the *W*14/15 gene, which was revealed from nucleotide sequences of more than 30 different cultivars/lines/species of *Gentiana*. Three restriction enzyme sites exhibiting restriction fragment length polymorphism (RFLP) (referred to as RFLP sites 1 and 2 for *Hae*III, and RFLP site 3 for *Hin*dIII) and three insertion/deletion sites (insertion sites 1, 2, and 3) have been identified. The lengths of insertions 1, 2, and 3 are 13, 11, and 9–20 bp C-stretch, respectively. Nucleotide lengths of introns 1 and 2 also showed polymorphism. These polymorphic sites/regions can be markers for examining the taxonomic status of *Gentiana* species and their variants.

According to the variation and combination of the above polymorphic sites, the *W*14/15 gene was classified into more than 10 variant types or haplotypes (Hikage et al. 2011). Basically, the genes carrying insertion 1 were referred to as the *W*14 series, and those without it as the *W*15 series. Genes that carry insertion 2 were marked "b" following the gene name (e.g., *W*14*b* and *W*15*b*), and those carrying both insertions 2 and 3 were termed *W*14*c* or *W*15*c*. Further classification was done with the sequential use of numbers and "dashes," such as *W*14*b*1, *W*14*b*2, *W*15*a*′, *W*15*b*, and *W*15*b*′, according to variation at the RFLP sites 1 to 3, other nucleotide substitutions (SNPs), and short nucleotide insertion/deletions (mono- to tetranucleotides) other than insertions 1, 2, and 3 (Table 9.2).

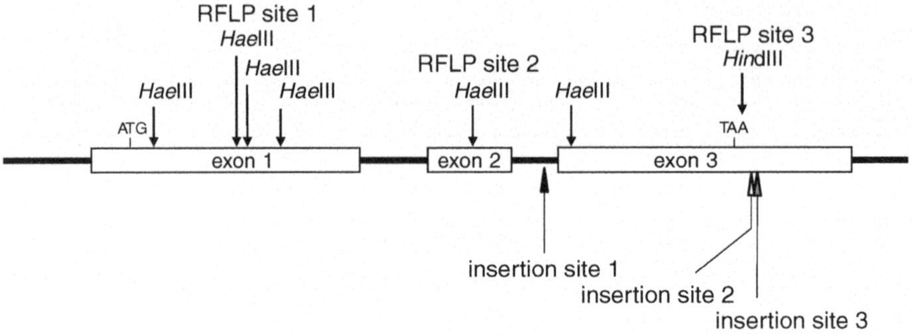

Fig. 9.2 Polymorphic features of the *W14/15* gene. Recognition sites of *Hae*III and *Hin*dIII are shown. Two of the *Hae*III sites (RFLP sites 1 and 2) and *Hin*dIII site (RFLP site 3) exhibited restriction fragment length polymorphism (RFLP). In addition, three insertion sites (insertion sites 1, 2, and 3) were found, one, two, or all of which were deleted in some *W14/15* gene variants in *Gentiana* species (see also Table 9.2)

Table 9.2 Variation of the *W14/15* gene in different gentian species

Haplotype	RFLP[a] 1 2 3	Insertion site[b] 1 2 3	Species
W14a	+ + +	+ − −	*G. triflora*
W14a′	+ − +	+ − −	*G. triflora*
W14b1	+ + −	+ − −	*G. scabra*
W14b1′	− + −	+ − −	*G. scabra*
W14b2	+ + −	+ + −	*G. scabra*
W14b2′	− + −	+ + −	*G. scabra*
W14b3	+ + +	+ + −	*G. pneumonanthe*
W14c1	− + +	+ + +	cv. Showtime starlet
W14c2	− + +	+ + +	cv. Showtime starlet
W15a	− + +	− − −	*G. triflora*
W15a′	− − +	− − −	*G. triflora*
W15a″	+ + +	− − −	*G. triflora*
W15b	+ + +	− + −	*G. septemfida*
W15b′	− + +	− + −	*G. paradoxa*

[a] Plus (+) denotes the presence of intact restriction enzyme sites. Minus (−), sites mutated
[b] Plus (+) denotes presence, minus (−), absence

These classifications revealed structural features that are characteristic of *Gentiana* species (Fig. 9.3), namely, the nucleotide length of introns 1 and 2 differed among the five species of *Gentiana* examined. The length of intron 1 was 116 bp in *G. triflora*, *Gentiana scabra*, *Gentiana pneumonanthe*, and *Gentiana septemfida*, while that in *Gentiana paradoxa* was 106 bp. Intron 2 was found to be more divergent in its length among the species. The *W14a*, *W14b*, and *W14c* type genes in *G. triflora*, *G. scabra*, and *G. pneumonanthe* had the same nucleotide length (82 bp). However, those of the *W15a* group, *W15b*, and *W15b′*, in *G. triflora*,

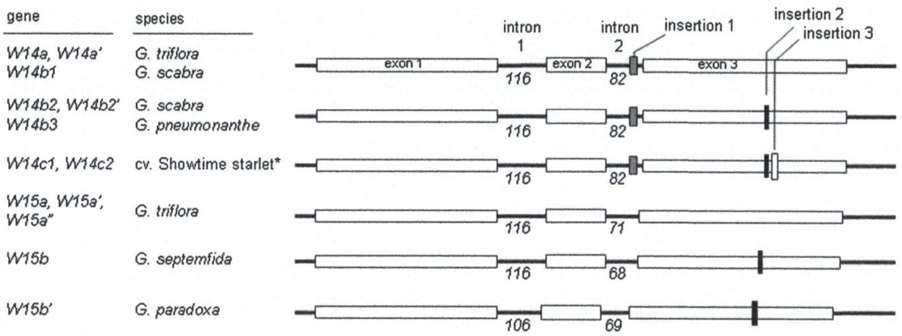

Fig. 9.3 Structure of the *W*14/15 gene variants in different *Gentiana* species. Organization of the *W*14/15 gene variants are shown. Nucleotide lengths of introns 1 and 2 are shown below each variant. The presence of insertion 1, 2, and 3 is indicated by *gray*, *black*, and open boxes, respectively

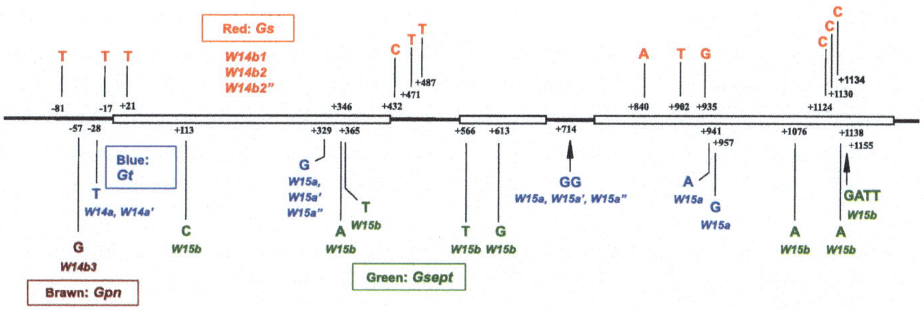

Fig. 9.4 Species- or gene-specific single nucleotide substitutions and short nucleotide insertions (*arrows*) in the *W*14/15 gene variants. Nucleotide positions are relative to the transcription start site in the *W*14*a* gene. Gs, *G. scabra*; Gt, *G. triflora*: Gpn, *G. pneumonanthe*; Gsept, *G. septemfida*

G. septemfida, and *G. paradoxa* had lengths of 71, 68, and 69 bp, respectively. The difference in intron 2 was mainly due to the presence or absence of insertion 1. In addition, several species carried insertion 2 just 3′ to the TAA codon. The *W*14*c* and *W*14*c*′ genes had both insertions 2 and 3. These *W*14*c* genes are found in the cv. "Showtime Starlet," a commercial cultivar (triploid; Rhindo New Zealand Ltd., New Zealand), which was developed using *G. scabra* and an unknown *Gentiana* species other than those listed in (Fig. 9.3). Insertion 3 was also found in the *W*15*c* type genes in native *G. triflora* var. *japonica* (details are described elsewhere). Other single nucleotide polymorphism (SNP) and short (1–4 bp) insertion/deletions were also found. Some of those, which are characteristic of the species, are illustrated in (Fig. 9.4) For instance, nucleotides at position −28 (T in *W*14*a* and *W*14*a*′ genes), +329 (G in *W*15*a* group genes), +941 (A in *W*15*a*), and +957 (G in *W*15*a*) are specific to *G. triflora*. The insertion of GG at position +714 is also characteristic of *W15a* type genes in *G. triflora*. Collectively, these structural differences can discriminate particular *Gentiana* species from others. The *W14/15* genes must have diverged along with the evolution of *Gentiana* species, and thus they can be utilized as markers for identification/confirmation/reconfirmation of the taxonomic status of *Gentiana* species (discussed later).

9.3.3 Variant W14/15 Proteins in Different Gentiana Species and Their Correlation with Winter Hardiness of Overwinter Buds

In contrast to the diverged nucleotide sequences among the species, the encoding amino acid sequences of the *W14/15* gene variants are relatively conserved, indicating the importance of the W14/15 protein for survival or evolutionary adaptation. Figure 9.5 shows the amino acids altered in different *Gentiana* species as compared to those in *G. triflora* W14a protein. When compared among *G. triflora*, *G. scabra*, *G. pneumonanthe*, *G. septemfida*, and *G. paradoxa*, only five residues at the most in 259 amino acids were found to be different. As an outgroup, the W14/15 protein in *Gentiana thunbergii* (termed W14Thb, details described elsewhere) was aligned in Fig. 9.5. This gentian is an annual/biennual plant, and belongs to a different section (*Chondrophyllae*) from *G. triflora* and other species described earlier (Section *Pneumonanthe*). This protein differed from the W14a protein at 14 positions distributed widely over the entire sequence; the difference was about three times greater than those among the species in the Section *Pneumonanthe*.

It is worth mentioning that the different amino acids in some species locate around the amino acid positions of so called "catalytic triad" i.e., S^{81}, D^{209}, and H^{237}, which are architects of a substrate-binding pocket in three-dimensional structure, and are thus critical to the catalytic activity of the esterases/hydrolases in the alpha/beta-hydrolase fold superfamily (Heikinheimo et al. 1999). The difference around the catalytic triad in the W14/15 proteins may indicate some differences in their catalytic properties among the species, such as substrate recognition, specificity, and reaction kinetics. In this regard, amino acid substitution of the W14/15 protein seems to reflect the winter hardiness of overwinter buds. As reported earlier, winter hardiness of the buds differs among cultivars (Hikage et al. 2007). Later, it was revealed that cultivars carrying specific allele types have less tolerance to overwintering. For example, the cv. "Ashiro-no-Aki" (AA) (*W14a*/*W15a*) and "Aki-6Ps" (*W15a′*/*W15a′*), which do not carry *W14b* group genes, survive winter better than those carrying *W14b*1 or *W14b*2 alleles, e.g., cv. "PW" and cv. "GSW". It is suggested that either or both of the two alleles carrying the *W14b*1 and/or *W14b*2 gene exhibit somehow weaker cold tolerance (Hikage et al., personal communication). The encoding proteins W14b1 and W14b2 in *G. scabra* differ only in four amino acid residues from W14a and W15a (Fig. 9.5). However,

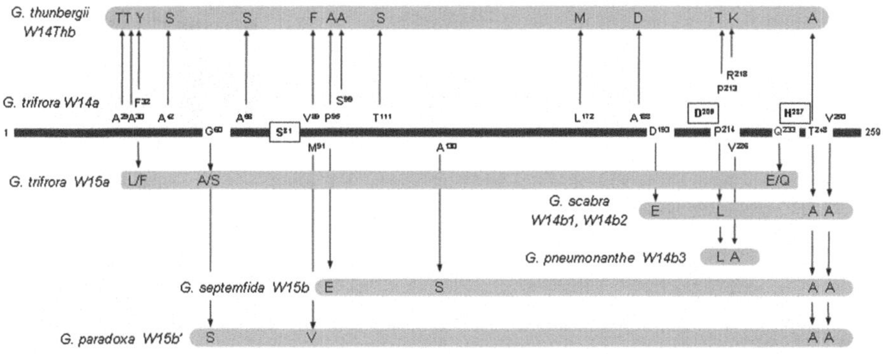

Fig. 9.5 Comparison of amino acid sequences of the W14/15 proteins from different species. Boxed amino acids in *G. triflora* W14a are those that constitute the catalytic triad (active center). Numbers are positions from the N-terminus

three of them are juxtaposing, or close to those amino acids constituting part of the catalytic triad (D^{209} and H^{237}). Thus, enzymatic function of W14b1 and W14b2 seems to differ from those of other W14/15 proteins, i.e., W14a and W15a groups. Although causal relationship remains to be elucidated, variant forms of the W14/15 protein/gene might reflect cold tolerance or winter hardiness. Thus, the *W14/15* gene variation could be a genotypic marker for winter hardiness, regardless of its direct involvement.

9.4 Practical Use of the *W14/15* Allele for Taxonomic, Pedigree, and Phenotypic Markers

As described in Sect. 9.3.3, variant forms of the W14/15 protein/gene can be utilized as markers for specific phenotypes, such as winter hardiness and cold tolerance. Expected structural relationships have been discussed between the W14/15 variant proteins and phenotype. In addition to phenotypic markers, allelic variants of the *W14/15* gene can be attributed to the pedigree and genetic relationship of gentian cultivars, since the variant forms of the gene, i.e., haplotypes, are of parental origin. Various cultivars have been analyzed for their *W14/15* gene variation, and several SNP sites have been found that are unique to particular cultivars and lines. Nucleotide substitutions occurred specifically in the genes in the cv. "Maerchen Ashiro" (MA; *G. triflora* x *G. scabra*), cv. "Shine Blue" (SB; *G. triflora* x *G. scabra*), cv. "*Gentiana Scabra* White" (GSW; *G. scabra*), and cv. "Lovely Ashiro" (LA; *G. triflora* x *G. scabra*) which are shown in (Fig. 9.6). For instance, nucleotides T^{+288} and T^{+413} were found only in the *W14b2* gene in SB. C^{+328}, T^{+457}, and C^{+937} were specific to the *W14b2* gene in MA, while nucleotides A^{+245} and G^{+1024} were found only in *W14b2* gene in GSW. Several other sites

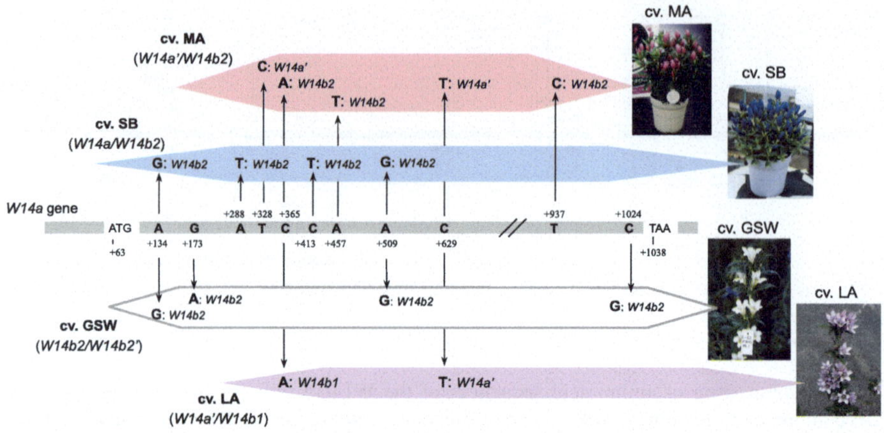

Fig. 9.6 Single-nucleotide polymorphism (SNP) found in the *W*14/15 genes, which is characteristic of some cultivars. *MA* cv. Maerchen Ashiro, *SB* cv. Shine Blue, *GSW* cv. *Gentiana Scabra* White, *LA* cv. Lovely Ashiro.

carried nucleotides unique to more than one cultivar/line, e.g., G^{+134} and G^{+469} were found only in the *W*14*b*2 gene in SB and GSW, A^{+365} occurred in *W*14*b*1 and *W*14*b*2 genes in LA and MA, while T^{+629} was present in the *W*14*a*′ gene in LA and MA. These sites can be exploited for practical use to identify cultivars/lines by a nonsequencing method, i.e., dot-blot hybridization using allele-specific oligonucleotide probes (Hikage et al. 2011).

The *W*14/15 variants can also be utilized to verify pedigree and breeding history. A typical example may be the cv. LA. This cultivar was derived from backcrossing between an unknown parental line/cultivar and an F_1 individual, the latter being a cross between *G. triflora* and the unknown species. The cultivar LA was bred by open pollination by honeybees. Consequently, paternal origin was unknown until our study. Nucleotide sequencing identified two types of genes, *W*14*a*′ and *W*14*b*1, in this cultivar, the former being typical of *G. triflora* and the latter of *G. scabra*. Thus, paternal origin should be *G. scabra*. In addition, LA might share common paternal origin with line OZ, because LA and OZ possessed almost the same *W*14*b*1 allele. Cultivar AA ("Ashiro-no-Aki") is another example. This cultivar is an F_1 hybrid generated from line Y8 (*G. triflora*, *W*15*a*/*W*15*a*′) and line N27 (*G. triflora*, *W*14*a*/*W*15*a*). Thus, the genotype segregates into four different types in the F_1 individuals, *W*14*a*/*W*15*a*, *W*14*a*/*W*15*a*′, *W*15*a*/*W*15*a*′ , and *W*15*a*/*W*15*a*. This genetic flow can be identified in AA individuals. In this cv., the *W*15*a*′ allele could be one parental marker. The *W*15*a*′ gene is characteristic of the native species Y8 (Yahaba), and therefore one parent of AA can be traced to be native Yahaba (Y8).

9.5 Phylogenetic and Phylogeographical Analyses of the *W14/15* Gene Variants

9.5.1 Taxonomic Relationships of the W14/15 Genes in the Section Pneumonanthe

A phylogenetic tree was constructed by the maximum parsimony method using nucleotide sequences of the *W*14/15 gene variants so far determined, to see taxonomic relationships of *Gentiana* species. As shown in (Fig. 9.7), genes from plants at different taxonomic positions clustered into different groups. Two major clades, I and II, except for a *G. thunbergii* clade, constituted the tree. Clade I consists of *G. pneumonanthe* and *G. triflora*, and clade II diverged to *G. septemfida*, *G. paradoxa*, and *G. scabra*. A similar tree constructed by morphological phylogenetic analysis has been reported for *G. septemfida*, *G. paradoxa*, *G. pneumonanthe*, and *G. scabra*, where *G. septemfida* and *G. paradoxa* constitute a clade (Septemphida Clade) that was independent of a clade that consists of *G. pneumonanthe* and *G. scabra* (Davitashvili and Karrer 2006). In contrast to this analysis, our results showed that *G. scabra* is included together with *G. septemfida* and *G. paradoxa* in the same clade (Clade II). This result indicates that *G. scabra* is closer to *G. septemfida* and *G. paradoxa*, rather than to *G. pneumonanthe*. All these species belong to the Section *Pneumonanthe*. *G. thunbergii* has been classified in a different Section *Chondrophyllae*, and it diverged far from other species examined here. Thus, the phylogenetic tree inferred by the *W*14/15 gene variants can classify gentian species in the Section *Pneumonanthe*, and may also discriminate the Section *Pneumonanthe* from *Chondrophyllae*.

9.5.2 Phylogeographical Divergence in Japan

In Japan, 13–18 native *Gentiana* species are recognized currently (Toyokuni 1963; Yonezawa and Kawano 1989; Ho and Liu 2001). Among those, *G. triflora* var. *japonica* is known to have several variants, including *G. triflora* var. *japonica* f. *montana*, an alpine form of *G. triflora* var. *japonica*. Classification of those plants of Japanese habitats has been based on gross morphological differences. However, it seems rather obscure whether or not the differences are due to genetic (genomic) backgrounds, because gentians sometimes differ in their appearance depending on growth conditions under different geographical environments (Koehlein 1991). In this context, *G. triflola* var. *japonica* grown in different mountainous areas (shown in red in Fig. 9.7) were examined for variation of the *W*14/15 gene. As shown in (Fig. 9.7), the *W*14/15 genes from different habitats were separated into several subfamilies on the tree. Subfamilies i and ii constitute the genotype of the habitat

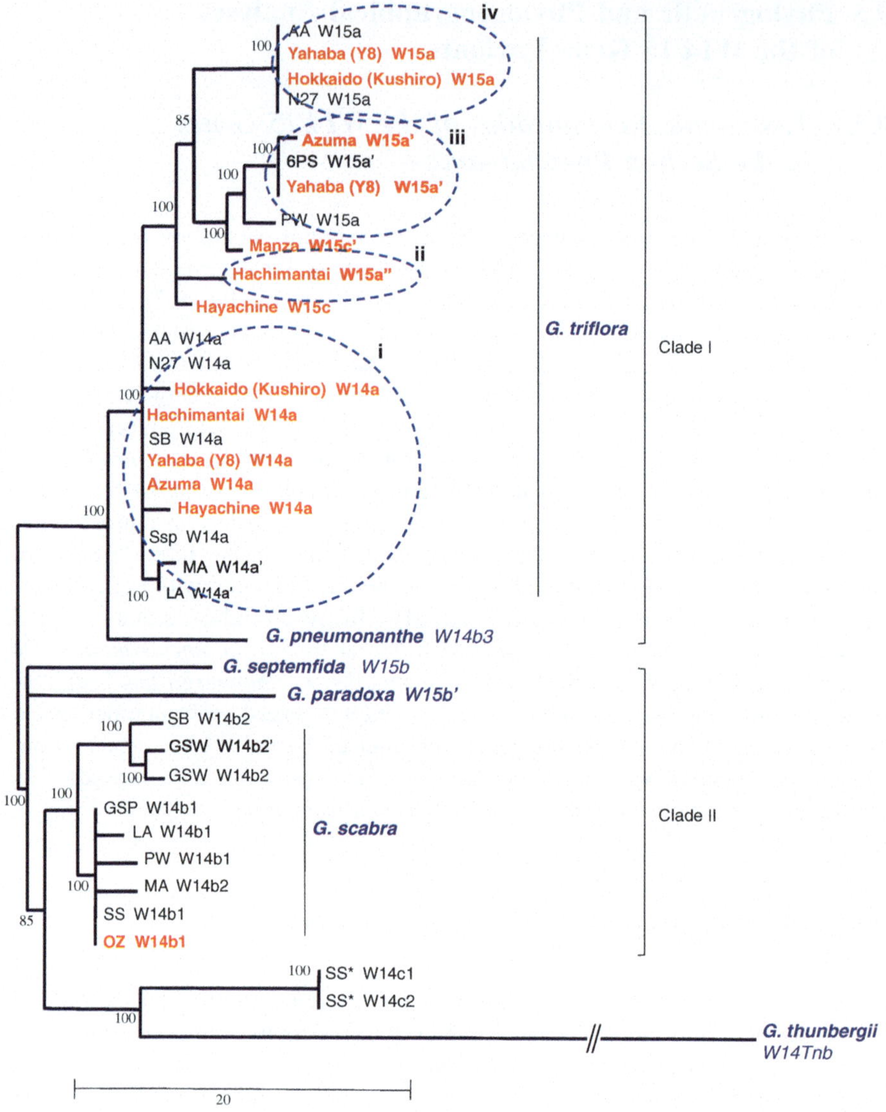

Fig. 9.7 Phylogenetic tree showing evolutionary distance. The tree was constructed by the maximum parsimony method with more than 35 *W14/15* genes from six different species. Nucleotide sequences subjected to this analysis were those corresponding to the region from position +63 to +1,184 of the AA *W14a* gene. Gentian species are shown on the right. Asterisk (cv. SS) indicates unknown species, other than *G. scabra*, *G. triflora*, *G. pneumonanthe*, *G. paradoxa*, *G. septemfida* and *G. thunbergii*. Genes from native gentians of different habitat in Japan are shown in *red*, and others (commercial cultivars) in *black*. Genes in *G. triflora* are classified tentatively into four subfamilies, 1 to 4

at Mt. Hachimantai, while subfamilies 1 and 3, and 1 and 4 constitute genotypes at Mt. Azuma and Hokkaido (Kushiro), respectively. These features imply phylogeographical divergence of *G. triflora* var. *japonica*. For example, the *W15a′* gene in Hachimantai (subfamily 2) was not found in Kushiro, Azuma, and Yahaba, and the *W15a* gene in Kushiro (4) was not found in Hachimantai. This geographical difference was further revealed from the genes in Yahaba (line Y8, 4) and Y29 (not shown) that were established from native species grown in Yahaba, Iwate Prefecture, Japan. These lines carried genotypes of *W15a*/*W15a′* and *W14a*/*W15a*, respectively. Among those genes, the *W15a′* gene was not found in Hachimantai and Kushiro. In contrast to the *W15* type genes, the *W14a* gene (1) was found to have diverged less.

9.6 Conclusions

The *W14/15* gene variants do not help simplification of the taxonomic classification of *G. triflora*. As mentioned above, native species of *G. triflora* var. *japonica* in four different mountainous areas (Hachimantai, Azuma, Kushiro and Yahaba) have different haplotypes and allelic pairs. Among those, gentians in Yahaba were classified into *G. triflora* var. *japonica*, and other three into *G. triflora* var. *japonica* f. *montana*. However, a difference in the genotype/haplotype between the two taxa is not clear from the *W14/15* gene variation. These observations may indicate the occurrence of intraspecies variation in those native gentians that are classified formerly into the same taxon *G. triflora* var. *japonica*, rather than different classification into subvarieties or forma. Precise phylogeographical analysis using *W14/15* gene variation is in progress.

Acknowledgments This study was supported in part by "Research and Development Projects for Application in Promoting New Policy of Agriculture, Forestry and Fisheries" from the Ministry of Agriculture, Forestry and Fisheries of Japan, and by a grant for scientific research from the Ministry of Education, Culture, Sports, Science and Technology of Japan. The authors thank Miho Takahashi, Claire Saito, Kiyomi Kogusuri, Shuji Watanabe, Sakura Chiba, Miyuki Yamazaki, Drs. Yoshitaka Shimotai, Hisako Doi, and Kohei Kume for collaboration, support, and discussion, and Drs. Yoshihito Takahata, Tetsuro Yamashita, Abidur Rahman, and Yasushi Saitoh for suggestions and discussions.

References

Amtmann A (2009) Learning from evolution: *Thellungiella* generates new knowledge on essential and crucial components of abiotic stress tolerance in plants. Mol Plant 2:3–12

Davitashvili N, Karrer G (2006) Taxonomic relationships of the western Asian taxa of entiana sect. *Pneumonanthe* Bot J Linn Soc 152:197–208

Doi H, Takahashi R, Hikage T, Takahata Y (2010) Embryogenesis and doubled haploid production from anther culture in gentian (*Gentiana triflora*). Plant Cell Tiss Org Cult 102:27–33

Doi H, Yokoi S, Hikage T, Nishihara M, Tsutsumi K, Takahata Y (2011) Gynogenesis in Gentians (*Gentiana triflora, G.* Scabra): production of haploids and doubled haploids. Plant Cell Rep 30:1099–1106

Espartero J, Sanchez-Aguayo I, Pardo JM (1995) Molecular characterization of glyoxalase I from a higher plant; upregulation by stress. Plant Mol Biol 29:1223–1233

Ganeshan S, Sharma P, Young L, Kumar A, Fowler DB, Chibbar RN (2011) Contrasting cDNA-AFLP profiles between crown and leaf tissues of cold-acclimated wheat plants indicate differing regulatory circuitries for low temperature tolerance. Plant Mol Biol 75:379–398

Heikinheimo P, Goldman A, Jeffries C, Ollis DL (1999) Of barn owls and bankers: a lush variety of a/b hydrolases. Struct Fold Des 7:R141–R146

Hikage T, Saitoh Y, Tanaka-Saitoh C, Hagami H, Satou F, Shimotai Y, Nakano Y, Takahashi M, Takahata Y, Tsutsumi K (2007) Structure and allele-specific expression variation of novel alpha/beta hydrolase fold proteins in gentian plants. Mol Genet Genomics 278:95–104

Hikage T, Kogusuri K, Tanaka-Saito C, Watanabe S, Chiba S, Kume K, Doi H, Saitoh Y, Takahata Y, Tsutsumi K (2011) W14/15 esterase gene haplotype can be a genetic landmark of cultivars and species of the genus *Gentiana* L. Mol Genet Genomics 285:47–56

Ho TN, Liu SW (2001) A worldwide monograph of *Gentiana*. Science Press, Beijing

Jensen SR, Schripsema J (2002) Chemotaxonomy and pharmacology of gentianaceae. In: Struwe L, Albert VA (eds) Gentianaceae-systematics and natural history. Cambridge University Press, UK, pp 563–573

Koehlein F (1991) Gentians. Christopher Helm Ltd., London

Kumar D, Klessig DF (2003) High-affinity salicylic acid-binding protein 2 is required for plant innate immunity and has salicylic acid-stimulated lipase activity. Proc Natl Acad Sci USA 100:16101–16106

Lee H, Guo Y, Ohta M, Xiong L, Stevenson B, Zhu JK (2002) LOS2, a genetic locus required for cold-responsive gene transcription encodes a bi-functional enolase. EMBO J 21:2692–2702

Lee D, Ahsan N, Lee S, Lee J, Bahk J, Kang K, Lee B (2009) Chilling stress-induced proteomic changes in rice roots. J Plant Physiol 166:1–11

Matsukawa K, Ogata M, Hikage T, Minami H, Shimotai Y, Saitoh Y, Yamashita T, Ouchi A, Tsutsumi R, Fujioka T, Tsutsumi K (2006) Antiproliferative activity of root extract from gentian plant (*Gentiana triflora*) on cultured and implanted tumor cells. Biosci Biotechnol Biochem 70:1046–1048

Ogata M, Matsukawa K, Kogusuri K, Yamashita T, Hikage T, Ito K, Saitoh Y, Tsutsumi K (2011) Gentian extract induces caspase-independent and mitochondria-modulated cell death. Adv Biol Chem 1:49–57

Shigeoka S, Ishikawa T, Tamoi M, Miyagawa Y, Takeda T, Yabuta Y, Yoshimura K (2002) Regulation and function of ascorbate peroxidase isoenzymes. J Exp Bot 53:1305–1319

Smirnoff N (1998) Plant resistance to environmental stress. Curr Opin Biotech 9:214–219

Struwe L, Kadereit JW, Klackenberg J, Nilsson S, Thiv M, von Hagen KB, Albert VA (2002) Systematics, character evolution, and biogeography of Gentianaceae, including a new tribal and subtribal classification. In: Struwe L, Albert VA (eds) Gentianaceae-systematics and natural history. Cambridge University Press, UK, pp 21–309

Stuhlfelder C, Mueller MJ, Warzecha H (2004) Cloning and expression of a tomato cDNA encoding a methyl jasmonate cleaving esterase. Eur J Biochem 271:2976–2983

Taji T, Seki M, Satou M, Sakurai T, Kobayashi M, Ishiyama K, Narusaka Y, Narusaka M, Zhu JK, Shinozaki K (2004) Comparative genomics in salt tolerance between *Arabidopsis* and *Arabidopsis*-related halophyte salt cress using *Arabidopsis* microarray. Plant Physiol 135:1697–1709

Takahashi M, Hikage T, Yamashita T, Saitoh Y, Endou M, Tsutsumi K (2006) Stress-related proteins are specifically expressed under non-stress conditions in the overwinter buds of the gentian plant *Gentiana triflora*. Breed Sci 56:39–46

Toyokuni H (1963) Conspectus Gentianacearum Japonicarum: a general view of the Gentianaceae indigenous to Japan. J Fac Sci Hokkaido Univ Ser 7:137–259

Yang Y, Xu R, Ma C-J, Vlot AC, Klessig DF, Pichersky E (2008) Inactive methyl indole-3-acetic acid can be hydrolyzed and activated by several esterases belonging to the AtMES esterase family of *Arabidopsis*. Plant Physiol 147:1034–1045

Yonezawa N, Kawano S (1989) Variations in Japanese *Gentiana* Sect. *Pneumonanthe* (Gentianaceae) with special reference to their taxonomic status. Acta Phytotax Geobot 40:13–30 (Japanese with English abstract)

Yoshiike T (1992) Rindou (*Gentiana*). Seibunndou Shinkousha, Tokyo, Japan (in Japanese)

Zang X, Komatsu S (2007) A proteomics approach for identifying osmotic-stress-related proteins in rice. Phytochemistry 68:426–437

Zong GY, Goren R, Riov J, Sisler EC, Holland D (2001) Characterization of an ethylene-induced esterase from *Citrus sinensis* by competitive hybridization. Physiol Plant 113:267–274

10.1 Introduction

Exacum bicolor Roxb., commonly known as country kreat, is an exquisite ornamental flowering plant found wild in the humid tropics. The plant is an erect herb with a sturdy stem that does not lodge even during heavy rain. It bears numerous attractive flowers that retain their beauty for a considerable time. Attempts to domesticate this attractive flowering plant have not been successful. Reports in the ancient literature state the medicinal uses of the plant for various ailments. Severe destruction of the natural habitats has now endangered the plant in the Western Ghats and dry grasslands of Southern India.

10.2 Systematics

E. bicolor Roxb. is a member of the tribe Exacaceae in the family Gentianaceae. Six genera are included in this tribe. According to Struwe et al. (2002), the genus *Exacum* was introduced by Linnaeus in 1747, with two species, *E. pedunculatum* and *E. sessile,* being added later (Linnaeus 1753). Roxburg (1832) included another five species in this genus. Klackenberg (1985) revised the genus describing 65 species. Members of the genus have a wide paleotropical distribution occurring from Africa to Madagascar and Socotra Island, and also from India to New Guinea to the Northern part of Australia. Broadly, the genus *Exacum* is divided into two Sections; Section *Exacum* has 21 species distributed in India, Sri Lanka, and South East Asia, while section *Africana* with 44 species is distributed in Madagascar, Socotra, Oman, and tropical Africa (Struwe et al. 2002). John et al. (2001) reported 16 species of *Exacum* occurring in India. The majority of the *Exacum* species are characterized by restricted and local endemic distribution. Some of the species have been collected only a few times and hence not studied extensively. In contrast, there are species such as *E. tetragonum* which are widely distributed.

Many authors have reported *E. bicolor* as synonymous with *E. tetragonum.* Roxburg (1832) allotted separate species status to the above two species in his classification. Bentham (1876) in *Genera Plantarum* described *E. bicolor* as only a form of *E. tetragonum.* Hooker (1885) gave separate species status to *E. bicolor* and *E. tetragonum,* but Klackenberg (1985) described *E. bicolor* as a synonym of *E. tetragonum.* Sivarajan and Philip (1997) highlighted the need for further studies before describing the Himalayan *E. tetragonum* and South Indian *E. bicolor* as a single species. A difference of opinion still exists about separate species status for *E. bicolor* and *E. tetragonum,* and more studies may be required to explain the following, namely,

1. The wide variation in chromosome numbers of the two species viz. *E. tetragonum* is $2n = 18$ and *E. bicolor* $2n = 62$ (Struwe et al. 2002).
2. Anthers of *E. bicolor* are curved (Gamble 1915; Hooker 1885) whereas *E. tetragonum* has straight anthers.

3. *E. bicolor* is not seen naturally in sandy soils or marshy meadows or swamps of Kerala's South Western Ghats. The plant has restricted occurrence in Kerala. There are places with less than five plants in total. These findings do not conform to the studies of Klackenberg (1985), in which he reported a wide distribution for *E. tetragonum* in clay to sandy soils or waterlogged places, such as marshy meadows and swamps.
4. All the authors who have assigned separate species status to *bicolor* and *tetragonum* have mentioned the variation in flower color in the two species, *E. bicolor* having white flowers with blue corolla tips, while *E. tetragonum* has a blue/azure-colored corolla.
5. The present authors also feel that the choice of the epithet *E. bicolor* is more apt than *E. tetragonum*, as there are many species with quadrangular stems and tetramerous flowers in the genus *Exacum*.

10.3 Habitat Characterization

E. bicolor is a species endemic to Peninsular India. Natural habitats of the plant are grasslands of both high and low altitudes, as well as scrub savannah. The plant has been reported in the plateau region from the Konkan coast to the Southern tip of India, including Western Ghats. The species has wide adaptability to the climate prevailing in the plains of the tropics and high ranges in Western Ghats. A survey was conducted representing various habitats in the South Indian State of Kerala, the latter being a major portion of Southern Western Ghats, a hot spot of biodiversity. Floristic studies were undertaken initially to obtain a preliminary idea about the distribution of the plant. The majority of these studies represented forest areas such as Wild Life Sanctuaries, National Parks, and Bioreserves occurring at high altitudes. In the plains, where dry grasslands are present, survey work was carried out based on local enquiry. Dry grasslands usually occur in hillocks at 50–200 m elevation. Of a total of 125 sampling units surveyed, *E. bicolor* was cited only in 41 units with a frequency of 32.8 %. Nearly 70 % of these holdings are in private possession. Again, among the 41 units where the plant was recorded, 20 or less than 20 plants were recorded in 10 units during 2005. Among the 10 units with less than 20 plants, 6 had only 5 or less than 5 plants in the entire area (approximately half km^2). The unique feature of the plant is its restricted occurrence in small areas even in a vast stretch of grassland. Sometimes, plants are seen only on one side of a hillock. During a revisit to 17 habitats in 2006, the plant disappeared from three of the sites due to complete destruction of the habitats. The status of various locations where *E. bicolor* was cited in Kerala is represented in Table 10.1. All the privately owned habitats are likely to be exploited at any time as decided by the owners for various reasons including laterite mining, soil excavation for various development activities, establishment of plantations of rubber and coconut, and also as industrial sites.

Table 10.1 Status of habitats of occurrence of *E. bicolor* in Kerala

District	Number of habitats				Total habitats
	Forest—shola grassland	Revenue—shola grassland	Revenue—dry grassland	Private—dry grassland	
Thiruvananthapuram	0	0	0	0	0
Kollam	0	0	0	0	0
Pathanamthitta	3	0	0	0	3
Idukki	1	0	0	0	1
Kottayam	0	1	0	0	1
Thrissur	0	0	1	6	7
Palakkad	1	0	0	2	3
Malappuram	1	0	0	9	10
Kozhikode	0	0	0	2	2
Wayanad	5	0	0	0	5
Kannur	0	0	0	7	7
Kasargode	0	0	0	2	2
Total	**11**	**1**	**1**	**28**	**41**

Table 10.2 Comparison of the altitude and occurrence of habitats of *E. bicolor* in Kerala

Sl.No.	Site particulars (Altitude from mean sea level)	No. of sites visited	No. of sites of plant occurrence	Frequency (%)	Density	Abundance
1	<500 m	80	29	36.3	2.2	3.9
2	>500 m	45	12	26.7	1.1	2.8

Altitudinal differences in plant distribution were also investigated (Sreelatha et al. 2009). For the study, sites were grouped mainly into two: those with an altitude less than 500 m, and those with an altitude exceeding 500 m. This grouping was done because above 500 m, the main associating flora, such as grasses, are very tall. Other associating plants were also similar from 500 m above mean sea level. Table 10.2 shows the frequency and numerical status of the plants above and below 500 m from mean sea level. It is evident that among the sites surveyed, plants were more frequent (36.3 %) at altitudes below 500 m. Greater density (2.2) and abundance (3.9) of *E. bicolor* were also recorded at these sites, indicating that the plant population is more concentrated in these areas than in high ranges (>500 m above mean sea level) of Kerala.

10.4 Associated Flora

The major associating flora in low altitude grasslands are grasses such as *Heteropogon contortus, Dimeria bialata, Ischaemum indicum*, ground orchids like *Habenaria diphylla*, and *H. heyneana* together with other herbs such as *Sopubia delphinifolia*, *Striga densiflora*, *Eusteralis quadrifolia*, *Sesamum laciniatum*, *Spermacoce pusilla* and *Oldenlandia corymbosa*. In high altitude grasslands, *E. bicolor* is seen in association with the grasses *Themeda cymbaria*, *Chrysopogon zeylanicus*, *Cymbopogon flexuous*, and *Arundinella purpurea*, and with other herbs including *Ageratum conyzoides*, *Osbeckia sp.*, *Tephrosia purpurea*, *Anaphalis sp.*, *Hypericum mysorensis*, *Swertia sp.* and *Phoenix lourieii*.

10.5 Mycorrhizal Association

Soil samples collected from four different natural habitats of the plant showed the presence of mycorrhizae. Preliminary studies identified *Glomus* and *Acaulospora* as the two associating fungi. Spores of the fungi were grown on maize roots and mycorrhizae infected roots of maize were added to the potting medium while transplanting seedlings of *E. bicolor*. There was considerable difference in the number of flowers per plant. Mycorrhizae-treated plants produced almost twofold more flowers than the control plants. This particular aspect needs detailed investigation for further confirmation.

10.6 Characteristics of Exacum bicolor

Exacum bicolor is a herbaceous perennial (Fig. 10.1), and in dry grasslands grows to a height of 25–80 cm. In high altitude grasslands attached to shola forests, where grasses grow thick and tall, the height of the plants ranges from 40 to 120 cm. The stem is quadrangular in shape and hard at the base when mature. It is nonlodging even in severe rains. Leaves are sessile, and linear lanceolate with five prominent veins, and are arranged in an opposite fashion. Variations such as oblong and cordate type leaves are also observed. In the plain regions, plants flower during September to November, while in high altitude areas flowering lasts up to January. The inflorescence is a dichasial cyme; the number of flowers is 2–125 per plant. On average, there may be 40 flowers per plant, with 10–20 buds in bloom at any time. The flowering period lasts for 30–45 days, with a single flower remaining for 8–10 days in the field. Flowers open from 6.00 to 8.00 am, close in the evening, and reopen the next morning. This pattern continues for 2–3 days after which flowers remain open.

Flowers are bisexual and tetramerous, with four winged sepals and a four-lobed rotate corolla. Petal length varies from 1.0 to 3.8 cm and width from 0.5 to 1.9 cm. The androecium consists of four epipetalous stamens alternating with the petals; the filaments are short, the anthers oblong, curved, and dehiscing by apical pores. The ovary is two celled, with many ovules on an axile placenta. The style is long, the stigma capitate, and well exerted above the anthers. The fruit is a dehiscent capsule with numerous minute seeds. The size of individual flowers ranges from

Fig. 10.1 Habit of *Exacum bicolor*

2.2 to 7.8 cm, with an average of 5.0 cm. The petals are white with dark purple tips; these are the most attractive part of the flower. Petal tip color varies in intensity from dark purple to blue and sometimes even violet. Five to six days after flower opening, the purple pigmentation spreads gradually downward with the whole petal finally turning light purple. The bright yellow stamens add to the beauty of the flowers. A bunch of flowers is usually a colorful mix of white, dark purple, light purple, and yellow (Fig. 10.2). Pollination is assisted by insects such as stingless bees, bumble bees, and carpenter bees. Fruit set is not observed when flowers are bagged or plants are kept in cloth covered cages. Nearly 4 months are required from flowering to fruit maturity. The capsules dehisce to release the minute seeds.

Each capsule weighs 130–150 mg with a seed weight of 25–30 mg per capsule. On average, 1 mg of seed contains 140 seeds, the average seed size being 298 μm. Considering an average of 40 flowers per plant, the number of seeds produced per plant ranges from 140,000 to 170,000. Even with this large quantity of seed production per plant, the spread of the plant to adjacent areas is not usually observed. This may be due to the fact that seeds dispersed by wind may not be deposited in suitable open places to germinate and also seeds may become buried deeply in low lands where there is water logging. Other supporting factors for the isolated occurrence of *E. bicolor* may be the discontinuous genus status reported by Good (1953) and the mycorrhizal association characteristic of the Gentianaceae (Meszaros et al. 2002). Natural seed germination is also less than 5 %, as in the case of other gentians.

Fig. 10.2 A bunch of flowers of *Exacum bicolor*

10.7 Propagation

In natural habitats, the plant is propagated from tuberous roots that perennate underground. This was confirmed from the sprouting of new plants from dry stumps of the previous year (Fig. 10.3) and the same pattern was observed throughout a study period by the present authors. In ex situ conditions, sprouting is not common and plants that sprout do not show vigorous growth. Hence, propagation through seed has been standardized. A compost of sand and coir pith in equal proportions is ideal for seed germination. Seeds germinate within 2 weeks in this medium provided high humidity is maintained. Propagation by cuttings has been unsuccessful.

The protocol for *in vitro* seed germination, shoot multiplication, and plantlet production has also been standardized. For *in vitro* seed germination, the optimum medium is that of Murashige and Skoog (1962) at one quarter strength, supplemented with 1.0 mg/l 6-benzylamino purine (BA). Darkness induced earlier increased seed germination. Multiple shoots from in vitro seedlings were best induced on MS medium at half strength supplemented with 0.5 mg/l BA with 0.5 mg/l indole-3-acetic acid (IAA). Optimum rooting was MS-based medium at one quarter strength supplemented with 1.0 mg/l indole-3-butyric acid (IBA), 20 g/l sucrose, and 0.125 % activated charcoal.

Fig. 10.3 Sprouting of *Exacum bicolor* from dry stumps

10.8 Variation in Seedling Progeny

Wide variation occurs in seedling populations of *E. bicolor*. A seedling population was raised from seeds obtained from wild plants collected from a natural habitat. A total of 350 seedlings were planted in beds and grown under the same management practices. Observations were recorded on various quantitative characters including plant height, internodal length, number of branches, number of leaves, leaf length, leaf width, number of flowers, flower size, petal length, petal width, and length of the dark purple patch at each petal tip. Qualitative characters were also observed such as leaf shape, flower shape, intensity of the purple color at the petal tip and flower color variants. All the quantitative characters showed wide variability (Table 10.3); maximum variability was expressed for the number of flowers per plant, from 15 to 215. The wide variation observed in seedling progeny offers immense scope for selection and further improvement of *E. bicolor*.

Qualitative characters also differed extensively especially in leaf shape, petal shape, and petal tip color. Leaf shape ranged from lanceolate, linear to cordate and oblong. Petal shapes, which ultimately decide the flower shape, also showed considerable variation. Petal tip color varied from low to high intensities of purple,

Table 10.3 Variability in the seedling population of *E. bicolor*

Sl.No.	Plant characters	Mean	SD	CV	Range
1	Height (cm)	48.4	12.6	26.1	23–100
2	Internodal length (cm)	7.3	2.1	28.7	3–14.5
3	Number of branches	8.0	3.1	39.1	3–18
4	Number of leaves	12.7	5.1	40.0	4–49
5	Leaf length (cm)	8.0	2.5	31.1	2.4–14.5
6	Leaf width (cm)	2.8	1.0	37.1	1–7.5
7	Number of flowers	66.4	36.1	54.4	15–215
8	Flower size (cm)	4.4	1.8	40.9	3–5.6
9	Petal length (cm)	2.1	0.4	17.1	1.2–4.0
10	Petal width (cm)	0.9	0.2	24.4	0.2–1.3
11	Length of purple patch at petal tip (cm)	0.7	0.2	30.7	0–1.2

with occasionally a blue color. A rare occurrence of flowers with fully white petals was also observed. Unlike the usual pattern of petals turning light purple on fading, some plants had flowers with petals that become dark purple during this period.

10.9 Ornamental Uses

The ornamental value of the plant was reported earlier by Woodrow (1910) in his book "*Gardening in the Tropics.*" Although beautiful, the plant was reported by the author as being most difficult to cultivate. The ornamental value of *E. bicolor* has been reported by several scientists (Rao 1986; John et al. 2001; Sreelatha et al. 2006). The most valued character of the plant is its tolerance to heavy rain and high humidity, but it needs these conditions for optimum growth. The plant could be utilized as a potted ornamental as well as for flower borders. Annuals that are nonlodging and suitable for cultivation are rare in tropical, high rainfall areas. A long flowering period of 30–45 days, pleasingly large individual flowers and long field life of the flowers, add to the aesthetic value of this plant. In Kerala, the plant was very popular about 5–8 decades ago and flowers of *E. bicolor* were given considerable religious importance in certain regions of the State. It was one of the choice flowers to adorn the earthen deity "Trikkakarayappan," worshipped during the important regional festival "Onam" in Kerala. The vernacular name, *Kannamthali,* very often finds a place in local linguistic poems and film songs.

10.10 Medicinal Uses

Like any other gentian, *E. bicolor* is also reported to have many medicinal uses. Dried stalks of the plant were sold in South Indian markets during the early twentieth century (Rao 1914) and were widely used for their febrifuge properties.

Table 10.4 Quantitative analysis of dry plant samples of *E. bicolor*

Sl. No.	Plant organ	Seedling stage			Bud initiation stage			Flowering stage		
		Starch (%)	Soluble sugar (%)	Phenol (%)	Starch (%)	Soluble sugar (%)	Phenol (%)	Starch (%)	Soluble sugar (%)	Phenol (%)
1	Shoot	6.7	7.5	2.3	7.3	5.5	1.9	6.0	7.1	1.8
2	Root	6.5	6.1	0.7	4.0	4.6	0.9	6.9	4.7	0.6
3	Flower	–	–	–	–	–	–	8.5	9.8	3.0

The tonic and stomachic properties of the plant were reported as a substitute for *Gentiana* and *Swertia* (Anon, 1952). Chopra et al. (1956) and Srivastava (1989) quoted similar medicinal properties for the plant. However, the plant is not in commercial trade anywhere in India. Local enquiry during studies by the present authors revealed that the plant was used for treatment of fever, eye, and skin diseases and urinary disorders by traditional medical practitioners. However, substitutes are being used today due to lack of availability of the plant.

Primary metabolites like starch and sugars are essential for plant growth; many primary metabolites act as precursors for pharmacologically active compounds. Preliminary phytochemical screening has been done to analyze the content of primary and secondary metabolites. Quantitative analysis of dry plant samples for starch, soluble sugar, and phenol are given in Table 10.4. During the flowering stage, the content of primary metabolites reached a maximum which could support the medicinal property of the whole plant. Identification tests for various compounds such as alkaloids, flavonoids, tannins and saponins in dry plant samples have given positive results. Jeeshna and Paulsamy (2011) isolated six phytochemical compounds of medicinal importance from powdered whole plant material of *E. bicolor*. Of these compounds, two were polyphenolic, two were alkaloidal, one was glycosidal, and one was steroidal in nature.

10.11 Conservation Strategies

In Kerala, nearly 70 % of the natural habitats are in private ownership. Severe habitat destruction continues in all these areas causing depletion of the plant at an alarming rate. Habitat destruction is also reported from the Western Ghats regions of other States. *Ex situ* conservation is the only solution in this context. For any *ex situ* conservation programme, plants have to be domesticated, and for this to be achieved, the habitat, plant habit, propagation, floral biology and phenology must be studied in depth. This needs urgent attention in the prevailing social situations in Kerala and other South Indian States.

10.12 Conclusions

Wide adaptability to the climate of plains and high ranges, nonlodging stems even in severe rain, improved flower morphology, a long flowering period, and better adaptations to field conditions give potential for *E. bicolor* to be domesticated as an ornamental plant. Future research should concentrate on standardization of production technologies for popularizing *E. bicolor* as a garden plant. Although there are several reports about the medicinal value of this plant, scientific studies are required for commercial exploitation of its medicinal properties.

References

Anon (1952) Wealth of India: raw materials, vol 3. Publication and Information Directorate CSIR, New Delhi

Bentham G (1876) Gentianaceae. In: Bentham G, Hooker J (eds) Genera Plantarum, Part 2, vol 2. L. Reeve & Co. Williams and Norgate, London, pp 799–820

Chopra RN, Nayar SL, Chopra IC (1956) Glossary of Indian medicinal plants. National Institute Of Science Communication, New Delhi

Gamble JS (1915) Flora of the presidency of Madras, vol 2. 1986. Repr Bishen Singh Mahendra Pal Singh, Dehra Dun

Good R (1953) Geography of flowering plants. Longman, Great Britain

Hooker JD (1885) The flora of British India: Asclepidaceae to Amarantaceae, vol 4. 1973. Repr Bishen Singh Mahendra Pal Singh, Dehra Dun

Jeeshna MV, Paulsamy S (2011) Phytochemistry and bioinformatics approach for the evaluation of medicinal properties of the herb, *Exacum bicolor* Roxb. Int Res J Pharm 2:163–168

John JK, Nizar AM, Velayudhan KC (2001) *Exacum bicolor*: a beautiful wild flowering herb. Indian Hortic 46:25

Klackenberg J (1985) Genus *Exacum* (Gentianaceae). Opera Botanica 84:1–144

Linnaeus C (1753) Species Plantarum, vol 1. L Salvii, Stockholm

Meszaros S, De Laet J, Goethals V, Smets E, Nilsson S (2002) Cladistics of Gentianaceae: a morphological approach. In: Struwe L, Albert VA (eds) Gentianaceae. Systematics and natural history. Cambridge University Press, UK, pp 310–376

Murashige T, Skoog F (1962) A revised medium for rapid growth and bioassays with tobacco tissue cultures. Physiol Plant 15:473–497

Rao MR (1914) Flowering plants of Travancore, Bishen Singh Mahendra Pal Singh, Dehra Dun

Rao RS (1986) Flora of Goa, Diue, Daman, Dadra and Nagarhaveli, vol 2. Botanical Survey of India, Calcutta

Roxburg W (1832) Flora indica: 1974 reproduction from carey's edition. Today & Tomorrow Printers and Publishers, New Delhi

Sivarajan VV, Philip M (1997) Flora of Nilambur (Western Ghats, Kerala). Bishen Singh Mahendra Pal Singh, Dehra Dun

Sreelatha U, Baburaj TS, Narajanankutty C, Mini Raj N, Nazeem PA (2006) Exacum bicolor: An exquisite, endemic flowering herb from the wild. In: Abstracts of national symposium on under utilized horticultural crops, IIHR Bangalore, p 137

Sreelatha U, Baburaj TS, Mini Raj N, Narajanankutty C (2009) Distribution pattern of *Exacum bicolor* Roxb. A Gentian endemic to Peninsular India. Indian J Trop Biodivers 17:105–110

Srivastava RC (1989) Drug plant resources of central India (an inventory). Today & Tomorrow's Printers and Publishers, New Delhi

Struwe L, Kadereit JW, Klackenberg J, Nilsson S, Thiv M, Von Hagen KB, Albert VA (2002) Systematics, character evolution and biogeography of Gentianaceae, including a new tribal and subtribal classification. In: Struwe L, Albert VA (eds) Gentianaceae: systematics and natural history. Cambridge University Press, UK, pp 21–309
Woodrow GM, Paisley (1910) Gardening in the Tropics. Alexander Gardener, London

Chapter 11
Indian *Swertia* from Eastern Himalaya: Strategies of Conservation and Biotechnological Improvements

Tapojita Samaddar, Sumita Jha and Timir Baran Jha

Abstract The Eastern Indian area of the Himalaya is highly acclaimed for its rich bioresources and traditional repository of medicinal plants. This region is considered to be the natural habitat of many species and populations of *Swertia*, an important and diverse genus of the Gentianaceae. The genus *Swertia*, popularly known as chirata, is one of the most important indigenous medicinal plants of India. Habitat destruction, other human-caused stresses, and impacts of climate change have resulted in considerable loss of genetic diversity and necessitate reassessment, documentation, molecular and biochemical screening, and well-formulated biotechnological strategies for sustainable use and improvement of these high-value medicinal crops. *Swertia chirata* is considered the prized Indian species because of its well-characterized bioactive molecules known as iridoid and secoiridoid glycosides, xanthones and xanthone derivatives, which are effective against many conventional and unconventional ailments. The increasing demand of chirata in national and international markets, the paucity of agricultural practices, and correct exomorphological and molecular screening have resulted in intentional and unintentional adulteration. The prized species, *Swertia chirata,* has been labeled as being critically endangered and most of the other important species may face similar threats in the near future. The present review discusses the status of *Swertia* species in the Eastern Himalayas, together with biotechnological approaches for their conservation and future improvement.

T. Samaddar · S. Jha
Department of Botany, University of Calcutta, 35 B. C. Road, Kolkata 700019, India

T. B. Jha (✉)
Barasat Govt. College, 10 KMC Road, Kolkata 700024, India
e-mail: tbjha2000@yahoo.co.in

J. J. Rybczyński et al. (eds.), *The Gentianaceae - Volume 1: Characterization and Ecology*, DOI: 10.1007/978-3-642-54010-3_11,

11.1 General Account of the Family Gentianaceae

The family Gentianaceae consists of 87 genera and about 1,650 species. Twenty six genera are indigenous to Asia, of which 11 are considered to be endemic (Struwe et al. 2002). The family belongs to the order *Gentianales* under the subclass *Asteridae* and class *Magnoliopsida* (Cronquist 1988; Judd et al. 1999). According to Struwe et al. (2002), the Gentianaceae is the third largest family of the *Gentianales*. Plants belonging to this family are primarily native to Northern temperate regions of the world. The *Gentianales* are characterized by leaves that are opposite, with interpetiolar stipules or stipular lines. Colleters (small glandular structures) are found in almost all species, on the insides of leaf bases and/or calyces (Struwe et al.1994).

11.2 The Genus *Swertia*

The genus *Swertia* of the family Gentianaceae was founded by Linnaeus in 1753 in honor of Emanuel Swert (also spelt Sweert), an eminent Dutch gardener and botanical author of the seventeenth century (Chassot et al. 2001). *Swertia* is one of the most diverse genera of the family, comprising about 170 species that are found mostly in the temperate regions of the Northern Hemisphere (Brahmachari et al. 2004).

Gilg (1895) dealt with *Swertia* as a primitive genus of the temperate subtribe *Gentianinae*, on the basis of the morphological phylogeny of the family (Mészáros and Höhn 2002). *Swertia* has been described in the literature as a highly polyphyletic genus (von Hagen and Kadereit 2001). There are several disagreements among researchers regarding the circumscription of *Swertia*. The genus has caused several debates owing to some superficial morphological similarities among the species of *Swertia* and some related genera like *Halenia*, *Lomatogonium* and *Veratrilla* (Xue et al. 2007). Despite similarities in gross morphological characters, the diverse embryology existing within this group supports the view that it is a loose group. According to Xue et al. (2007), *Swertia* should be considered as a polyphyletic rather than a monophyletic group. Chassot et al. (2001) suggested *Swertia* to be a taxon that acts as a stem group, giving rise to diverse lineages on the basis of cladistic analyses relying on DNA sequence data.

11.3 Distribution of *Swertia*

The genus *Swertia* is distributed in the mountains of Tropical Asia, Europe, America, and Africa. Regionalization of world flora by Takhtajan (1978) suggested *Swertia* to be present in 14 regions of the world, with the major distribution center being in the Himalayas and South-West China.

In India, about 40 species of *Swertia* are found at high altitude (1,200–3,000 m) in the temperate Himalayan region ranging from Kashmir to Bhutan, and also in the Khasia hills and Western Ghats, of which 16 are reported from the North West Himalayas (Garg 1987; Ho et al. 1994; Misra et al. 2010; Chopra et al. 1956) and about 13–14 species from the north Eastern states of India (Pradhan and Badola 2010). Clarke (1883), in J. D. Hooker's *Flora of British India*, reported 30 species from British India (including, Afghanistan, Bangladesh, Bhutan, Myanmar, Nepal, Pakistan, and India). There are some species that have been reported to be present in undivided India (Clarke 1883), but have not been reported since then, to the best of the authors' knowledge.

The distribution of different species of *Swertia* as reported until the present time, are listed in Table 11.1. Several species have been stated as having been found in the Eastern Himalayas by different authors. Collection and documentation have been conducted of different populations of *S. bimaculata*, *S. chirata*, *S. dilatata*, *S. nervosa*, and *S. paniculata* from different areas of Darjeeling District (West Bengal), Sikkim, and Arunachal Pradesh (Pandey et al. 2012). Further efforts to explore regions of Eastern Himalayas are being attempted by the authors in search of *Swertia* species.

11.4 Morphology of *Swertia*

Rijal (2009) described *Swertia* L. as annual, biennial, or perennial herbs, with fibrous or woody roots and opposite leaves, the latter being rarely whorled or alternate. The inflorescence is cymose, usually grouped into simple or paniculate thyrses, sometimes reduced to single flowers with the inflorescences raceme-like, or with flowers that are solitary and terminal. The flowers are tetra or pentamerous and pedicellate. Bracts are sessile and leaf-like. The calyx and corolla both rotate. There are one or two nectaries per corolla lobe, either naked or covered by a scale or flaps. Fruit is in the form of oblong capsules, enclosed in a persistent calyx and corolla.

In the course of our study on the conservation of *Swertia* species in the Eastern Himalayas, the following five species of *Swertia* collected are described below with a description of identifying floral characters (Pandey et al. 2012) in Table 11.2 and Figures (a–e).

11.4.1 Swertia chirata Buch.-Ham. Ex Wall (Fig. 11.1a)

Habit: Up to 90 cm in height. Roots light yellow and fibrous. Stem terete to winged; twigs rarely quadrangular, but branched.

Leaf morphology: Basal leaves are sessile to shortly petiolate, ovate-elliptic to ovate-lanceolate, slightly wavy at their margins, and acuminate at their apex. The

Table 11.1 Distribution of *Swertia* species in India, as reported by Clarke (1883) and other authors

Species	Distribution in British India[a] (Clarke 1883)	Distribution in Western Himalayas	Distribution in Eastern Himalayas
S. affinis	Deccan Peninsula (Chotanagpur to Pulney Hills)	–*	–
S. alata	Temperate Western Himalayas (Kashmir to Kumaun)	Chamba, Kalamuni (Uttarakhand) (Misra et al. 2010; Gupta et al. 2011)	Darjeeling; West Bengal (Misra et al. 2010)
S. alternifolia	Western Himalayas (Uttarakhand)	–	–
S. angustifolia	Chenab to Bhutan	Kalamuni (Uttarakhand) (Gupta et al. 2011); Snow view, Kilbury, Jageshwar, Jalna, Deenapani, Majhkhali, Katarmal (Kumaun Himalayas) (Bhatt et al. 2007); Tehri Garhwal, Chamba, (Uttarakhand), (Misra et al. 2010)	Sukhia pokhari- Darjeeling; W.B (Misra et al. 2010); Singhalila National Park, Singhalila-Senchal (http://www.icimod.org)
S. beddomei	Deccan Peninsula, Mysore	–	–
S. bimaculata	Eastern Himalayas (Sikkim, Darjeeling and Bhutan)	–	Darjeeling; W.B (Misra et al. 2010); Khangchendzonga Biosphere Reserve, Senchal Wildlife Sanctuary, Singhalila National Park (http://www.icimod.org); Darjeeling; W.B, Sikkim, Arunachal Pradesh (Jha et al. Unpublished)
S. chirata	Temperate Himalayas (Kashmir to Bhutan) and Khasia Hills (Meghalaya)	–	Sukhia pokhari- Darjeeling; W.B., Sikkim (Misra et al. 2010); Singhalila National Park, Khangchendzonga Biosphere reserve, Senchal Wildlife Sanctuary (http://www.icimod.org); Darjeeling; WB, Sikkim, Arunachal Pradesh (Jha et al. Unpublished)

(continued)

Table 11.1 (continued)

Species	Distribution in British India[a] (Clarke 1883)	Distribution in Western Himalayas	Distribution in Eastern Himalayas
S. coerulea	Western Himalayas	–	–
S. cordata	Temperate Himalayas and Western Tibet (Kashmir and Karakoram to Bhutan); Khasia Hills (Meghalaya)	Bhatwari, Chamba, Gangotri (Uttarakhand), Manali (Himachal Pradesh) (Misra et al. 2010)	–
S. corymbosa	Deccan Peninsula	–	–
S. cuneata	Western Himalayas	–	Khangchendzonga Biosphere reserve (http://www.icimod.org)
S. decussata	West Deccan Peninsula (Konkan to Travancore)	–	–
S. dilatata	Sikkim and Nepal	–	Mahananda-Neora Valley, Senchal-Mahananda (http://www.icimod.org); Darjeeling; W. B., Arunachal Pradesh (Jha et al. Unpublished)
S. griffithi	Bhutan	–	–
S. hookeri	Sikkim	–	Khangchendzonga Biosphere Reserve (http://www.icimod.org)
S. lurida	NW Himalayas (Naini Tal)	–	–
S. macrosperma	Meghalaya (Khasia Hills)	–	–
S. multicaulis	Alpine Nepal and Sikkim Himalayas	–	Khangchendzonga biosphere reserve (http://www.icimod.org)
S. nervosa	Temperate Nepal and Sikkim	Mussoorie (Uttaranchal) (Bhatia et al. 2004)	Temperate Sikkim and Darjeeling Himalayas (Bhatia et al. 2004); Senchal Wildlife Sanctuary (http://www.icimod.org); Darjeeling; W. B., (Jha et al. Unpublished)

(continued)

Table 11.1 (continued)

Species	Distribution in British India[a] (Clarke 1883)	Distribution in Western Himalayas	Distribution in Eastern Himalayas
S. paniculata	Temperate Western Himalayas (Kashmir to Nepal)	Kalamuni (Uttarakhand) (Gupta et al. 2011)	Barsey Rhododendron Sanctuary, Khangchendzonga biosphere reserve, Senchal Wildlife Sanctuary (http://www.icimod.org); Darjeeling; W. B., (Jha et al. Unpublished)
S. petiolata	Western Himalayas (Kashmir)	Jammu and Kashmir, Himachal Pradesh (Garg 1988)	–
S. pulchella	Subtropical Himalayas (Kangra valley in Himachal Pradesh to Bhutan) and Meghalaya (Khasia Hills)	–	–
S. purpurascens	Temperate NW Himalayas (Kashmir to Kumaon)	–	–
S. racemosa	North India	–	Sikkim, Assam (Garg 1988)
S. rex	Sikkim Himalayas (Singhalila)	–	Endemic to Sikkim (Garg 1988)
S. speciosa	Western Himalayas (Kashmir to Kumaon)	Jammu and Kashmir, Himachal Pradesh (Garg 1988)	Sikkim (Garg 1988); Singhalila National Park (http://www.icimod.org)
S. tetragona	Temperate Himalayas (Kashmir to Simla)	Jammu and Kashmir, Himachal Pradesh (Garg 1988)	Arunachal Pradesh (Garg 1988)
S. thomsonii	Western Himalayas (Kashmir)	Jammu and Kashmir (Garg 1988)	–
S. trichotoma	Nilgheeries (Nilgiri Hills)	Endemic to Western Ghats (Garg 1988)	–
S. zeylanica	Sri Lanka	–	–

[a] Including Afghanistan, Bangladesh, Bhutan, Myanmar, Nepal, Pakistan, and present India
– not reported

Table 11.2 Comparative floral features for field identification of some important species of *Swertia* occurring in Eastern Himalayas

Species	Altitude	Characteristic identifying floral characters	Flowering time	Fruiting time	Figures a–e[a]
S. chirata	1,500–2,500 m	Strongly tetramerous. Dark violet corolla. Branched appendages developed from middle of each petal	Peak in mid September to early October	Late October to November	
S. nervosa	1,500–2,500 m	Strongly tetramerous. Calyx lobe larger than corolla lobe. Upper part of each corolla lobe contains irregular black-striated region. A single pocket like gland present in each corolla lobe	Peak in mid September to early October	Late October to November	
S. dilatata	2,200–3,200 m	Strongly pentamerous. Calyx lobes triangular, corolla rotate, and greenish white. Dark purple band encircling two green spots at the base of each petal. Anther connate white, filaments light purple	September to October	October to November	

(continued)

Table 11.2 (continued)

Species	Altitude	Characteristic identifying floral characters	Flowering time	Fruiting time	Figures a–e[a]
S. paniculata	1,800–2,800 m	Strongly pentamerous. Calyx bell shaped, corolla rotate, creamish yellow with light purple band encircling a green zone at the base of the corolla	Peak in mid October to November	Late October to December	
S. bimaculata	1,600–2,500 m	Pentamerous to heptamerous. Corolla greenish white. Presence of many black spots on upper part and two conspicuous green spots at the middle of each corolla lobe	Peak in September to early October	Late October to November	

[a] Flowers of different *Swertia* species: a. *S. chirata* (Bar = 0.15 cm), b. *S. nervosa* (Bar = 0.20 cm), c. *S. dilatata* (Bar = 0.25 cm), d. *S. paniculata* (Bar = 0.20 cm), e. *S. bimaculata* (Bar = 0.30 cm)

laminae of stem leaves are usually ovate-elliptic, ovate, elliptic to elliptic-lanceolate, finely serrulate-ciliolate, and slightly wavy at the margin (toothed at basal part), acuminate at their apex, broadly cuneate at the base, glabrous, dark green adaxially, and glabrous, light green abaxially.

Inflorescence: Panicles of cymes, each raceme 12–25 flowered, glabrous, winged, and 2-bracteate.

Fruit: Capsule, loculicidal.

11.4.2 Swertia nervosa (Wall. Ex G. Don) C. B. Clarke (Fig. 11.1b)

Habit: An erect dwarf shrub, up to 80 cm in height. Roots are light yellow and fibrous. Stem is four-lineolate or four-winged, often narrowly winged on angles, and branched.

Leaf morphology: Basal leaves are sessile to shortly petiolate, broadly elliptic, and acute at their apices. Laminae of stem leaves are usually elliptic to oblong-lanceolate, finely serrulate-ciliolate, acute at the apex, cuneate at base, glabrous, dark green adaxially, and glabrous, light green abaxially.

Inflorescence: Panicles of cymes, each raceme 8–15 flowered, glabrous, four-winged, and 2-bracteate.

Fruit: Capsule, loculicidal.

11.4.3 Swertia dilatata C. B. Clarke (Fig. 11.1c)

Habit: An erect dwarf shrub, up to 80 cm in height. Roots light yellow in color and fibrous. Stem terete with ridges, branched, and glabrous.

Leaf morphology: Basal leaves are shortly petiolate, oblong-lanceolate, finely serrulate-ciliolate at their margins, and acute to shortly acuminate at the apex. Stem leaves are mostly sessile to subsessile, lamina usually oblong-lanceolate to sometimes linear-lanceolate, finely serrulate-ciliolate at the margin, acuminate at the apex, cuneate at base, glabrous, dark green adaxially, and light green with minute glandular dots abaxially.

Inflorescence: Panicles of cymes, each raceme 6–15 flowered, glabrous, terete, and 2-bracteate.

Fruit: Capsule, loculicidal.

11.4.4 Swertia paniculata Wall. (Fig. 11.1d)

Habit: An erect dwarf shrub, up to 50 cm high. Roots are light yellow and fibrous. Stem four-lineolate to four-winged, branched, and pubescent near nodes.

Fig. 11.1 Habit of different species of *Swertia*: **a** *S. chirata* (Bar = 3.5 cm), **b** *S. nervosa* (Bar = 4.0 cm), **c** *S. dilatata* (Bar = 1.0 cm), **d** *S. paniculata* (Bar = 4.0 cm), **e** *S. bimaculata* (Bar = 3.0 cm)

Leaf morphology: Basal leaves are shortly petiolate, oblong-lanceolate, finely serrulate-ciliolate at the margin, acute to shortly acuminate at the apex, and base margin ciliate. The laminae of stem leaves are usually elliptic-lanceolate to oblong-lanceolate, finely serrulate-ciliolate at their margins, acute to shortly acuminate at the apex, cuneate at the base, glabrous, dark green adaxially, and light green with minute glandular dots abaxially.

Inflorescence: Panicles of cymes, each raceme 5–10 flowered, sparsely pubescent, quadrangular, and two-bracteate.

Fruit: Capsule, loculicidal.

11.4.5 Swertia bimaculata (Sieb. & Zucc.) Hook. f. & Thomson ex C. B. Clarke (Fig. 11.1e)

Habit: An erect dwarf shrub, 30–150 cm high. Roots are light yellow and fibrous. Stem terete, subquadrangular to winged and branched.

Leaf morphology: Basal leaves are shortly petiolate, broadly ovate to ovate-elliptic, and acuminate at their apeces. The laminae of stem leaves oval (Arunachal population), ovate-elliptic to elliptic (Darjeeling population), finely serrulate-ciliolate at the margin, acuminate, shortly acuminate to acute (Darjeeling population) at the apex, cuneate to rounded at the base, glabrous, dark green adaxially, and glabrous, light green abaxially.

Inflorescence: Panicles of cymes that are lax, spreading, each raceme 5–6 flowered, glabrous, winged, and two-bracteate.

Fruit: Capsule, loculicidal.

11.5 Medicinal Importance of *Swertia*

Various species of *Swertia* are important medicinally. In the indigenous system of medicine, different species of *Swertia* find their use as crude drugs for the treatment of various ailments and for primary health care purposes. *Swertia* is colloquially known as Chirata, Chiretta, Chiraito, and Chirayat with *S. chirata* being the most important species, owing to its extensive medicinal properties. The medicinal usage of this species has been reported in the Indian pharmaceutical codex, the British and American Pharmacopoeias, and in different traditional systems of medicines such as the Ayurveda, Unani, and Siddha (Joshi and Dhawan 2005). *S. chirata* is used as an antipyretic, anthelmintic, antiperiodic, cathartic, analeptic and stomachic. It is also used in the treatment of asthma and leucorrhoea in the Ayurveda system. The herb serves to mitigate inflammation and for relaxing the uterus during pregnancy (Kirtikar and Basu 1984). Traditionally, the herb's root is considered to be the most potent part and serves as a drug used for fever, cough, and cold (Joshi and Dhawan 2007a). The aqueous and amarogentin rich crude extract of *S. chirata* was reported to have anticarcinogenic activity (Saha et al. 2004). Other species, such as *S. alternifolia*, *S. angustifolia*, *S. cordata*, *S. lurida*, *S. nervosa*, *S. paniculata*, and *S. purpurascens*, collected from the Himalayas, showed anti-hepatotoxic activity in rat hepatocyte cultures (Reen et al. 2001). Reports on the medicinal importance of various *Swertia* species are summarized in Table 11.3.

11.6 Present Status of *Swertia*

11.6.1 Status of Wild Populations

"Chirata" represents very valuable indigenous medicinal plants of the Eastern Himalayas. The destruction of natural habitats through climatic changes, as well as human activities, is the major threat toward loss of natural diversity. *S. chirata* is

Table 11.3 Species of *Swertia* reported from Eastern Himalayas that are used commercially and in traditional medicine

Species	Plant part used	Uses in traditional medicine	References
S. alata	Whole plant	Infusion of plants used as tonic and febrifuge	Bhatt et al. (2007)
S. angustifolia	Whole plant, roots	Cough and cold relief	Joshi (2008)
S. chirata	Whole plant, roots	Malarial fever, cough and cold, asthma, headache, liver disorders, anthelmintic, blood purifier, skin disease, bitter tonic for fever, indigestion, laxative, antidiarrhoetic, and bronchial asthma	Bhatt et al. (2007), Joshi (2008)
S. cordata	Whole plant	Febrifuge, anthelmintic, and bitter tonic. Used as substitute of *S. chirata*	Rahman et al. (1994), Bhatt et al. (2007)
S. hookeri	Whole plant	Treatment of microbial infections in man, in hypertension and as a mood enhancer	Ghosal et al. (1980)
S. multicaulis	Whole plant	Fever, cough and cold, anthelmintic, wound healing	Joshi (2008)
S. nervosa	Whole plant, roots	Malarial fever, skin disease, stomach problems. Used as substitute for *S. chirata*	Bhatt et al. (2007), Joshi (2008)
S. paniculata	Whole plant	Fever, malaria, used as substitute of *S. chirata*	Bhatt et al. (2007), Joshi (2008)
S. petiolata	Whole plant	Medicinal for gall disorder, antimalarial, and laxative	Bhatt et al. (2007), Khetwal et al. (1990)
S. racemosa	Whole plant	Fever, cough and cold, eczema, pimples, jaundice	Joshi (2008)
S. speciosa	Roots	Used as substitute for *S. chirata*	Bhatt et al. (2007)
S. tetragona	Whole plant	Not known but due to bitter taste used as substitute for *S. chirata*	Bhatt et al. (2007)
S. thomsonii	Whole plant	Decoction used in head ache and fever	Bhatt et al. (2007)

conventionally propagated by seed. Low seed viability and percentage of germination act as impediments in large-scale commercial plantations of this herb (Wang et al. 2009). The plant is on the list of the Natural Habitats of Prioritized Species in the Western and Eastern Himalayan region by the National Medicinal Plants Board, Government of India (Sastry and Chatterjee 2000). This priority plant has now been designated as being critically endangered, according to new criteria of the International Union for Conservation of Nature and Natural resources (IUCN), as a result of excessive human exploitation, both in terms of over-harvesting and loss of habitat (Anonymous 1997; Joshi and Dhawan 2007a). *S. chirata* is considered to be virtually endemic to the Himalayas (Samant et al. 1998). Balick et al. (1996) considered that extinction of even one population of the plant can cause loss of all its unique germplasm, phytochemicals and associated properties.

Since the entire plant of chirata is used for its medicinal properties, individuals are uprooted indiscriminately resulting in a decrease in the density of natural populations in the wild. Removal of entire plants before seed maturation also reduces the possibility of seed development for future regeneration (Sheldon et al. 1997). The raw materials for the herbal industry are collected consistently from the wild. This has lead to a decline in the number and, in some places, extinction of natural populations (Joshi and Dhawan 2007a). Chirata is cultivated only sparsely and negligible effort has gone into developing agrotechniques for the plant (Joshi and Dhawan 2005).

S. angustifolia is considered near endemic to Western Himalayas and has also become an endangered medicinal plant (Samant et al. 1998) due to adverse conditions, including habitat destruction and overgrazing that have affected the growth and survival of the species. Similar threats may also be applicable to other species of *Swertia* that are used commercially. In April 2002 at the Convention on Biological Diversity, reassessment, identification, documentation, and conservation was suggested as part of the Global Strategy for Plant Conservation. In order to conserve its genetic diversity, there is a need to maintain sufficiently large populations of *Swertia* in natural habitats. If appropriate conservation measures are not adopted, a major part of the genetic diversity of this genus may be lost in the near future.

11.6.2 Trade in Herbal and Pharmaceutical Industries

In the Indian raw drug markets, most of the medicinal plants are sourced entirely or largely from the wild, the remaining traded species being obtained from either cultivation or imports. *S. chirata* is one such prized species whose demand is increasing steadily in domestic and pharmaceutical industries within and outside India. Most of the collected plant materials come primarily from forests. However discrepancy is noted in trade and consumption of chirata in India. While estimated annual trade of the species suggests 500–1,000 MT during the year 2005–2006, official import of *chirata* was 97.4 MT (Ved and Goraya 2007) and annual demand in India was around 37.3 MT (Negi et al. 2011). Lack of proper identification of plants in trade and the high commercial demand of *S. chirata* has resulted in a trend of intentional or unintentional mixing of the authentic drug with some of its less effective substitutes. Allied species such as *Andrographis paniculata*, *Exacum pedunculatum*, *E. tetragonum*, and *Slevolgia orientalis* as well as *Swertia alata*, *S. angustifolia*, *S. bimaculata*, *S. ciliata*, *S. cordata*, *S. densifolia*, *S. elegans*, *S. lawii*, *S. minor*, *S. multiflora* and *S. paniculata* are used as substitutes or adulterants in the *S. chirata* trade (Anonymous 1982). Most of these adulterants/substitutes have inferior biological activity in comparison to *S. chirata*. Molecular profiling of the "Chirayat" complex has been reported recently using inter simple sequence repeat (ISSR) markers (Tamhankar et al. 2009). Studies on amplified fragment length polymorphism (AFLP)-based DNA markers were undertaken to generate markers for six species of *Swertia*, namely *S. alata*, *S. angustifolia*, *S. bimaculata*, *S. chirata*, *S. ciliata*, and *S. cordata*, that are most commonly used in the herbal industry. AFLP analysis can play a useful role in identification and authentication of drugs (Misra et al. 2010). Extensive chemical analysis of the wide array of phytochemicals must be studied in various species of *Swertia* in order to screen out the medicinally and commercially important species.

11.7 Conservational Strategies and Biotechnological Improvements

11.7.1 Conservation and Resource Management

Low seed viability, a poor germination rate, and a long gestation period have contributed toward the endangered status of *S. chirata*, together with extensive habitat loss and over-harvesting from the wild beyond a sustainable limit.

As the demand for chirata is extensive in the herbal and pharmaceutical industries, appropriate conservation measures are essential to maintain a supply of the herb. The development of conservation strategies for any threatened species require details of its occurrence, availability, and distribution in natural habitats. Assessments of natural populations are important for the development of in situ conservational strategies (Bhatt et al. 2007). Ex situ conservation methods need to be adopted for sustainable maintenance of various species of *Swertia*. Since, the Eastern Himalayan region harbors many species of *Swertia*, local community initiatives of in situ and ex situ conservation need to be encouraged. The cultivation of *Swertia* species by local inhabitants on their private land, as well as conservation action by different Himalayan Institutes and Botanical Gardens, can be adopted as conservation measures.

11.7.2 Biotechnological Approaches for Improvement and Conservation

Germplasm preservation through conventional approaches alone cannot guarantee re-establishment of this endangered plant in its natural habitat. If alternatives for production of the crude drug are not found soon, there may be a severe effect on the chirata trade.

Tissue culture and molecular techniques are useful tools that can complement conventional breeding in order to meet market demand. Biotechnological approaches show considerable potential in the management of valuable plant resources in a sustainable way. Plant tissue culture can be employed for conservation of commercially exploited medicinal plants. Micropropagation techniques, cell and organ culture are some of the efficient methods for multiplication and ex situ conservation of plant biodiversity from a minimum of available plant material. Such techniques can also serve as alternative means of secondary metabolite production (Joshi and Dhawan 2007a).

Genetic engineering is a potent tool for phytochemical production from simple, readily available, and inexpensive starting materials. Comparison of nontransformed and *Agrobacterium rhizogenes*-transformed root cultures have been studied with respect to growth and amarogentin content, following culture in various growth media by Keil et al. (2000). A 15-fold enhancement of amarogentin content in the medium was achieved by root permeabilization using the surfactant Tween 20. The generation of somatic hybrids by protoplast fusion may result in the introgression of nuclear and cytoplasmic DNA from one species to another. This approach may be used in the future to alter metabolite content through genetic engineering. Studies on somatic hybrids between *Bupleurum scorzonerifolium* and *S. mussotii* have been reported by Wang et al. (2011).

Several tissue culture studies in *S. bimaculata* and *S. chirata* have been reported, although there are limited reports on the culture of Eastern Himalayan *Swertia*. The reports are summarized in Table 11.4 and are categorized on the basis of the explants used for initiating cultures.

11.7.3 Chromosome Analysis

Chromosome studies play a vital role in the improvement of medicinally important species, since such assessments can aid determination of taxonomic relationships and information about past evolutionary events. They are also important for understanding the nature of plant variation, particularly from the level of population to genus. Differences in absolute chromosome size can indicate evolutionary changes in nuclear DNA (Govindarajan and Subramanian 1983). Very few investigations on the cytological analysis and documentation of chromosome morphology in Eastern Himalayan *Swertia* have been performed in the recent past,

Table 11.4 Tissue culture studies on different explants from species of *Swertia*

Species	Explant	Mode of regeneration	References
S. bimaculata	Leaf, stem, and stem with bud	Micropropagation	Long et al. (2009)
S. chirata	Root	Shoot organogenesis	Wawrosch et al. (1999)
S. chirata	Nodal explants	Shoot multiplication	Ahuja et al. (2003)
S. chirata	Nodal explants	Shoot multiplication	Chaudhuri et al. (2007)
S. chirata	Nodal explants	Shoot regeneration	Joshi and Dhawan (2007a)
S. chirata	Shoot tip explants from in vitro grown seedlings	Micropropagation	Balaraju et al. (2009)
S. chirata	Leaf explants	Shoot organogenesis	Chaudhuri et al. (2008)
S. chirata	Leaf explants from field grown plants	Shoot regeneration via somatic embryogenesis	Wang et al. (2009)
S. chirata	Leaf explants	Shoot regeneration via somatic embryogenesis	Jha et al. (2011)
S. chirata	Immature seeds	Indirect organogenesis	Chaudhuri et al. (2009)

and there are several species which have not yet been evaluated. Most of the reports date back almost two decades. Hence, reassessment of *Swertia* species at the chromosomal level is recommended in order to obtain complete information relating to somatic and gametic chromosomes. This information can contribute to the development of a chromosome data base of Eastern Himalayan *Swertia* species that can be used as a ready source of information for future studies. Preparation of chromosomal databases has been initiated in different species of *Swertia*, with reports of somatic chromosome numbers of *S. chirata* (Chaudhuri et al. 2007) and *S. bimaculata* (Dafadar and Jha 2012) from plants collected from the Eastern Himalayas. Existing reports on chromosome numbers of different species of *Swertia* are summarized in Table 11.5.

11.7.4 DNA Fingerprinting

Since *Swertia* occupies a range of different niches, there is a possibility of the existence of several ecotypes and/or chemotypes in nature, even in a single species such as *S. chirata* (Joshi and Dhawan 2005). Thus, identification of high yielding chemotypes is very important for extraction of the maximum concentration of the therapeutically significant secondary metabolites.

Molecular markers that are unique to a relevant plant and are stable under different environmental conditions, such as plant age, can act as a potent tool for identification. DNA markers are best suited for this purpose. In a study using ISSR markers, 98.7 % polymorphism was found among 19 genotypes of *Swertia* species

Table 11.5 Chromosome numbers of different Eastern Himalayan species of *Swertia*

Species	Chromosome number	
	Diploid	Haploid
S. alata	$2n = 24$ (Mehra and Gill 1968), $2n = 26$ (Khoshoo and Tandon 1963)	$n = 13$ (Vasudevan 1975; Khatoon and Ali 1993)
S. angustifolia	$2n = 24$ (Mehra and Gill 1968), $2n = 26$ (Khoshoo and Tandon 1963; Mallikarjuna 1985)	$n = 13$ (Vasudevan 1975)
S. beddomei	$2n = 26$ (Mallikarjuna et al. 1987)	–*
S. bimaculata	$2n = 18$ (Wada 1954), $2n = 24$ (Wada 1954), $2n = 26$ (Wada and Tanaka 1968; Shigenobu 1982; Dafadar and Jha 2012)	$n = 11$ (Roy et al. 1988), $n = 13$ (Sharma 1970; Dafadar and Jha 2012)
S. chirata	$2n = 20$ (Wada 1955), $2n = 24$ (Wada 1955), $2n = 26$ (Khoshoo and Tandon 1963; Chaudhuri et al. 2007)	$n = 13$ (Jha et al. Unpublished)
S. cordata	$2n = 26$ (Khoshoo and Tandon 1963; Mehra and Gill 1968)	$n = 13$ (Vasudevan 1975)
S. corymbosa	$2n = 26$ (Mallikarjuna 1985)	–
S. dilatata	$2n = 20$ (Khoshoo and Tandon 1963)	–
S. lurida	–	$n = 13$ (Mehra and Vasudevan 1972; Vasudevan 1975)
S. nervosa	$2n = 26$ (Sharma 1970)	–
S. paniculata	$2n = 16$ (Khoshoo and Tandon 1963)	$n = 8$ (Khatoon and Ali 1993; Vasudevan 1975)
S. petiolata	$2n = 26$ (Khoshoo and Tandon 1963)	$n = 13$ (Jee et al. 1989)
S. purpurascens	$2n = 20$ (Khoshoo and Tandon 1963)	–
S. speciosa	$2n = 26$ (Khoshoo and Tandon 1963)	$n = 13$ (Vasudevan 1975)
S. tetragona	$2n = 18$ (Khoshoo and Tandon 1963)	–
S. thomsonii	$2n = 26$ (Khoshoo and Tandon 1963)	–
S. trichotoma	$2n = 26$ (Mallikarjuna 1985)	–

* not reported

(13 of *S. chirata* and 2 each of *S. cordata*, *S. paniculata*, and *S. purpurascens*; Joshi and Dhawan 2007b). Species-specific markers were identified for six *Swertia* species (272, 19, 181, 131, 47, and 94 markers for *S. alata*, *S. angustifolia*, *S. bimaculata*, *S. chirata*, *S. ciliata* and *S. cordata*, respectively) in another study (Misra et al. 2010). Isozyme markers have also been used to differentiate between different species of *Swertia* (Verma and Kumar 2001).

Molecular markers are one of the most desirable tools for studies of genetic uniformity on micropropagated plantlets. ISSR markers were used to validate the genetic (clonal) fidelity of *S. chirata* plantlets derived from axillary multiplication in vitro in order to screen tissue culture-induced genetic variation (Joshi and Dhawan 2007b). DNA fingerprinting using the Random Amplified Polymorphic DNA (RAPD) technique, has also been applied to determine clonal fidelity and genetic stability in micropropagated *S. chirata* plantlets and their donor plants (Chaudhuri et al. 2008).

11.7.5 Biochemical Fingerprinting

Xanthones and bitter principles (secoiridoid glycosides) are widely distributed in plants of the Gentianaceae (Singh 2008). Xanthones (9H-xanthen-9-ones) are heterocyclic compounds with the framework of dibenzo-γ-pyrone (Pedro et al. 2002). Xanthones contain antidepressant and antitubercular activity. Some xanthones are also known to exhibit choleretic, diuretic, antimicrobial, antiviral, and cardiotonic actions (Peres et al. 2000). Bellidifolin, gentisin, norswertianolin, swerchirin, gentiacaulein and gentiakochianin are some of the xanthone compounds found in members of the Gentianaceae (Singh 2008). The iridoid glucosides found in Gentianaceae are mostly secoiridoids. The carbocyclic iridoids found in Gentian herbs are almost exclusively derivatives of loganin, which is the obligatory biosynthetic precursor of secoiridoids (Jensen and Schripsema 2002). Amarogentin (chirantin), amaroswerin, sweroside, and swertiamarin are secoiridoid glycosides commonly found in Gentianaceae. Amarogentin is the most bitter substance known. Gentianaceae plants are also known to contain gentianine, a bitter, crystalline monoterpene alkaloid, and gentiopicrin, a crystalline monoterpene gentipicroside (Singh 2008). The genus *Swertia* is a rich source of xanthone compounds and some species are known to contain secoiridoids and terpenoids (Chopra et al. 1956).

The microanalytical high pressure liquid chromatography (HPLC) technique for determination of swertiamarin in *Swertia* species, and High Performance Thin Layer Chromatography (HPTLC) detection of swertiamarin and amarogentin content in different samples of *S. chirata* and *S. cordata*, have been reported by Bhandari et al. (2006a, b) and Vishwakarma et al. (Vishwakarma et al. 2004). Gupta et al. (2011) determined ursolic acid by HPTLC in *S. alata*, *S. angustifolia* and *S. paniculata* collected from Uttarakhand, India. Three tetroxygenated xanthones have been reported to be present in *S. nervosa* collected from Mussoorie,

India by Bhatia et al. (2004). Xanthones in *S. chirata*, *S. paniculata*, and *S. speciosa* were determined by HPLC (Negi et al. 2009a, 2010a, b). Different concentrations of mineral elements can play different roles in human and plant health. Nine elements (Zn, Cu, Mn, Fe, Co, Na, K, Ca, and Li) have been analyzed in *S. chirata* and *S. speciosa* by atomic absorption spectrometry (Negi et al. 2009b, 2010c).

Biochemical fingerprinting is essential for screening elite populations within and between *Swertia* species. This approach can play a vital role in reducing intentional and unintentional mixing of different species in the commercial trade.

11.8 Conclusions

Sustainable management of medicinal plants plays a crucial role in the conservation of biodiversity. Conservation of medicinal plants must involve scientifically planned strategies for sustainable utilization and harvesting. Due to their high demand in the domestic as well as the international market, collectors face extensive competition leading to indiscriminate, unmanaged exploitation of different species of *Swertia*. This has resulted in a decrease in natural populations. Also, there are no established agrotechniques for promoting the cultivation of *Swertia*. This has necessitated proper resource assessment and establishment of sustainable harvesting techniques.

Resource assessment is essential to estimate the quantity of any plant that can be harvested without harming its natural resource base. This means estimating the number of plants that can be collected from a particular area on a sustainable basis. The North Eastern Himalayan region is a natural repository of several species of *Swertia*. Habitat conservation plays a primary role in maintaining the natural populations found in this region. The most widely accepted scientific techniques of biodiversity conservation involve in situ and ex situ approaches. In situ conservation of *Swertia* can be achieved through active participation and awareness of people living in and around the source of natural populations.

Botanical Gardens can play a key role in ex situ conservation of *Swertia*, especially where species face a serious threat of extinction. In India, a negligible number of Botanical Gardens have an active programme on conservation. Participation of different Himalayan Institutes and Botanical Gardens will be essential in germplasm conservation of *Swertia* species.

Biodiversity studies at the morphological and genetic levels are also essential to estimate the extent of variability within the existing germplasm of the genus. Considering climatic changes over the years, reassessment of species at the chromosomal level has become essential for identification of variation. There is a strong need to screen different chemotypes of chirata growing at different phytogeographical locations, for optimal resource utilization, causing minimal deterioration of natural populations. Identification of elite populations by extensive biochemical screening may reveal a significant role of species, other than *S. chirata*, in meeting high commercial demand. Correct resource management, along with the use of the latest biotechnological techniques for resource development, is essential to maintain the diverse populations and species of *Swertia* found in the Eastern Himalayas.

Acknowledgments T.B. Jha acknowledges financial support from the Council of Scientific and Industrial Research, New Delhi, India and the assistance of Dr. S. Panda, Dr. A. Ghorai, and S. Basu for collection of materials.

References

Ahuja A, Koul S, Koul BL, Verma NK, Kaul MK, Raina RK, Qazi GN (2003) Media composition for faster propagation of *Swertia chirayita*. WO 03/045132 AL. US Patent 7238527

Anonymous (1982) The wealth of India: raw materials. Publication and Information Directorate, CSIR, New Delhi

Anonymous (1997) Biodiversity conservation prioritization project, conservation assessment and management plan (CAMP) for endemic medicinal plants in India. Central Institute of Medicinal and Aromatic Plants, Lucknow

Balaraju K, Agastian P, Ignacimuthu S (2009) Micropropagation of *Swertia chirata* Buch.-Hams. ex Wall.: a critically endangered medicinal herb. Acta Physiol Plant 31:487–494

Balick MJ, Elinabotsky E, Laird SA (1996) Medicinal resources of the tropical forest. Columbia University Press, New York

Bhandari P, Gupta AP, Singh B, Kaul VK (2006a) HPTLC determination of swertiamarin and amarogentin in *Swertia* species from the Western Himalayas. J Planar Chromatogr 19:212–215

Bhandari P, Kumar N, Gupta AP, Singh B, Kaul VK (2006b) Micro-LC determination of swertiamarin in *Swertia* species and bacoside-A in *Bacopa monnieri*. Chromatographia 64:599–602

Bhatia A, Karan M, Vasisht K (2004) Xanthones from *Swertia nervosa* Wall. Indian J Pharm Sci 66:708–709

Bhatt A, Bisht AK, Rawal RS, Dhar U (2007) Assessment of status and biomass of *Swertia angustifolia*: a high value Himalayan medicinal plant. Afr J Plant Sci 1:1–6

Brahmachari G, Mondal S, Gangopadhyay A, Gorai D, Mukhopadhyay B, Saha S, Brahmachari AK (2004) *Swertia* (Gentianaceae): chemical and pharmacological aspects. Chem Biodivers 1:1627–1651

Chassot P, Nemomissa S, Yuan YM, Küpfer P (2001) High paraphyly of *Swertia* L. (Gentianaceae) in the *Gentianella*-lineage as revealed by nuclear and chloroplast DNA sequence variation. Plant Syst Evol 229:1–21

Chaudhuri RK, Pal A, Jha TB (2007) Production of genetically uniform plants from nodal explants of *Swertia chirata* Buch. Ham. ex Wall—an endangered medicinal herb. In Vitro Cell Dev Biol-Plant 43:467–472

Chaudhuri RK, Pal A, Jha TB (2008) Conservation of *Swertia chirata* through direct shoot multiplication from leaf explants. Plant Biotech Rep 2:213–218

Chaudhuri RK, Pal A, Jha TB (2009) Regeneration and characterization of *Swertia chirata* Buch.-Ham. ex Wall. plants from immature seed cultures. Sci Hortic 120:107–114

Chopra RN, Nayar SL, Chopra IC (1956) Glossary of Indian medicinal plants. CSIR, New Delhi

Clarke CB (1883) Gentianaceae. In: Hooker JD (ed) The Flora of British India, vol IV. L Reeve and Co., Ltd., London, pp 121–130

Cronquist A (1988) The evolution and classification of flowering plants. The New York Botanical Garden, Bronx

Dafadar A, Jha TB (2012) In vitro propagation and conservation of *Swertia bimaculata* Hook F. & Thoms. Indian J Biotech 11:295–299
Garg S (1987) Gentianaceae of the North West Himalaya (a revision). International Bioscience Monograph 17. Today and Tomorrow's Printers and Publishers, New Delhi
Garg S (1988) Indian Gentianaceae. A Check-List. Northern Book Centre, New Delhi
Ghosal S, Biswas K, Jaiswal D (1980) Xanthone and flavonol constituents of *Swertia hookeri*. Phytochemistry 19:123–126
Gilg E (1895) Gentianaceae. In: Engler A, Prantl K (eds) Die Natürlichen Pflanzenfamilien, vol 4. Engelmann, Leipzig, pp 50–108
Govindarajan T, Subramanian D (1983) Karyomorphological studies in South Indian Acanthaceae. Cytologia 48:491–504
Gupta M, Bisht D, Khatoon S, Srivastava S, Rawat AKS (2011) Determination of ursolic acid a biomarker in different *Swertia* species through high performance thin layer chromatography. Chin Med 2:121–124
Ho TN, Xue CY, Wang W (1994) The origin, dispersal and formation of the distribution pattern of *Swertia* L. (Gentianaceae). Acta Phytotaxon Sin 32:525–537
Jee V, Dhar U, Kachroo P (1989) Cytogeography of some endemic taxa of Kashmir Himalaya. Proc Indian Natl Sci Acad 55:177–184
Jensen SR, Schripsema J (2002) Chemotaxonomy and pharmacology of Gentianaceae. In: Struwe L, Albert VA (eds) Gentianaceae: systematics and natural history. Cambridge University Press, Cambridge, pp 573–626
Jha TB, Dafadar A, Chaudhuri RK (2011) Somatic embryogenesis in *Swertia chirata* Buch. Ham ex Wall-A multipotent medicinal herb. Asian J Biotechnol 3:186–193
Joshi K (2008) *Swertia* L. (Gentianaceae) in Nepal: ethnobotany and agenda for sustainable management. Ethnobot Leafl 12:1–6
Joshi P, Dhawan V (2005) *Swertia chirayita*—an overview. Curr Sci 89:635–640
Joshi P, Dhawan V (2007a) Axillary multiplication of *Swertia chirayita* (Roxb. Ex Fleming) H. Karst., a critically endangered medicinal herb of temperate Himalayas. In Vitro Cell Dev Biol-Plant 43:631–638
Joshi P, Dhawan V (2007b) Assessment of genetic fidelity of micropropagated *Swertia chirayita* plantlets by ISSR marker assay. Biol Plant 51:22–26
Judd WS, Campbell CS, Kellog EA, Stevens PF (1999) Plant systematics—a phylogenetic approach. Sinauer Associates Inc., Sunderland
Keil M, Hartle B, Guillaume A, Psiorz M (2000) Production of amarogentin in root cultures of *Swertia chirata*. Planta Med 66:452–457
Khatoon S, Ali SI (1993) Chromosome atlas of the angiosperms of Pakistan. Department of Botany, University of Karachi, Karachi
Khetwal KS, Joshi B, Bisht RS (1990) Tri- and tetraoxygenated xanthones from *Swertia petiolata*. Phytochem 29:1265–1267
Khoshoo TN, Tandon SR (1963) Cytological and pollination studies on some Himalayan species of *Swertia*. Caryologia 16:445–477
Kirtikar KR, Basu BD (1984) Indian medicinal plants. LM Basu Publishers, Dehradun
Long H, Hu X, Huang H (2009) Tissue culture of *Swertia bimaculata*. Chin Tradit Herb Drugs 3:004
Mallikarjuna MB (1985) Karyomorphological and cytotaxonomic studies in the family Gentianaceae. Ph.D. Thesis, Bangalore University
Mallikarjuna MB, Sheriff A, Krishnappa DG (1987) Chromosome number reports 97. Taxon 36:766–767
Mehra PN, Gill LS (1968) In IOPB chromosome number reports XVI. Taxon 17:199–204
Mehra PN, Vasudevan KN (1972) In IOPB chromosome number reports XXXVI. Taxon 21:333–346
Mészáros S, Höhn M (2002) Species diversity and advancement of *Swertia* (Gentianaceae): ecological and morphological correlates. Acta Bot Hung 44:317–334

Misra A, Shasany AK, Shukla AK, Darokar MP, Singh SC, Sundaresan V, Singh J, Bagchi GD, Jain SP, Saikia D, Khanuja SPS (2010) AFLP markers for identification of *Swertia* species (Gentianaceae). Genet Mol Res 9:1535–1544

Negi JS, Singh P, Pant GJ, Rawat MSM (2009a) Quantitative assessment of xanthone derivatives in *Swertia chirata* (Wall.) by RP-HPLC with UV detection. Med Plants Int J Phytomed Relat Ind 1:97–100

Negi JS, Singh P, Rawat MSM, Pant GJ (2009b) Study on trace elements in *Swertia chirayita* (Roxb.) H. Karsten. Biol Trace Elements Res 133:350–356

Negi JS, Singh P, Pant GJ, Rawat MSM (2010a) RP-HPLC analysis and antidiabetic activity of *Swertia paniculata*. Nat Prod Comm 5:907–910

Negi JS, Singh P, Rawat MSM, Pant GJ (2010b) Qualitative and quantitative determination of major xanthones in *Swertia speciosa* by high performance liquid chromatography. Med Plants Int J Phytomed Relat Ind 2:45–50

Negi JS, Singh P, Pant GJ, Rawat MSM (2010c) Study on the variations of mineral elements in *Swertia speciosa* (G. Don). Biol Trace Elements Res 138:300–306

Negi JS, Singh P, Rawat B (2011) Chemical constituents and biological importance of *Swertia*: a review. Curr Res Chem 3:1–15

Pandey DK, Basu S, Jha TB (2012) Screening of different East Himalayan species and populations of *Swertia* L. based on exomorphology and mangiferin content. Asian Pac J Tropical Biomed 2:1450–1456

Pedro M, Cerqueira F, Sousa ME, Nascimento MSJ, Pinto M (2002) Xanthones as inhibitors of growth of human cancer cell lines and their effects on the proliferation of human lymphocytes in vitro. Bioorganic Med Chem 10:3725–3730

Peres V, Nagem TJ, de Oliveira FF (2000) Tetraoxygenated naturally occurring xanthones. Phytochem 55:683–710

Pradhan BK, Badola HK (2010) Chemical stimulation of seed germination in ex situ produced seeds in *Swertia chirayita*, a critically endangered medicinal herb. Res J Seed Sci 3:139–149

Rahman AU, Pervin A, Feroz M, Choudhary MI (1994) Phytochemical studies on *Swertia cordata*. J Nat Prod 57:134–137

Reen RK, Karan M, Singh K, Karan V, Johri RK, Singh J (2001) Screening of various *Swertia* species extracts in primary monolayer cultures of rat hepatocytes against carbon tetrachloride- and paracetamol-induced toxicity. J Ethnopharmacol 75:239–247

Rijal DP (2009) Taxonomic study of some medicinally important species of *Swertia* L. (Gentianaceae) in Nepal. Bot Orient: J Plant Sci 6:18–24

Roy SC, Ghosh S, Chatterjee A (1988) A cytological survey of eastern Himalayan plants II. Cell Chromosome Res 11:93–97

Saha P, Mandal S, Das A, Das PC, Das S (2004) Evaluation of the anticarcinogenic activity of Swertia chirata Buch. Ham, an Indian medicinal plant on DMBA-induced mouse skin carcinogenesis model. Phytother Res 18:373–378

Samant SS, Dhar U, Palni LMS (1998) Medicinal plants of Himalaya: diversity, distribution and potential values. Gyanodaya Prakashan, Nainital

Sastry ARK, Chatterjee S (2000) Prioritization of medicinal plants of India. In: Singh S, Sastry ARK, Mehta R, Uppal V (eds). Setting biodiversity conservation priorities for India. World Wide Fund for Nature, India, p 707

Sharma AK (1970) Annual report, 1967–1968. Res Bull Univ Calcutta Cytogenetics Lab 2:1–50

Sheldon JW, Balick MJ, Laris SA (1997) Medicinal plants can utilization and conservation coexist? Adv Econ Bot 12:1–104

Shigenobu Y (1982) Cytological relationship between *Swertia bimaculata* and *S. swertopsis*. J Jpn Bot 57:353–357

Singh A (2008) Phytochemicals of Gentianaceae: a review of pharmacological properties. Int J Pharm Sci Nanotechnol 1:33–36

Struwe L, Albert VA, Bremer B (1994) Cladistics and family level classification of the *Gentianales*. Cladistics 10:175–206

Struwe L, Hagen KB, Kadereit JW, Klackenberg J, Nilsson JS, Thiv M, Albert VA (2002) Systematics, character evolution, and biogeography of Gentianaceae, including a new tribal and subtribal classification. In: Struwe L, Albert VA (eds) Gentianaceae: systematics and natural history. Cambridge University Press, Cambridge, pp 21–309

Takhtajan AL (1978) The floristic regions of the world. Soviet Sciences Press, Moscow

Tamhankar S, Ghate V, Raut A, Rajput B (2009) Molecular profiling of "Chirayat" complex using inter simple sequence repeat (ISSR) markers. Planta Med 75:1266–1270

The International Centre for Integrated Mountain Development. http://www.icimod.org

Vasudevan KN (1975) Contribution to the cytotaxonomy and cytogeography of the flora of the western Himalayas (with an attempt to compare it with the flora of the Alps). Ber Schweiz Bot Ges 85:210–252

Ved DK, Goraya GS (2007) Demand and supply of medicinal plants in India. NMPB (National Medicinal Plants Board), New Delhi & FRLHT, Bangalore, India

Verma NK, Kumar A (2001) Isozyme polymorphism and genetic diversity among *Swertia* species-endangered medicinal plants of Himalayas. Indian J Plant Gen Res 14:74–77

Vishwakarma SL, Baqul MS, Ranjani M, Goyal RK (2004) A sensitive HPTLC method for estimation of swertiamarin in *Enicostemma littorale* blume, *Swertia chirata* (wall) clarke and in formulations containing *E. littorale*. J Planar Chromatogr 17:128–131

von Hagen KB, Kadereit JW (2001) The phylogeny of *Gentianella* (Gentianaceae) and its colonization of the southern hemisphere as revealed by nuclear and chloroplast DNA sequence variation. Org Divers Evol 1:61–79

Wada Z (1954) Cytological studies in some species of *Swertia*. Jpn J Genet 29:180

Wada Z (1955) Cytological studies of four species of *Swertia* and one species of *Halenia*. Jpn J Genet 30:191

Wada Z, Tanaka N (1968) Cytological studies in Gentianaceae (V). Karyotype analysis of some species with evolutionary view. In: Proceedings of 12th International Congress of Genetics, vol 1, Tokyo, p 186

Wang L, An L, Hu Y, Wei L, Li Y (2009) Influence of phytohormones and medium on the shoot regeneration from leaf of *Swertia chirata* Buch.-Ham. Ex wall. in vitro. Afr J Biotechnol 8:2513–2517

Wang J, Zhao C, Liu C, Xia G, Xiang F (2011) Introgression of *Swertia mussotii* gene into *Bupleurum scorzonerifolium* via somatic hybridization. BMC Plant Biol 11:71

Wawrosch C, Maskay N, Kopp B (1999) Micropropagation of the threatened Nepalese medicinal plant *Swertia chirata* Buch.-Ham. ex. Wall. Plant Cell Rep 18:997–1001

Xue CY, Ho TN, Li DZ (2007) Embryology of *Swertia* (Gentianaceae) relative to taxonomy. Bot J Linn Soc 155:383–400

Chapter 12
The Role of Arbuscular Mycorrhiza in the Growth and Development of Plants in the Family Gentianaceae

Zuzana Sýkorová

Abstract Arbuscular mycorrhiza (AM) is an ancient symbiosis between land plants and those fungi from the Phylum Glomeromycota. It is estimated to occur in a large majority of plant species, including plants from the family Gentianaceae. Arbuscular mycorrhizal fungi (AMF) colonize plant roots and in exchange for carbohydrates, they contribute to plant mineral nutrient uptake, increased water absorption, tolerance to pathogens, drought, high soil temperatures, toxic heavy metals, extremes in pH and transplant shock. AMF were also reported to have a decisive influence on plant diversity and community productivity. A large plant–fungal network may be established, because more than one plant can be colonized by the same fungal "individual" and each plant may host several fungal species. Non-photosynthetic plants, known as mycoheterotrophs, may parasitize their green neighboring plants by taking up carbohydrates from the shared fungal network. This chapter summarizes research on mycorrhiza and gentians published to date. In brief, all gentians investigated were colonized by AMF and this colonization always had a specific morphology. AMF may play an important role in the life cycle of gentians. For example, several lineages of achlorophyllous mycoheterotrophic gentians showed greatest host specificity known so far in AM. The role of AMF in the growth and development of Gentianaceae is discussed in detail.

12.1 Arbuscular Mycorrhiza: General Definition, Life Cycle and Morphology

Arbuscular mycorrhiza (AM) is an ancient symbiosis (Redecker et al. 2000) between the large majority of land plants and fungi from the Phylum Glomeromycota (Schüßler et al. 2001). Arbuscular mycorrhizal fungi (AMF) colonize

Z. Sýkorová (✉)
Department of Mycorrhizal Symbioses, Institute of Botany, Academy of Sciences of the Czech Republic, CZ-252 43 Průhonice, Czech Republic
e-mail: Zuzana.sykorova@ibot.cas.cz

J. J. Rybczyński et al. (eds.), *The Gentianaceae - Volume 1: Characterization and Ecology*, DOI: 10.1007/978-3-642-54010-3_12,

plant roots, produce an extensive network of extraradical mycelium in the soil and contribute to the mineral nutrient uptake of the hosts in exchange for carbohydrates (Smith and Read 2008). Such fungi are present in most terrestrial ecosystems and crop production systems, and mediate plant root and soil interactions. The roots are colonized regularly by AMF and in most habitats it is almost impossible to find non-mycorrhizal individuals. AMF are obligate symbionts and cannot grow without their plant hosts.

AMF spores (diameter ca. 40–800 μm) are commonly present in soils. They possess layered walls and contain several hundreds to thousands of nuclei. Under favorable environmental conditions, the spores germinate and the hyphae colonize susceptible roots. Signaling molecules and genes are involved in this process (Parniske 2008). After the recognition of the host plant root, the hyphae form apressoria on the root epidermis and penetrate the cell lumen, which is followed by proliferation of the intraradical hyphae in the upper, middle and inner cell layers of the root cortex (Genre and Bonfante 2010). However, the fungus never enters meristems or vascular cylinders. Although the fungal hyphae may grow inside the host cells, they always remain surrounded by the intact plasma membrane of the plant host. In the deeper cortical layers, the fungal hyphae may form arbuscules, highly branched tree-like structures, in the plant cells. A plant periarbuscular membrane surrounds the thin fungal branches of the arbuscules forming an extremely large contact surface between the two symbionts. Here the nutrients are transferred between the plant and the fungus, mediated through a series of specific proteins (Parniske 2008). Despite their central role in nutrient exchange, arbuscules have a relatively short life, collapsing and degrading after a few days or weeks, leaving the host plant cell intact. Some AMF taxa form also intra or intercellular vesicles in the host plant roots (Schüßler and Walker 2010). Vesicles are lipid-filled storage structures with an ovoid shape. Once fungal colonization is established in the root, extensive growth of the extraradical mycelium (ERM) begins. Its finely branched fans are well adapted to the exploration of soil pores, mineral nutrient uptake and association with soil particles, and stabilization of soil aggregates. The ERM forms a complex network, which can link roots of plants of the same or different species (Barto et al. 2011). The life cycle of the AMF is completed by the formation of spores or sporocarps on the ERM or, rarely, inside the roots. The AMF form a coenocytic mycelium lacking hyphal septa.

Two types of AM symbiosis have been defined according to the morphology of the intraradical colonization. They were named after the plants in which they were first described, notably the *Arum* and the *Paris* type (Gallaud 1905). In the *Arum* type, the AMF produce long intercellular hyphae whose branches form typical arbuscules inside the cortical cells. The *Paris* type (Figs. 12.1 and 12.2) is characterized by predominantly intracellular hyphal growth and the formation of extensive intracellular hyphal swellings, coils and/or arbusculate coils in the root cortex (Smith and Smith 1997; Dickson et al. 2007). In addition to arbuscules, coils are also believed to participate in nutrient transfer between the symbionts. Both arbuscules and coils increase the contact area between the fungus and the plant cell. The two AM types differ not only in their morphology, but also in the kinetics of

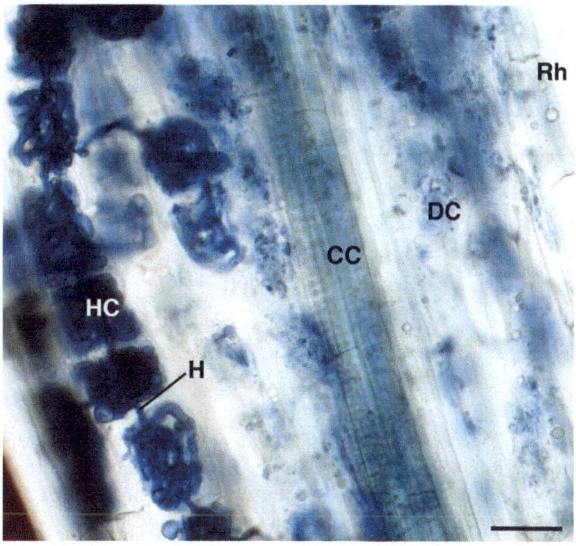

Fig. 12.1 Root of *Gentiana acaulis* collected in Swiss Alps, stained with Trypan blue and observed under a light microscope. Fungal structures were stained in *dark blue* and showed a *Paris* type of morphology of the arbuscular mycorrhiza. *HC*—intracellular hyphal coils, *H*—hypha growing directly from cell to cell, *DC*—degraded intracellular hyphal coils. *Rh*—rhizodermis, *CC*—central cylinder. Bar = 50 μm

root colonization. The development of *Paris* type colonization within the root is much slower than in the *Arum* type (Azcon-Aguilar et al. 1994). The type of AM colonization seems to be determined largely by the host plant genotype, as the same AMF species known to form the *Arum* type in some host plant species produces the *Paris* type in other hosts (Sýkorová et al. 2003; Ahulu et al. 2006).

12.2 Arbuscular Mycorrhiza: Its Beneficial Features

The symbiosis between AMF and autotrophic plants is regarded as mutualistic. With the exception of a few achlorophyllous species, AM host plants are autotrophic and, although normally colonized by mycorrhizal fungi in the field, they are usually capable of satisfactory growth in the absence of mycorrhizal colonization, provided that mineral nutrient supplies are adequate. In contrast, the glomeromycotan fungi are obligate symbionts, as their extraradical hyphae are unable to take up carbohydrates. Thus, they depend on photosynthates supplied by the plant and utilize a considerable proportion of its assimilated carbon (probably ca. 20 % of net photosynthate; Bago et al. 2000). It is predicted that nutrient transfer occurs at the arbuscule or the hyphal coil/root cortical cell interface (Parniske 2008; Helber et al. 2011). Hyphae from colonized roots extend into the soil and aid root hairs in absorbing water and mineral nutrients. Next to the increasing mineral nutrients and water absorption, AM associations have been shown to benefit the plants by root health and longevity (resistance to pathogens

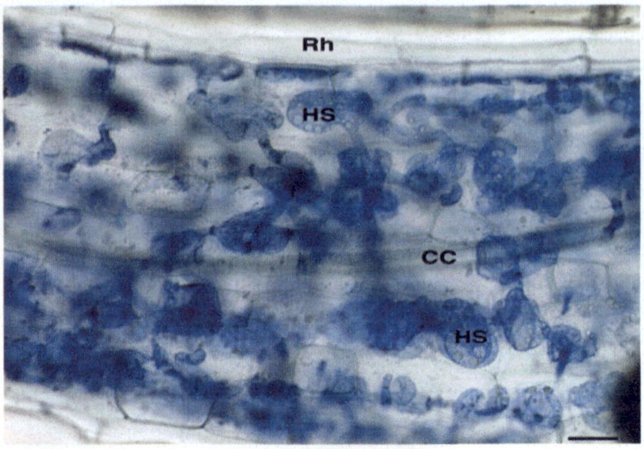

Fig. 12.2 Root of *Gentiana verna* collected in Swiss Alps, stained with Trypan blue and observed under a light microscope. Fungal structures were stained in *dark blue* and showed a *Paris* type of morphology of the arbuscular mycorrhiza. *HS*—intracellular hyphal swellings. *Rh*—rhizodermis, *CC*—central cylinder. Bar = 50 μm

and insect herbivores), tolerance to drought, high soil temperatures, toxic heavy metals, extremes in pH and transplant shock (Smith and Read 2008).

Many studies have shown that mycorrhizal colonization is followed by a considerable stimulation of plant growth. Phosphorus (P) is a major limiting nutrient in many ecosystems and enhanced phosphorus supply by AMF plays a key role in plant growth and the functioning of ecosystems (Klironomos et al. 2000; Lekberg and Koide 2005). The beneficial effects of AM are most apparent under conditions of limited nutrient availability, with the amount of root colonization and its benefits typically decreasing when nutrients are abundant. However, plants respond not always positively to colonization in terms of growth or P content; there is a range of plant growth responses from strongly positive, through neutral to negative (Smith et al. 2009). Interestingly, recent research has demonstrated that AM fungi play a significant role in plant P uptake, regardless of whether the plant responds positively to colonization in terms of growth or P content. In such plants, direct P uptake by the roots is reduced and the fungi deliver P as well as use C. Poor plant growth may be the result of P deficiency because AM fungi reduce the amount of P taken up directly by roots, but the AM uptake of P does not compensate for the reduction (Smith and Smith 2012). Moreover, it was demonstrated that the impacts of mycorrhizal association also differ according to the plant species and fungus involved (Munkvold et al. 2004). Recent investigations showed that partners maintain a fair, two-way transfer of resources; plants can detect, discriminate, and reward the best fungal partners with more carbohydrates. In turn, their fungal partners enforce cooperation by increasing phosphorus transfer only to those roots providing more carbohydrates (Kiers et al. 2011).

Additionally, hyphae of mycorrhizal fungi are important stabilizing agents in the soil. In many soils, the roots and hyphae, together with other organic components, play a major role in stabilization soil aggregates, erosion control and water management (Rillig and Mummey 2006). Because a single fungus can associate simultaneously with multiple unrelated host plants, this allows for the formation of mycorrhizal networks, linking plants of the same or different species (Barto et al. 2011). The below-ground diversity of AMF is an important factor contributing to the maintenance of plant biodiversity and to ecosystem functioning (van der Heijden et al. 1998). Thus, a healthy AM community with intact soil structure is of high importance for agriculture, as it can, to a certain extent, alleviate the necessity for high input of agrochemicals in agricultural practice (Gianinazzi and Vosátka 2004).

12.3 Arbuscular Mycorrhiza in the Family Gentianaceae

12.3.1 History of Mycorrhizal Research in Gentians

Neumann (1934) studied mycorrhiza of several species of Gentianaceae and linked them to the mycorrhiza of Orchidaceae, as their morphology was superficially similar to orchid mycorrhizas. Intracellular coils occurred within the roots of both these groups. Later, Jacquelinet-Jeanmougin and Gianinazzi-Pearson (1983) investigated the identity of mycorrhizal fungi in the Gentianaceae by inoculating *Gentiana lutea* with fungi from orchid, ericoid, and AMs. The results clearly demonstrated that mycorrhiza in this plant species is formed by AM fungi, but that it has an unusual morphology. The infection spread directly from cell to cell with a single hypha penetrating the cell wall, and intercellular hyphae were never observed. This mycorrhizal structure is typical for the *Paris* type of AM. Demuth and Weber (1990) showed in their experiments that roots of several flowering plants, such as *Plantago lanceolata* and *Cirsium vulgare*, cultivated under axenic conditions can be inoculated successfully with AMF from gentians (*Gentiana acaulis, G. purpurea*). Colonization showed structures typical for the *Arum* type of AM. In addition, the authors used the infected roots of the non-gentians as inoculum for several species of gentians. In all gentians, all AM fungi formed structures typical for the *Paris* type.

12.3.2 Life Cycle of Arbuscular Mycorrhiza in Gentians

As already mentioned, the Gentianaceae have a specific morphology of AM, which is known as the *Paris* type. Jacquelinet-Jeanmougin et al. (1987) described the AM colonization in *G. lutea*; intracellular hyphae in the outer layers of cells crossed these cells linearly without branching and did not spread laterally into the root tissue. Demuth et al. (1991) and Demuth and Weber (1990) described characteristic "S" shaped hyphae in rhizodermal cells prior to penetration of the cortex in several gentian species. In contrast, when the fungus reached the large parenchymal cortical cells, the infection spread intensely from cell to cell, and as many as 40 % of parenchymal cells in any section were colonized (Jacquelinet-Jeanmougin et al. 1987). In this tissue, two intracellular hyphal formations were distinguished in the light microscope, namely coils (spirals of thick hyphae, Fig. 12.1) and arbuscules (highly branched hyphae of different diameters). Hyphae never developed in the intercellular spaces of the roots. Demuth et al. (1991) suggested that, for this reason, the spread of the fungus is limited to small areas, primarily along the long axis of the root. Large areas of the root cortex may be colonized by repeated infections of the external infecting hyphae. However, these results were not in agreement with the results of Kühn and Weber (1986), who described some intercellular hyphae in roots of *Gentiana asclepiadea.*

Hyphal walls in coils become thinner (0.1–0.2 μm; Jacquelinet-Jeanmougin et al. 1987). These coiled hyphae contain cytoplasm, nuclei, mitochondria, glycogen particles, and numerous small vacuoles, which often enclose alkaline phosphatase and osmiophilic granules described as polyphosphate by many authors. As the fungus continues to develop in cortical host cells, the arbuscules can be observed, consisting of many fine hyphal branches arising by repeated division from the hyphal coil. The diameter of the smallest branches is about 0.4 μm, with the fungal wall being only 0.02 μm in thickness (Jacquelinet-Jeanmougin et al. 1987). The decrease in wall thickness is accompanied by an important simplification in its texture, composition and molecular organization, which results in an exceptionally large surface of contact and a greatly reduced apoplast between the AM fungus and the host protoplast. Hyphae fill up the host cell, which completely loses its initial aspect. Very little remains from the vacuole, and cytoplasm scattered among the fungal branches is rich in organelles, especially mitochondria and plastids, endoplasmatic reticulum and dictyosomes. These features indicate enhanced metabolic activity in the host cell following fungal invasion and is particularly favorable to solute movement between the symbionts. It contrasts with the weak host reaction observed in the outer cell layers of the root.

The first stages of fungal degradation, both in outer root cells and in parenchymal cortical cells, are characterized by deformation of the fungal wall and change in the hyphal contents, including accumulation of lipids. Degradation of arbuscules begins in the smallest branches and continues to the arbuscule trunk. The metabolic activity of the host cell appears greatly enhanced during the early stages of fungal senescence. The host cytoplasm contains numerous mitochondria, plastids, dictyosomes and an extensive endoplasmic reticulum. As fungal senescence in a cell advances, all hyphae become empty and the collapsed fungal walls clump together (Fig. 12.1). Again, changes in host cells occur, which result in uninfected cells with large vacuoles and a thin layer of peripheral cytoplasm (Jacquelinet-Jeanmougin et al. 1987).

12.3.3 Factors Shaping the Morphology of Mycorrhiza in Gentians

Interestingly, morphological differences in AM between some gentian–fungus combinations can be also under the control of the fungus. The roots of *Gentiana verna* collected in two different altitudes in the Savoy mountains in France were studied by Knöbel and Weber (1988). During secondary growth, the cortical cells died and colonization was restricted to thin lateral roots. The occurrence of root hairs was very low. Mycorrhizal colonization was characterized by the presence of intracellular loops, coils or swellings. Moreover, composite arbuscules (more arbuscules in one cell) and thick-walled vesicles were formed; the coiled hyphae became swollen and filled the cortical cells of *G. verna* (Fig. 12.2). In contrast to the vesicles, which had walls thickened with their increasing age, the wall of hyphal swellings was very thin. Hyphal swellings were found mainly in roots collected from locations of high altitude (2,000 m), whereas with advanced development of vegetation, intercellular hyphae and vesicles were also found at a lower altitude (1,400 m). It is possible that these morphological differences are produced by different endophytes at different locations (Jacquelinet-Jeanmougin and Gianinazzi-Pearson 1983). Differences in the morphology of arbuscules between various endophyte–host combinations were also observed by Demuth et al. (1991). Some arbuscules were larger and had an oblong-ovate profile, while others were smaller with a rounded profile. It also depended on the species of the fungus, if the coils and loops were loosely or tightly coiled and whether the mycorrhiza spread laterally. Kühn and Weber (1986) studied subterranean organs of *G. asclepiadea* in plants of different biotopes, from humid moorland and dry woodland. The intercellular hyphae were present only in plants from humid moorland as the second form in addition to the intracellular hyphae.

Gay et al. (1982) investigated the density of mycorrhizae of some biennial species of gentians (such as *Gentianella amarella*). They found marked seasonal variation with the highest values in the autumn of the first year and in the summer of the second year at the period of maximum growth and maximum accumulation of nutrients. These authors reported their observation on the morphology of mycorrhizae. The external hyphae penetrated the root at numerous places, giving rise to many small, dense infection units, which sometimes fused together. The colonization was sometimes so dense that it filled nearly all the cortical cells. However, the post-flowering mycorrhizal densities did not increase and widespread lysis of the mycorrhizal structures occurred.

Knöbel and Weber (1988) studied roots of an achlorophyllous gentian, *Voyria truncata*, from the Savoy mountains. The roots retained their cortex and, therefore, mycorrhizal colonization during secondary root growth. The authors proposed that this fact could be an essential adaptation of the plant to its special life form. Initially, the endophyte of *V. truncata* formed hyphal swellings with compositae arbuscules in the root tips, while later, development of arbuscules at lateral branches on hyphae coils decreased, so that coils and some vesicles were found in old roots. Since this plant does not contain chlorophyll, all carbohydrates as well as mineral nutrients must be delivered by the symbiotic fungus. This symbiotic relationship is called mycoheterotrophy. In principle, the plants hook themselves into mycorrhizal networks and benefit from carbon that they receive via the photosynthetic activity of interlinked green plants (Merckx et al. 2009). Several authors use the term "epiparasitism" to describe this kind of mycoheterotrophy (Bidartondo et al. 2002). Imhof (1997) studied the same plant species, but did not observe any arbuscules in its roots, although, hyphal bridges and direct root connections with roots of neighboring living or dead plants were common. Cortical tissues often enveloped attached roots. This direct contact accelerated the decay of a neighboring plant root. Less well understood and described within the family are the partially mycoheterotrophic and green members of the Gentianaceae. The small North American genera, namely *Bartonia* and *Obolaria,* are green, but have leaves reduced to scales or foliose bracts. The natural abundant stable isotope profiles strongly suggest that both species are partially mycoheterotrophic (Cameron and Bolin 2010).

12.3.4 Diversity and Host Specificity of AMF in Gentians

The diversity of AMF can be evaluated either using microscopy of spore morphology or molecular methods. The production of spores is highly dependent on environmental conditions, the physiological status and life strategy of the particular mycorrhizal fungus. In contrast, molecular methods allow identification of the symbiotic community currently colonizing the roots of an individual plant at any given time. The majority of recent molecular studies have used PCR with AMF-specific primers for nuclear-encoded ribosomal RNA genes (rDNA; Krüger et al. 2009), cloning and sequencing. However, identification of AMF at the species or even isolate level is complicated by the heterogeneity of rDNA within glomeromycotan spores and isolates. Several authors have shown that different variants of rDNA genes coexist within single AMF spores (Sanders et al. 1995; Jansa et al. 2002). Hence, a conservative approach in the evaluation of phylogenetic analyses is advisable, for example, by defining a well-supported monophyletic sequence cluster as an AMF phylotype (the molecular equivalent for species).

On the basis of the morphological features and, in some cases, also DNA sequences of their spores, only about 200 species of AMF have been described globally so far (Schüßler and Walker 2010). This small number of species was originally thought to colonize the majority of higher plants, and under laboratory conditions, most plant species can be colonized by most AMF species. As a consequence, their host specificity or preference was thought to be very low (Smith and Read 2008). However, molecular studies of AMF field communities from the last decade (Öpik et al. 2008; Wubet et al. 2009) revealed numerous previously unknown phylotypes. In several cases, significant differences were observed between the AMF communities inhabiting roots of different host plant species in the same habitat, mainly in grasslands (e.g., Helgason et al. 2002; Gollotte et al. 2004). In contrast, an apparent lack of host specificity was reported by other authors (Santos et al. 2006; Öpik et al. 2009).

Achlorophyllous mycoheterotrophic members of the Gentianaceae were among the plants showing the strongest host specificity known so far in AM (Bidartondo et al. 2002). These authors investigated the AMF communities of five *Voyria* and one *Voyriella* species, and their neighboring green plants, using molecular techniques. They discovered that all species are highly specialized to a few narrow lineages within a clade of *Glomus* (*Glomus*-group A as defined by Schwarzott et al. 2001). These results were confirmed by Franke et al. (2006), who analyzed an AMF community of *Seabaea oligantha*, a mycoheterotrophic gentian from Africa, and by Merckx et al. (2010) who investigated AMF community of *Voyria aphylla* from French Guiana. Such specificity is the characteristic of epiparasites. However, Merckx et al. (2012) analyzed mycoheterotrophic plants from nine families and found that while some of them were growing with only a single phylotype, many species are able to associate with a wide range of AMF. Interestingly, mycoheterotrophs from different plant lineages and different geographical regions tend to be dependent on lineages of AMF that are phylogenetically related.

There has been just a single molecular investigation of AMF diversity in the roots of green gentians, *G. acaulis* and *G. verna*. Despite similar morphology of their mycorrhizae, they did not show a similar restriction to such a narrow clade of fungal symbionts like the mycoheterotrophic gentians. These species occur together in Swiss upper montane meadows (Figs. 12.1 and 12.2), and harbored AMF communities significantly different from each other, but comprising multiple phylotypes from different fungal lineages, some of which were also found in neighboring non-gentians (Sýkorová et al. 2007). Hence, they showed host preference but not host specificity as strictly as defined earlier. In this respect, *G. acaulis* and *G. verna* are more similar to other green plants from other families than to their mycoheterotrophic relatives.

12.3.5 Cultivation of Gentians and Inoculation with AMF

Inoculation of AMF in general is used increasingly in commercial applications as a biofertilization tool for agriculture and horticulture (Gianinazzi and Vosátka 2004). Promising areas for these large-scale inoculations are the reclamation of anthropogenically disturbed ecosystems, horticulture, and sustainable plant production. The purpose is to increase crop yield, accelerate the process of soil revitalization and plant establishment, increase the soil inoculum potential, and reduce the input of industrial fertilizers and cultivation costs. A meta-analysis showed that plant biomass production and P uptake in the field are usually correlated positively with AMF root colonization, which can even be improved by inoculation (Lekberg and Koide 2005).

Under glasshouse as well as field conditions, plants can be inoculated either with AMF spores, ERM, or fragments of colonized roots, or they can be planted into a network of ERM established by mycorrhizal individuals of the same or different plant species. Although AMF inoculations have been used in many field experiments and their results in terms of plant growth response, for example, have been attributed to the inoculation treatments, it has been impossible so far to detect and trace inoculated AMF isolates in roots under field conditions and thus prove the inoculation efficiency. The AMF inoculants are generally applied to soils already containing a resident AMF community and, using only microscopy, the introduced strains are often not recognizable from the native ones. However, several recent molecular studies verified the success and effectiveness of inoculated AMF in the field, where conspecific native strains also occurred (Pellegrino et al. 2012; Sýkorová et al. 2012). Nevertheless, the possible exclusion of native AM fungi from colonizing roots raises concerns about the bioinvasion potential of non-native inoculants (Schwartz et al. 2006). Further studies are necessary of possible biodiversity losses and homogenization as a result of anthropogenic translocations of AMF inocula between biogeographic regions (Rosendahl et al. 2009).

There have been several reports on the cultivation of biennial as well as perennial gentians and their inoculation with AMF under glasshouse conditions. Such reports showed that the presence of AMF is not necessary for the germination of gentian seeds (Gay et al. 1982; Demuth et al. 1989). However, most plants easily develop a mycorrhizal association when inoculated (Heymons et al. 1986; Demuth et al. 1991). When seed is sown in the field, these plant species become infected within about two weeks after seed germination. These findings, together with observations of field-collected root samples, show that mycorrhiza is generally the rule rather than the exception in gentians in nature. Demuth et al. (1991) cultivated *G. asclepiadea*, *G. clusii* and *G. nivalis* under glasshouse conditions with three different AMF species. When spores were used as inoculum, all AMF tested colonized gentian roots and more than 90 % of the gentians tested were infected. The percentage of colonization varied according to the plant and fungus species. These authors proposed that inoculation of gentians with spores of AMF represents good cultural practice.

However, in several cases the conventional method of inoculation of gentians, using spores or colonized root segments as inoculum source, resulted in none or negligible root colonization. The experiments of Sýkorová et al. (2003) demonstrated that mycorrhizal colonization of *G. verna* microplants was very weak after their direct inoculation with spores or root fragments of AMF, being either indigenous to a specific field site or originating from an AMF collection. Successful mycorrhization of this plant in experimental conditions occurred only via an extraradical mycelial network established by a neighboring mycorrhizal plant of another species (nurse plant effect), or after transplantation of the plants to the field. Another example of the need for such a nurse plant for gentians is *Centaurium erythraea* (McGee 1985), which required *Trifolium dubium* for successful establishment of its colonization. In low nutrient soil, seedlings of *Centaurium* even died when cultivated without a mycorrhizal nurse plant. These observations highlight the importance of the ERM network or root-to-root contact mainly in grasslands, where the density of AM spores is often very low. The importance of such ERM network for biennial gentians is discussed by Gay et al. (1982).

An extraradical mycelium is also important for achlorophyllous members of the Gentianaceae, such as *V. truncata*. Imhof and Weber (1997) observed hyphal bridges connecting mycorrhizal roots of this species with roots of neighboring plants. Although the achlorophyllous plants cannot provide carbohydrates to the symbiotic fungi and, moreover, AMF had to receive all carbohydrates and other nutrients from the host plants, Imhof (1997) did not observe any negative effects on the symbiotic fungus. The authors proposed that the ERM linked to roots of many other plant species may be extremely important for nutrition of the fungus.

There have been several studies investigating the growth response of gentians following inoculation with AMF. As examples, no evidence was found that either indigenous or non-indigenous AMF induced a positive growth response in *G. verna* under two levels of inoculation (Sýkorová et al. 2003). Conversely, Heymons et al. (1986) inoculated *G. lutea* grown under glasshouse conditions in a low nutrient soil, and found that mycorrhizal plants developed better as compared to non-mycorrhizal ones, the latter showing nutrient deficiency symptoms. However, it is known that the extent of AM colonization and plant growth response to AM colonization can depend on various factors, such as identity of the fungal symbionts involved, propagule density, nutrient availability, the season, climate, soil type, and other environmental influences. Therefore, no general conclusions can be drawn about the growth response of gentians to AM colonization. Moreover, it must be kept in mind that without an obvious positive plant growth response, increased plant success may be due to hidden AM benefits related to nutrition, together with other non-nutritional advantages. Last but not least, there are also other ecosystem services delivered by the symbiosis, which are less obviously related to plant or fungal success (Smith and Smith 2012).

12.4 Conclusions

AM is a common feature of plants from the family Gentianaceae. It can be anticipated that under field conditions, almost all gentians will be colonized by AMF and it is highly probable that these mycorrhizae would have a morphology typical for the *Paris* type of AM. Though only limited physiological investigations of AM were carried out within this morphotype, it can be predicted that nutrient exchange between both partners takes place on hyphal coils or swellings. Achlorophyllous mycoheterotrophic plants of the Gentianaceae are specialized to a few narrow lineages of Glomeromycota and are entirely dependent on these fungi for their nutrition with carbohydrates as well as mineral nutrients. However, a considerable amount of work is required to clarify the importance of environment, host plant physiology and fungal species on the morphology, diversity and physiological roles of mycorrhizae in the Gentianaceae.

Acknowledgments The author thanks Miroslav Kolařík for critically reading this chapter, the Academy of Sciences of the Czech Republic (The Institutional Research Program, Grant No. AV0Z60050516) and the Czech Science Foundation (Grant No. P504/10/P021) for financial support.

References

Ahulu EM, Gollotte A, Gianinazzi-Pearson V, Nonaka M (2006) Cooccurring plants forming distinct arbuscular mycorrhizal morphologies harbor similar AM fungal species. Mycorrhiza 17:37–49

Azcon-Aguilar C, Encina CL, Azcon R, Barea JM (1994) Mycotrophy of *Annona cherimola* and the morphology of its mycorrhizae. Mycorrhiza 4:161–168

Bago B, Pfeffer PE, Shachar-Hill Y (2000) Carbon metabolism and transport in arbuscular mycorrhizas. Plant Physiol 124:949–957

Barto EK, Hilker M, Muller F, Mohney BK, Weidenhamer JD, Rillig MC (2011) The fungal fast lane: common mycorrhizal networks extend bioactive zones of allelochemicals in soils. PLoS ONE 6(11):1–7. doi:ARTNe27195

Bidartondo MI, Redecker D, Hijri I, Wiemken A, Bruns TD, Dominguez L, Sersic A, Leake JR, Read DJ (2002) Epiparasitic plants specialized on arbuscular mycorrhizal fungi. Nature 419:389–392

Cameron DD, Bolin JF (2010) Isotopic evidence of partial mycoheterotrophy in the Gentianaceae: *Bartonia virginica* and *Obolaria virginica* as case studies. Amer J Bot 97:1272–1277

Demuth K, Weber HC (1990) Structural incompatibility of VAM fungi in Gentianaceae. Angew Bot 64:247–252

Demuth K, Forstreuter W, Weber HC (1989) Cultivation of gentians and inoculation with VAM-fungi. Angew Bot 63:551–557

Demuth K, Forstreuter W, Weber HC (1991) Morphological differences in vesicular-arbuscular mycorrhizae of Gentianaceae produced by different endophytes. Flora 185:127–132

Dickson S, Smith FA, Smith SE (2007) Structural differences in arbuscular mycorrhizal symbioses: more than 100 years after Gallaud, where next? Mycorrhiza 17:375–393

Franke T, Beenken L, Doring M, Kocyan A, Agerer R (2006) Arbuscular mycorrhizal fungi of the *Glomus*-group A lineage (Glomerales; Glomeromycota) detected in myco-heterotrophic plants from tropical Africa. Mycol Prog 5:24–31

Gallaud I (1905) Études sur les mycorrhizes endotrophes. Revue Générale de Botanique 17:5–48, 66–83, 123–136, 23–239, 313–325, 425–433, 479–500

Gay PE, Grubb PJ, Hudson HJ (1982) Seasonal changes in the concentrations of nitrogen, phosphorus and potassium, and in the density of mycorrhiza, in biennial and matrix-forming perennial species of closed chalkland turf. J Ecol 70:571–593

Genre A, Bonfante P (2010) The making of symbiotic cells in arbuscular mycorrhizal roots. In: Koltai H, Kapulnik Y (eds) Arbuscular mycorrhizas: physiology and function. Springer, Dordrecht, pp 57–71

Gianinazzi S, Vosátka M (2004) Inoculum of arbuscular mycorrhizal fungi for production systems: science meets business. Can J Bot 82:1264–1271

Gollotte A, van Tuinen D, Atkinson D (2004) Diversity of arbuscular mycorrhizal fungi colonising roots of the grass species *Agrostis capillaris* and *Lolium perenne* in a field experiment. Mycorrhiza 14:111–117

Helber N, Wippel K, Sauer N, Schaarschmidt S, Hause B, Requena N (2011) A versatile monosaccharide transporter that operates in the arbuscular mycorrhizal fungus *Glomus* sp. is crucial for the symbiotic relationship with plants. Plant Cell 23:3812–3823

Helgason T, Merryweather JW, Denison J, Wilson P, Young JPW, Fitter AH (2002) Selectivity and functional diversity in arbuscular mycorrhizas of co-occurring fungi and plants from a temperate deciduous woodland. J Ecol 90:371–384
Heymons S, Holzl J, Weber HC (1986) VA-mycorrhiza in *Gentiana lutea*, the importance of cultivation and influence on constituents. Planta Med 6:510
Imhof S (1997) Root anatomy and mycotrophy of the achlorophyllous *Voyria tenella* Hook (Gentianaceae). Bot Acta 110:298–305
Imhof S, Weber HC (1997) Root anatomy and mycotrophy (AM) of the achlorophyllous *Voyria truncata* (Standley) Standley & Steyermark (Gentianaceae). Bot Acta 110:127–134
Jacquelinet-Jeanmougin S, Gianinazzi-Pearson V (1983) Endomycorrhizas in the Gentianaceae. 1. The fungi associated with *Gentiana lutea* L. New Phytol 95:663–665
Jacquelinet-Jeanmougin J, Gianinazzi-Pearson V, Gianinazzi S (1987) Endomycorrhizas in the Gentianaceae. 2. Ultrastructural aspects of symbiont relationships in *Gentiana lutea* L. Symbiosis 3:269–286
Jansa J, Mozafar A, Banke S, McDonald BA, Frossard E (2002) Intra- and intersporal diversity of ITS rDNA sequences in *Glomus intraradices* assessed by cloning and sequencing, and by SSCP analysis. Mycol Res 106:670–681
Kiers ET, Duhamel M, Beesetty Y, Mensah JA, Franken O, Verbruggen E, Fellbaum CR, Kowalchuk GA, Hart MM, Bago A, Palmer TM, West SA, Vandenkoornhuyse P, Jansa J, Bucking H (2011) Reciprocal rewards stabilize cooperation in the mycorrhizal symbiosis. Science 333:880–882
Klironomos JN, McCune J, Hart M, Neville J (2000) The influence of arbuscular mycorrhizae on the relationship between plant diversity and productivity. Ecol Lett 3:137–141
Knöbel M, Weber HC (1988) Vergleichende Untersuchungen zur Mycotrophie bei *Gentiana verna* L. und *Voyria truncata* (Stand.) Stand. & Stey. (Gentianaceae). Beitr Biol Pflanzen 63:463–477
Krüger M, Stockinger H, Krüger C, Schüßler A (2009) DNA-based species level detection of Glomeromycota: one PCR primer set for all arbuscular mycorrhizal fungi. New Phytol 183:212–223
Kuhn KD, Weber HC (1986) Vesicular arbuscular mycorrhiza in *Gentiana asclepiadea* L (Gentianaceae) on natural habitats. Angew Bot 60:427–439
Lekberg Y, Koide RT (2005) Is plant performance limited by abundance of arbuscular mycorrhizal fungi? A meta-analysis of studies published between 1988 and 2003. New Phytol 168:189–204
Mcgee PA (1985) Lack of spread of endomycorrhizas of *Centaurium* (Gentianaceae). New Phytol 101:451–458
Merckx V, Bidartondo MI, Hynson NA (2009) Myco-heterotrophy: when fungi host plants. Ann Bot 104:1255–1261
Merckx V, Stockel M, Fleischmann A, Bruns TD, Gebauer G (2010) ^{15}N and ^{13}C natural abundance of two mycoheterotrophic and a putative partially mycoheterotrophic species associated with arbuscular mycorrhizal fungi. New Phytol 188:590–596
Merckx VSFT, Janssens SB, Hynson NA, Specht CD, Bruns TD, Smets EF (2012) Mycoheterotrophic interactions are not limited to a narrow phylogenetic range of arbuscular mycorrhizal fungi. Mol Ecol 21:524–1532
Munkvold L, Kjoller R, Vestberg M, Rosendahl S, Jakobsen I (2004) High functional diversity within species of arbuscular mycorrhizal fungi. New Phytol 164:357–364
Neumann G (1934) Über die Mykorrhiza in der Gattung *Gentiana*. Zlb Bakt 89:433–458
Öpik M, Moora M, Zobel M, Saks U, Wheatley R, Wright F, Daniell T (2008) High diversity of arbuscular mycorrhizal fungi in a boreal herb-rich coniferous forest. New Phytol 179:867–876
Öpik M, Metsis M, Daniell TJ, Zobel M, Moora M (2009) Large-scale parallel 454 sequencing reveals host ecological group specificity of arbuscular mycorrhizal fungi in a boreonemoral forest. New Phytol 184:424–437
Parniske M (2008) Arbuscular mycorrhiza: the mother of plant root endosymbioses. Nat Rev Microbiol 6:763–775

Pellegrino E, Turrini A, Gamper HA, Cafa G, Bonari E, Young JPW, Giovannetti M (2012) Establishment, persistence and effectiveness of arbuscular mycorrhizal fungal inoculants in the field revealed using molecular genetic tracing and measurement of yield components. New Phytol 194:810–822

Redecker D, Kodner R, Graham LE (2000) Glomalean fungi from the Ordovician. Science 289:1920–1921

Rillig MC, Mummey DL (2006) Mycorrhizas and soil structure. New Phytol 171:41–53

Rosendahl S, Mcgee P, Morton JB (2009) Lack of global population genetic differentiation in the arbuscular mycorrhizal fungus *Glomus mosseae* suggests a recent range expansion which may have coincided with the spread of agriculture. Mol Ecol 18:4316–4329

Sanders IR, Alt M, Groppe K, Boller T, Wiemken A (1995) Identification of ribosomal DNA polymorphisms among and within spores of the Glomales: application to studies on the genetic diversity of arbuscular mycorrhizal fungal communities. New Phytol 130:419–427

Santos JC, Finlay RD, Tehler A (2006) Molecular analysis of arbuscular mycorrhizal fungi colonising a semi-natural grassland along a fertilisation gradient. New Phytol 172:159–168

Schüßler A, Walker C (2010) The Glomeromycota: a species list with new families and new genera. Avaible at www.amf-phylogeny.com

Schüßler A, Schwarzott D, Walker C (2001) A new fungal phylum, the Glomeromycota: phylogeny and evolution. Mycol Res 105:1413–1421

Schwartz MW, Hoeksema JD, Gehring CA, Johnson NC, Klironomos JN, Abbott LK, Pringle A (2006) The promise and the potential consequences of the global transport of mycorrhizal fungal inoculum. Ecol Lett 9:501–515

Schwarzott D, Walker C, Schüßler A (2001) *Glomus*, the largest genus of the arbuscular mycorrhizal fungi (Glomales), is nonmonophyletic. Mol Phylogenet Evol 21:190–197

Smith SE, Read DJ (2008) Mycorrhizal symbiosis, 3rd edn. Academic Press, Elsevier, Oxford, pp 1–765

Smith FA, Smith SE (1997) Tansley review No. 96. Structural diversity in (vesicular)-arbuscular mycorrhizal symbioses. New Phytol 137:373–388

Smith SE, Smith FA (2012) Fresh perspectives on the roles of arbuscular mycorrhizal fungi in plant nutrition and growth. Mycologia 104:1–13

Smith FA, Grace EJ, Smith SE (2009) More than a carbon economy: nutrient trade and ecological sustainability in facultative arbuscular mycorrhizal symbioses. New Phytol 182:347–358

Sýkorová Z, Rydlová J, Vosátka M (2003) Establishment of mycorrhizal symbiosis in *Gentiana verna*. Folia Geobot 38:177–189

Sýkorová Z, Wiemken A, Redecker D (2007) Cooccurring *Gentiana verna* and *Gentiana acaulis* and their neighboring plants in two Swiss upper montane meadows harbor distinct arbuscular mycorrhizal fungal communities. Appl Environ Microb 73:5426–5434

Sýkorová Z, Börstler B, Zvolenská S, Fehrer J, Gryndler M, Vosátka M, Redecker D (2012) Long-term tracing of *Rhizophagus irregularis* isolate BEG140 inoculated on *Phalaris arundinacea* in a coal mine spoil bank, using mitochondrial large subunit rDNA markers. Mycorrhiza 22:69–80

van der Heijden MGA, Klironomos JN, Ursic M, Moutoglis P, Streitwolf-Engel R, Boller T, Wiemken A, Sanders IR (1998) Mycorrhizal fungal diversity determines plant biodiversity, ecosystem variability and productivity. Nature 396:69–72

Wubet T, Kottke I, Teketay D, Oberwinkler F (2009) Arbuscular mycorrhizal fungal community structures differ between co-occurring tree species of dry Afromontane tropical forest, and their seedlings exhibit potential to trap isolates suited for reforestation. Mycol Prog 8:317–328

Erratum to: A Monographic Revision of the Neotropical Genus *Macrocarpaea* (Gentianaceae) in Ecuador

Jason R. Grant

Erratum to:
Chapter 3 in: J. J. Rybczyński et al. (eds.),
***The Gentianaceae - Volume 1: Characterization and Ecology*, DOI 10.1007/978-3-642-54010-3_3**

The online version of the original chapter can be found under
DOI 10.1007/978-3-642-54010-3_3

J. R. Grant (✉)
Laboratoire de botanique évolutive, Institut de Biologie Université de Neuchâtel, Rue Émile-Argand 11, CH 2000 Neuchâtel, Switzerland
e-mail: jason.grant@unine.ch

J. J. Rybczyński et al. (eds.), *The Gentianaceae - Volume 1: Characterization and Ecology*, DOI: 10.1007/978-3-642-54010-3_13,

Figure 3.43 should be an illustration of *Macrocarpaea micrantha*, but an illustration of *M. xerantifulva* was erroneously placed here. The correct illustration of *M. micrantha* is:

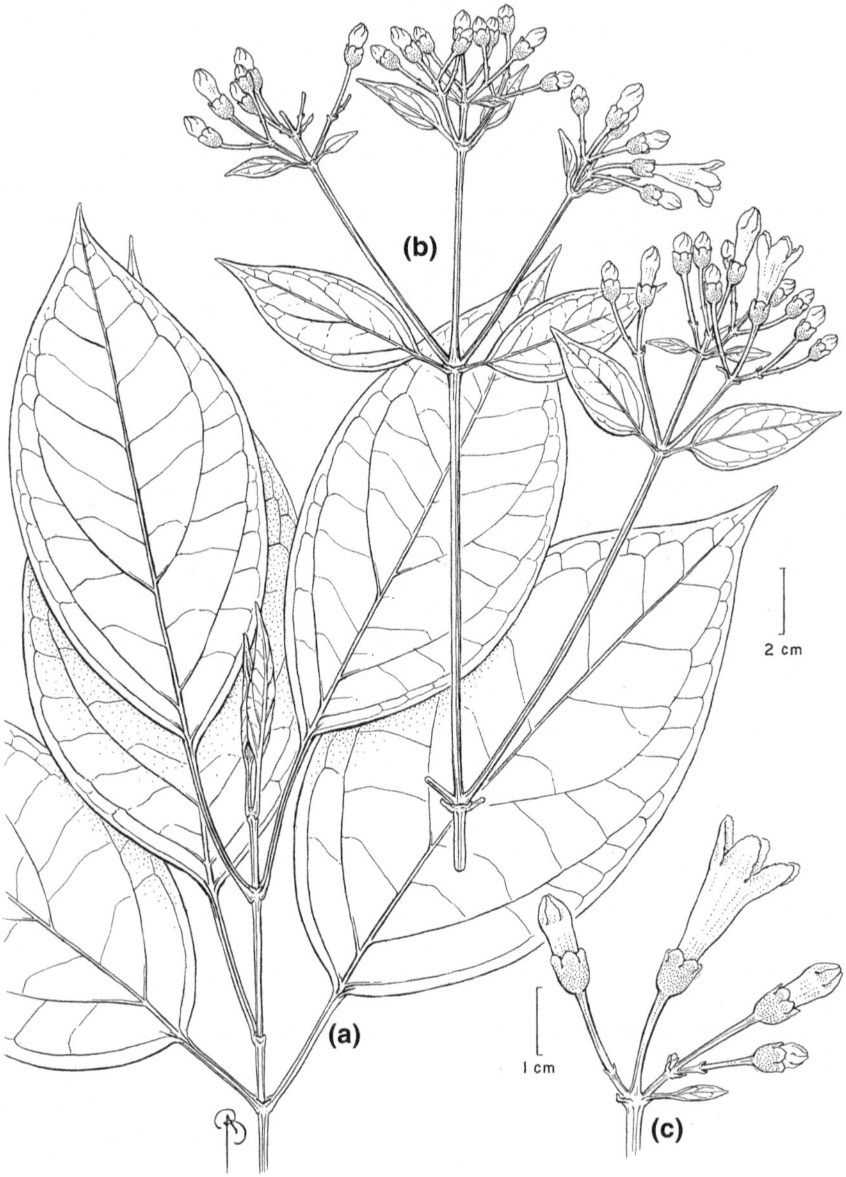

Figure 3.56 should be an illustration of *Macrocarpaea xerantifulva*, but it was erroneously included as Fig. 3.43. The correct illustration of *M. xerantifulva* is:

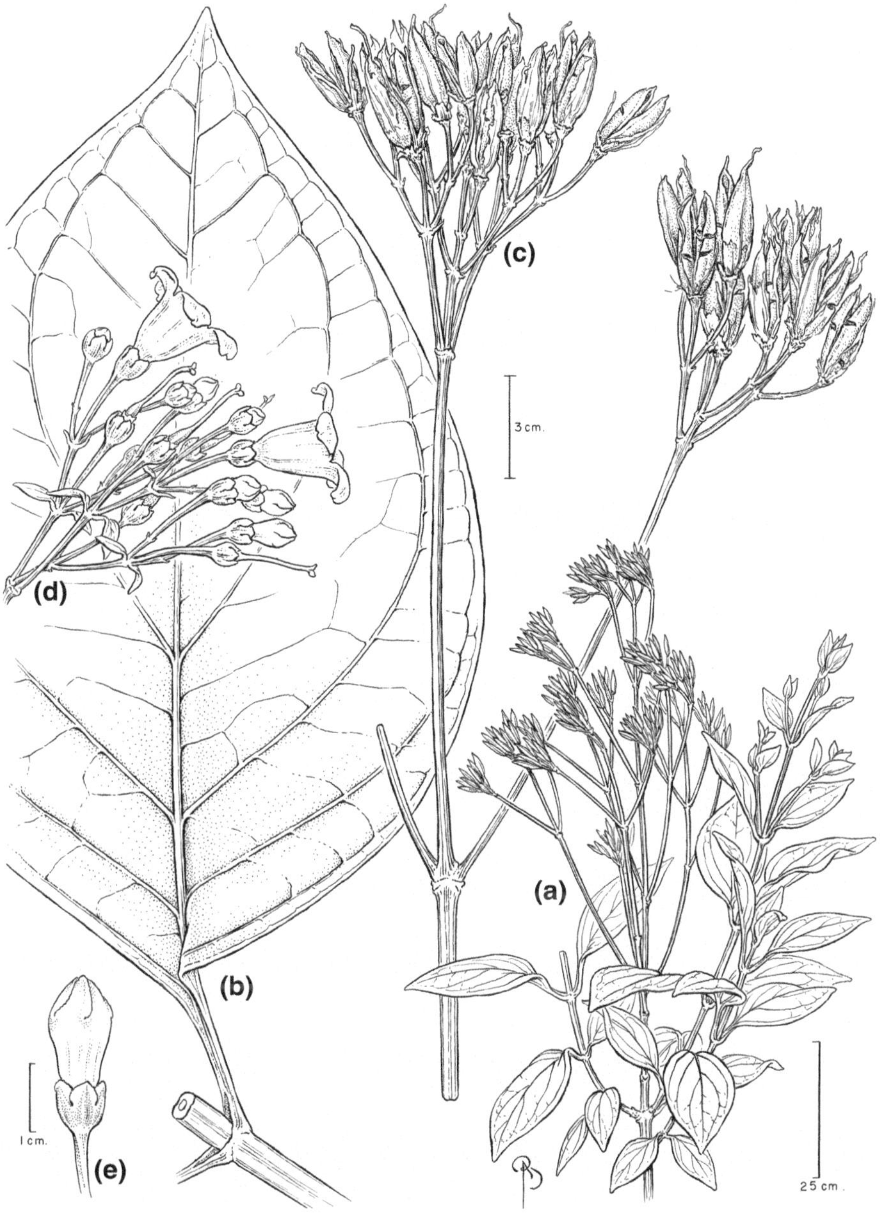

Latin Name Index

J. J. Rybczyński et al. (eds.), *The Gentianaceae - Volume 1: Characterization and Ecology*, DOI: 10.1007/978-3-642-54010-3,

Subject Index

J. J. Rybczyński et al. (eds.), *The Gentianaceae - Volume 1: Characterization and Ecology*, DOI: 10.1007/978-3-642-54010-3,